Modern Challenges and Approaches to Humanitarian Engineering

Yiannis Koumpouros
University of West Attica, Greece

Angelos Georgoulas
University of West Attica, Greece

Georgia Kremmyda
University of Warwick, UK

IGI Global
PUBLISHER of TIMELY KNOWLEDGE

A volume in the Advances in Human Services and Public Health (AHSPH) Book Series

Published in the United States of America by
IGI Global
Engineering Science Reference (an imprint of IGI Global)
701 E. Chocolate Avenue
Hershey PA, USA 17033
Tel: 717-533-8845
Fax: 717-533-8661
E-mail: cust@igi-global.com
Web site: http://www.igi-global.com

Library of Congress Cataloging-in-Publication Data

Names: Koumpouros, Yiannis, editor. | Georgoulas, Angelos, DATE- editor. | Kremmyda, Georgia, DATE- editor.
Title: Modern challenges and approaches to humanitarian engineering / Yiannis Koumpouros, Angelos Georgoulas, and Georgia Kremmyda, editors.
Description: Hershey, PA : Engineering Science Reference, [2022] | Includes bibliographical references and index. | Summary: "This book is those who want to improve their understanding of the strategic role of humanitarian engineering at different levels of the modern world and the various challenges facing our society, covering a wide variety of topics that fall under the domain of technology-based solutions that tackle humanitarian problems, such as pandemics, ageing of population, climate change, social inclusion, extreme poverty and hunger, maternal health and child mortality, and education for all"-- Provided by publisher.
Identifiers: LCCN 2021051158 (print) | LCCN 2021051159 (ebook) | ISBN 9781799891901 (hardcover) | ISBN 9781799891925 (ebook)
Subjects: LCSH: Technology--Social aspects. | Science--Social aspects. | Humanitarianism.
Classification: LCC T14.5 .H3536 2022 (print) | LCC T14.5 (ebook) | DDC 303.48/3--dc23/eng/20211118
LC record available at https://lccn.loc.gov/2021051158
LC ebook record available at https://lccn.loc.gov/2021051159

This book is published in the IGI Global book series Advances in Human Services and Public Health (AHSPH) (ISSN: 2475-6571; eISSN: 2475-658X)

British Cataloguing in Publication Data
A Cataloguing in Publication record for this book is available from the British Library.

Advances in Human Services and Public Health (AHSPH) Book Series

Jennifer Martin
RMIT University, Australia

ISSN:2475-6571
EISSN:2475-658X

Mission

The well-being of the general public should be a primary concern for any modern civilization. Ongoing research in the field of human services and public healthcare is necessary to evaluate, manage, and respond to the health and social needs of the global population.

The **Advances in Human Services and Public Health (AHSPH)** book series aims to publish high-quality reference publications focused on the latest methodologies, tools, issues, and strategies for managing the health and social welfare of the public. The AHSPH book series will be especially relevant for healthcare professionals, policy makers, government officials, and students seeking the latest research in this field.

Coverage

- Social Work
- Domestic Violence
- Poverty
- Healthcare Reform
- Public Funding
- Access to Healthcare Services
- Health Policy
- Medicare and Medicaid
- Social Welfare Policy
- Public Welfare

IGI Global is currently accepting manuscripts for publication within this series. To submit a proposal for a volume in this series, please contact our Acquisition Editors at Acquisitions@igi-global.com or visit: http://www.igi-global.com/publish/.

The Advances in Human Services and Public Health (AHSPH) Book Series (ISSN 2475-6571) is published by IGI Global, 701 E. Chocolate Avenue, Hershey, PA 17033-1240, USA, www.igi-global.com. This series is composed of titles available for purchase individually; each title is edited to be contextually exclusive from any other title within the series. For pricing and ordering information please visit http://www.igi-global.com/book-series/advances-human-services-public-health/102256. Postmaster: Send all address changes to above address.

Titles in this Series

For a list of additional titles in this series, please visit: www.igi-global.com/book-series/advances-human-services-public-health/102256

Tactical Sciences for Biosecurity in Animal and Plant Systems
Kitty F. Cardwell (Oklahoma State University, USA) and Keith L. Bailey (Charles River Laboratories, Inc. USA)
Information Science Reference • © 2022 • 454pp • H/C (ISBN: 9781799879350) • US $195.00

Handbook of Research on Interdisciplinary Perspectives on the Threats and Impacts of Pandemics
Şahver Omeraki Çekirdekci (Haliç University, Turkey) Özlem İngün Karkış (Doğuş University, Turkey) and Suna Gönültaş (Doğuş University, Turkey)
Information Science Reference • © 2022 • 556pp • H/C (ISBN: 9781799886747) • US $245.00

Smart Cities and Machine Learning in Urban Health
J. Joshua Thomas (UOW Malaysia KDU Penang University College, Malaysia) Vasiliki Geropanta (Technical University of Crete, Greece) Anna Karagianni (Technical Chamber of Greece, Greece) Vladimir Panchenko (Russian University of Transport, Russia) and Pandian Vasant (MERLIN Research Center, TDTU, Vietnam)
Information Science Reference • © 2022 • 269pp • H/C (ISBN: 9781799871767) • US $195.00

Handbook of Research on Digital Citizenship and Management During Crises
Erdem Öngün (Trakya University, Turkey) Nilüfer Pembecioğlu (Istanbul University, Turkey) and Uğur Gündüz (Istanbul University, Turkey)
Information Science Reference • © 2022 • 440pp • H/C (ISBN: 9781799884217) • US $275.00

Public Health and Economic Resiliency in the Post-COVID-19 Era
Kholoud Kahime (Cadi Ayyad University, Morocco) Mustapha Zahir (SAEDD, ESTE, Cadi Ayyad University, Morocco) Mohamed Hadach (SAEDD, ESTE, Cadi Ayyad University, Morocco) Moulay Abdelmonaim El Hidan (SAEDD, ESTE, Cadi Ayyad University, Morocco) and Blaid Bougadir (Cadi Ayyad University)
Information Science Reference • © 2022 • 276pp • H/C (ISBN: 9781799882022) • US $215.00

Handbook of Research on Historical Pandemic Analysis and the Social Implications of COVID-19
Antonio Cortijo Ocaña (University of California, Santa Barbara, USA) and Vicent Martines (University of Alicante, Spain)
Information Science Reference • © 2022 • 353pp • H/C (ISBN: 9781799879879) • US $245.00

IGI Global
PUBLISHER of TIMELY KNOWLEDGE

701 East Chocolate Avenue, Hershey, PA 17033, USA
Tel: 717-533-8845 x100 • Fax: 717-533-8661
E-Mail: cust@igi-global.com • www.igi-global.com

Table of Contents

Detailed Table of Contents

As humanitarian crises have become increasingly protracted, strengthening the collaboration between development and humanitarian aid actors has been identified as a priority by the donor community, popularized in the concept of the humanitarian-development nexus (HDN). In recent years, aid practices based on the HDN approach have emerged increasingly in response to humanitarian crises. The remarkable thing about these aid practices is that donors try to address humanitarian crises by developing projects and programs aimed at equipping humanitarian victims with specific technical and vocational skills and integrating them into the labor market through value chains. Incorporating humanitarian crisis victims into labor markets according to the needs and demands of capital has recently been the main concern of the international donor community. This chapter aims to provide insights into the role and relevance of the development-oriented humanitarian responses in the attempts to incorporate humanitarian victims into the exploitative global capitalist relations at the global level.

Bangladesh is going to be the worst-hit country from the climate change. The impacts of climate change at specific warming levels (SWLs) of 1.5°C, 2°C, and 4°C upon the natural and social systems of Bangladesh are assessed. Eleven regional climate models' findings show that the average Bangladesh summer temperature is likely to rise from 3.24°C to 5.77°C while the mean annual precipitation is likely to increase up to 25% by the end of the century. The yield trends of Bangladesh in terms of Boro and Aman rice are projected to gradually decrease from 2-15% at a specific warming level of 1.5°C, 5-20% at a specific warming level of 2°C, and 5-25% at a specific warming level of 4°C. A few adaptation options are proposed for different SWLs and sea-level rise. Managing climate for SWLs could be for 'short

and Central Vietnam's coastlines. Intensive socio-economic development, sea level rise, more frequent and strong intensity of disasters are predicted to occur in a future of global climate change, which poses an urgent need for coastal hazard management strategies. This chapter describes main causes and discusses the applied engineering interventions to reduce coastal erosion at these sites. Hard engineering structures are often constructed in these areas, but they seem to be merely short-term costly solutions and have a negative impact on the coastal environment and its residents. Nature-based solutions and soft engineering approaches are proposed, which seem to be sustainable and less expensive than hard engineering options. These possible future solutions can be applied in coastal settings to meet the principles of sustainable and humanitarian engineering with multiple benefits to reduce the risk and negative impacts on both humans and the environment.

Chapter 6
Community-Based Waste Management Model in Optimizing Waste Reduction: Waste Bank Practices in Indonesia 98
Christia Meidiana, Brawijaya University, Indonesia
Tonni Agustiono Kurniawan, Xiamen University, China
Adipandang Yudono, Brawijaya University, Indonesia
Surjono Surjono, Brawijaya University, Indonesia

The concept of collect-transport-and-dump of waste in Indonesia has changed into reduce, reuse, and recycle (3R) since it causes environmental consequences that the Government of Indonesia (GoI) has banned including the open dumping practices since 2013. GoI targets 30% waste reduction by 2030, and some policies are introduced to achieve the target including promoting waste bank. Waste bank is an implementation of 3R involving community. The number of waste banks are increasing since waste banks offer many benefits. Currently, there are 7600 waste banks nationwide. Through waste banks, the community learns to sort the waste. It also raises public awareness about waste issues and the importance of 3R. It becomes an innovation program at the grassroots level to increase income and support the local governments to improve the performance of waste management. In this chapter, some cases presenting waste bank implementation and their related aspects, such as waste bank performance, public acceptance, and participation, as well as determinants of public participation, are discussed.

Chapter 7
Lesson Learned From Public Participation in the Urban Infrastructure Projects for Engineering: Case Studies in Ho Chi Minh City, Vietnam 114
Pham Thi Anh, Ho Chi Minh City University of Transport, Vietnam
Tien Thuy Nguyen, Ho Chi Minh City University of Transport, Vietnam
Tuan Anh Nguyen, Ho Chi Minh City University of Transport, Vietnam
Dong Doan Van, Ho Chi Minh City University of Transport, Vietnam

Raising public awareness and participation of the local communities can have an important input for public involvement in the project design and implementation process, and it has become a major objective of sound development programs. This chapter focuses on (1) public participation in the design and adjust boundary of the underground station of the Metro line 2 in Ho Chi Minh City and (2) public awareness and participation in canal environmental protection. Research is mainly based on the results of community meetings on project information, communication to raise public awareness on environmental protection, and survey questionnaires. The study will add to several recent studies on the roles of public participation in infrastructure projects for sustainable development. There are three important lessons that stand out

from the experiences using public consultants in infrastructure projects.

About one million Rohingya Muslim minority people fled from their home on 25 August 2017 due to the recent violence in the Rakhine State of Myanmar and entered Bangladesh, a neighboring country. Many of them were the vulnerable people with 58% refugees under 18 children, 67% female, and 9% infants below one year. This was a great humanitarian challenge for the Government of Bangladesh and the local community. The mass influx of refugees caused a major humanitarian emergency, and the majority were reliant on humanitarian assistance including shelter, food, healthcare, clean water, sanitation, education, nutrition, safety, and security. They were given accommodation, food, emergency aids, etc. upon arrival, and finally, 890,276 Rohingya refugees settled in 34 refugee camps in Cox's Bazar and Teknaf area. Many houses, schools, shops, hospitals, and other service centers were built day by day in the camp area. Children are getting vaccines for different diseases and are going to learning centers for education. Still, there are several challenges for better management.

This chapter seeks to explore the key considerations that should be employed by camp coordinators and designers to mitigate the fundamental issues being faced by camps today and provide a safer and more durable living environment for refugees. It also aims to provide alternative camp design and material solutions and provide insight into the use of new technologies to aid these considerations. Specifically, this chapter will be exploring GIS technologies and their use in field operations. The need to rethink refugee camps is vital given the continuously increasing influx of refugees worldwide. However, within the context of the COVID-19 pandemic, this study becomes even more critical for the safety of refugees, camp workers, and host countries alike. This chapter will delve deeper into the topics of camp design and layout, camp shelter and materials, and employing technologies such as GIS in the process of redesigning camps.

Global population ageing is a pressing issue with significant humanitarian implications associated with compromised independence, social isolation, frailty, and diminished quality of life affecting millions of older individuals and their families worldwide as well as healthcare and social infrastructures. In this context, this chapter presents a personalized smart-living framework for promoting healthy and successful ageing, combining interdisciplinary knowledge and technologies including internet of things (IoT), artificial intelligence (AI), and affective computing (AC). The proposed intervention tackles the health-related, societal, and emotional impacts of ageing within a society with a rising mean global age aiming to mitigate the effects of decline and reduce the associated risks. The outlined architecture is based on sensor-derived information blended with non-sensor-based computerized modules and voice-based inputs for establishing a smart ecosystem that enables older people to live independently inside and outside their home, maintaining a high cognitive and physical status.

Contact tracing apps for the COVID-19 pandemic, from a personal data protection point of view, are discussed in this chapter. More precisely, having the relevant European legal framework as a basis, the authors investigate the underlying personal data processing that occurs from these apps in relation with the relevant privacy risks that are inherent in the smart mobile ecosystem. The analysis indicates that there exist several different approaches across the world, whereas even in countries with unified legal framework there exist some discrepancies in the design philosophy and the operation of these apps. Since user trust is essential for the effectiveness of such types of apps, they discuss the importance of respecting the fundamental right of privacy when developing and operating such applications. The ultimate goal is to illustrate the need to strike the proper balance between substantial public interests and the right to the protection of personal data.

The purpose of this chapter is to describe a possible best practice to teaching chemistry from a humanitarian engineering perspective. The interest in teaching chemistry by focusing on humanitarian engineering arises from the economic and environmental concerns that the country of this study faces, some of which are poverty, climatic changes, food crisis, inadequate healthcare, water crisis, and pollution. As an educator, there is an interest in educating future generations to be able to cope with environmental changes that face their countries and the world at large. This exposition of a possible new approach with appropriate pedagogies that is presented here may be an answer that underdeveloped, developed, and emergent economies may adopt to close the gap between themselves and other industrialised nations.

People with disabilities raise important and special humanitarian approaches. There is a strong necessity for an epidemiological and sociological study of disabled population with regard to disease distribution and the patients' social profile based on the bio-psycho-social model of health and illness. Adverse economic data, low social cohesion, and consequently, social exclusion are often observed within the specific population. According to the quantitative research conducted, the population's nosological profile has serious effects on patients' everyday lives and is directly related to social parameters. Fundamental social factors and access to health services affect people with a degree of disability of over 67% health in terms of disease prevention and treatment. At the same time, social determinants seem to negatively affect their health and treatment. There is a clear need for an information technology system that will monitor and produce appropriate key performance indicators to help design appropriate policies for the specific population.

Preface

The 21st century is associated to a number of environmental, social and economic challenges, spanning from globalization and migration to climate change, global health, urbanization and natural hazards.

Globalization, in addition to its positive aspects, has enabled the mobility of population, often combined with the transmission of infectious diseases and underpinning challenges in the detection and management of migrants. Conflicts waged in different parts of the world have led to a large influx of refugees impacting strongly other communities.

On the other hand, countries like those in sub-Saharan Africa are plagued by a multitude of problems, such as malnutrition, extreme poverty, violation of basic human rights, corruption, water and energy deficiencies, and similar. Combating climate change appears to be a high priority, and although discussed extensively in the recent years, has made relatively little progress. The interests of countries, multinationals and industries are lagging behind in finding comprehensive solutions. The requirement for conciliation cannot be met easily since the perspective of each country on this issue is influenced by different parameters. The phenomena, however, are now felt in every corner of the Earth. Emissions of pollutants to the environment and the greenhouse effect make their appearance in the form of unprecedented environmental conditions in every part of the planet. Extreme weather conditions are now common in many areas. Furthermore, the pollution of oceans and seas is becoming apparent with the reduction of marine wealth and life, with the unprecedented movement of species to other regions and countries or the extinction of species, at high rate. Commercial activities and the exploitation of natural wealth by few bring unique humanitarian threats, such as malnutrition in countries that traditionally had a dominant export character in agri-food products. In addition, millions of our fellow human beings are experiencing exploitation in the labor market in developing countries. The right to work is not self-evident in many such countries. The same goes for access to information and knowledge. This, in turn, has the effect that part of the world's population is entering deeper into poverty and despair. Natural disasters, earthquakes, volcanic eruptions, tsunamis, hurricanes, etc., are a daily occurrence in some countries with extremely large populations.

All of the above are just some of the challenges of the modern age which commands our immediate reaction towards an equal society, with equal opportunities for every human being. There is an urgent need for scientists, academics, researchers and politicians to take the reins by providing immediate solutions to tackle this harsh reality. The need for a more human-centered approach has relatively recently led to what we call humanitarian engineering.

Humanitarian engineering is to provide individuals and communities facing global challenges with solutions, tools and processes that improve lives and livelihoods. The provision of water, energy, food, shelter and information are some of the issues targeted by this "discipline". Humanitarian Engineers could be a catalyst towards achieving the Sustainable Development Goals. Currently, UNESCO is working to strengthen engineering education through curricula development and capacity building. It is also incorporating sustainability topics into engineering education by highlighting the need for green technology in engineering applications. The careful use of resources in a way that does not compromise the environment or deplete the materials for future generations is recognized as sustainable engineering. Both sustainable engineering and humanitarian engineering require a highly interdisciplinary approach since they aim to improve the quality of life for all.

This book aims to provide relevant frameworks and the latest research findings in the theme of humanitarian engineering. It is written from researchers and professionals who want to improve the strategic role of humanitarian engineering at different levels of the modern world and the various challenges facing our society.

The book investigates the most recent challenges and approaches in the field of humanitarian engineering. A wide variety of topics that fall under the domain are technology-based solutions that tackle humanitarian problems. Pandemics, ageing of population, climate change, social inclusion, education for all, are some indicative topics that could be addressed by technology. Early warning and alerting mechanisms for physical disasters, green engineering approaches, mobile health solutions for remote and underserved populations, are some paradigms that fall under the researched themes. The book is presenting research, real-life case studies, innovative models and approaches, and other work that lies in the presented theme. The interdisciplinarity of the authors combined with the chapters that analyze and focus on real events and conditions in the most vulnerable countries (e.g. Indonesia, Bangladesh, Vietnam, etc.) makes it a valuable tool for academics, researchers and students in the field.

The book's approach is highly interdisciplinary covering a wide range of topics, like: sustainable development, climate emergency, global health, water, sanitation and environmental management, natural disasters and risk reduction management, sustainable and resilient infrastructure, information and communication technologies in health, as well as some more general but also critical aspects of humanitarian challenges e.g. education, health status and socio-economic issues.

ORGANIZATION OF THE BOOK

Chapter 1: As humanitarian crises have become increasingly protracted, strengthening the collaboration between development and humanitarian aid actors has been identified as a priority by the donor community, popularized in the concept of the Humanitarian-Development Nexus (HDN). In recent years, aid practices based on the HDN approach have emerged increasingly in response to humanitarian crises. The remarkable thing about these aid practices is that donors try to address humanitarian crises by developing projects and programs aimed at equipping humanitarian victims with specific technical and vocational skills and integrating them into the labor market through value chains. Incorporating humanitarian crisis victims into labor markets according to the needs and demands of capital has recently been the main concern of the international donor community. This chapter aims to provide insights into the role and relevance of the development-oriented humanitarian responses in the attempts to incorporate humanitarian victims into the exploitative global capitalist relations at the global level.

Chapter 2: Bangladesh is going to be the worst-hit country from climate change. The impacts of climate change at specific warming levels (SWLs) of 1.5°C, 2°C, and 4°C upon the natural and social systems of Bangladesh are assessed. Eleven Regional Climate Models' findings show that the average

Bangladesh summer temperature is likely to rise from 3.24°C to 5.77°C while the mean annual precipitation is likely to increase up to 25% by the end of the century is predicted. The yield trends of Bangladesh in terms of Boro and Aman rice are projected to gradually decrease from 2 to 15% at a specific warming level of 1.5°C, 5 to 20% at a specific warming level of 2°C, and 5 to 25% at a specific warming level of 4°C. A few adaptation options are proposed for different SWLs and sea-level rise. Managing climate for SWLs could be for 'Short Term' up to 2020, 'Medium Term' up to 2050, and 'Long Term' up to 2080. The end goal is sustainable, resilient, and transformed Bangladesh, where additional policy documents, strategies, and action plans to mainstream adaptation to combat climate change are necessary.

Chapter 3: WASH interventions have not always been successful, despite the significant efforts to improve coordination and delivery of quality interventions. Assessments have shown that sustainable management of interventions must transcend the immediate intervention period to allow continuous infrastructural service delivery, social and behavioural change, civil participation, improved governance and more robust community and public sector systems. This chapter highlights the challenges of sustainable management, approaches and lessons learned from the humanitarian response, especially in providing WASH facilities in conflict zones and fragile communities. It would also focus on a case study of conflict-affected Pulka Town, Gwoza Local Government Area, Borno State, Nigeria.

Chapter 4: This chapter will discuss the issues of plastic, primarily microplastic pollution in freshwater and drinking water, with a focus on developing nations. Microplastics, generally defined as plastic particles with a size less than 5 mm, are beginning to gain attention as an emerging contaminant of concern. Whilst testing has recently begun on the contamination of freshwater and treated drinking water by microplastics in a number of developed regions, literature regarding microplastic pollution in the water of less economically developed countries is lacking. Microplastics pose a threat to human health and therefore it is important that cost effective methods for the testing, detection and removal of these plastic items from drinking water globally is considered with a higher level of urgency. It is argued that by achieving the aims laid out by the Sustainable Development Goals, particularly goals 6 and 12, the threats from microplastic pollution will subside.

Chapter 5: Both natural and anthropogenic forces could play significant roles in coastal erosion in Western Taiwan and Central Vietnam's coastlines. Intensive socio-economic development, sea level rise, more frequent and strong intensity of disasters are predicted to occur in a future of global climate change, which poses an urgent need for coastal hazard management strategies. This chapter describes main causes and discusses the applied engineering interventions to reduce coastal erosion at these sites. Hard engineering structures are often constructed in these areas, but they seem to be merely short-term costly solutions and have a negative impact on the coastal environment and its residents. Nature-based solutions and soft engineering approaches are proposed, which seem to be sustainable and less expensive than hard engineering options. These possible future solutions can be applied in coastal settings to meet the principles of sustainable and humanitarian engineering with multiple benefits to reduce the hazards risk and negative impacts on both humans and the environment.

Chapter 6: Concept of collect-transport-and dump of waste in Indonesia has changed into reduce, reuse and recycle (3R) since it causes environmental consequences that the Government of Indonesia (GoI) has banned the open dumping practices since 2013. GoI targets 30% waste reduction by 2030 and some policies are introduced to achieve the target including promoting waste bank. Waste bank is an implementation of 3R involving community. The number of waste banks are increasing since waste bank offers many benefits. Currently, there are 7600 waste banks national wide. Through waste bank, the community learn to sort the waste It raises also public awareness about waste issues and importance of 3R. It becomes an innovation program at the grassroots level to increase income and support the local governments to improve the performance of waste management. In this article, some cases presenting waste bank implementation and their related aspects, such as waste bank performance, public acceptance, and participation, as well as determinants of public participation are discussed.

Chapter 7: Raising public awareness and participation of the local communities can have an important input for public involvement in the project design and implementation process, and it has become a major objective of sound development programs. This paper focuses on (1) public participation in the design and adjust boundary of the underground station of the Metro line 2 - in Ho Chi Minh City, and (2) public awareness and participation in canal environmental protection. Research is mainly based on the results of community meetings on project information, communication to raise public awareness on environmental protection, and survey questionnaires. The study will add to several recent studies on the roles of public participation in infrastructure projects for sustainable development. There are three important lessons that stand out from the experiences using public consultants in infrastructure projects.

Chapter 8: About one million Rohingya Muslim minority people fled from their home on 25 August 2017 due to the recent violence in the Rakhine State of Myanmar and entered Bangladesh, a neighboring country. Many of them were the vulnerable people with 58% refugees under 18 children, 67% female and 9% infants below one year. This was a great humanitarian challenge for the Government of Bangladesh and the local community. The mass influx of refugees caused a major humanitarian emergency and the majority were reliant on humanitarian assistance including shelter, food, healthcare, clean water, sanitation, education, nutrition, safety and security. They were given accommodation, food, emergency aids etc. upon arrival and finally 890,276 Rohingya refugees settled down in 34 refugee camps in Cox's Bazar and Teknaf area. Many houses, schools, shops, hospitals and other service centers were built day by day in the camp area. Children are getting vaccines for different diseases and are going to learning centers for education. Still, there are several challenges for better management.

Chapter 9: This chapter seeks to explore the key considerations that should be employed by camp coordinators and designers to mitigate the fundamental issues being faced by camps today and provide a safer and more durable living environment for refugees. It also aims to provide alternative camp design and material solutions and provide insight into the use of new technologies to aid these considerations. Specifically, this chapter will be exploring GIS technologies and their use in field operations. The need to rethink refugee camps is vital given the continuously increasing influx of refugees worldwide. However, within the context of the Covid-19 pandemic, this study becomes even more critical for the safety of refugees, camp workers, and host countries alike. This chapter will delve deeper into the topics of camp design and layout, camp shelter and materials, and employing technologies such as GIS in the process of redesigning camps

Chapter 10: Global population ageing is a pressing issue with significant humanitarian implications, associated with comprised independence, social isolation, frailty and diminished quality of life, affecting millions of older individuals and their families worldwide, as well as healthcare and social infrastructures. In this context, this chapter presents a personalized smart-living framework for promoting healthy and successful ageing, combining interdisciplinary knowledge and technologies including Internet-of-Things (IoT), Artificial Intelligence (AI) and Affective Computing (AC). The proposed intervention tackles the health-related, societal and emotional impacts of ageing within a society with a rising mean global age, aiming to mitigate the effects of decline and reduce the associated risks. The outlined architecture is based on sensor-derived information blended with non-sensor-based computerized modules and voice-based inputs, for establishing a smart ecosystem that enables older people to live independently inside and outside their home, maintaining a high cognitive and physical status.

Chapter 11: Contact tracing apps for COVID-19 pandemic, from a personal data protection point of view, are being discussed in this chapter.More precisely, having the relevant European legal framework as a basis, we investigate the underlying personal data processing that occurs from these apps, in relation with the relevant privacy risks that are inherent in the smart mobile ecosystem. The analysis indicates that there exist several different approaches across the world, whereas even in countries with unified legal framework there exist some discrepancies in the design philosophy and the operation of these apps. Since users' trust is essential for the effectiveness of such types of apps, we discuss the importance of respecting the fundamental right of privacy when developing and operating such applications. Our ultimate goal is to illustrate the need to strike the proper balance between substantial public interests and the right to the protection of personal data.

Chapter 12: The purpose of this chapter is to describe a possible best practice to teaching chemistry from a humanitarian engineering perspective. The interest in teaching chemistry by focusing on humanitarian engineering arises from the economic and environmental concerns that the country of this study faces, some of which are poverty, climatic changes, food crisis, inadequate health care, water crisis and pollution. As an educator, there is an interest in educating future generations to be able to cope with environmental changes that face their countries and the world at large. This exposition of a possible new approach, with appropriate pedagogies that is presented here may be an answer that under-developed, developed and emergent economies may adopt to close the gap between themselves and other industrialised nations.

Chapter 13: People with disabilities raise important and special humanitarian approaches. There is a strong necessity for an epidemiological and sociological study of disabled population, with regard to disease distribution and the patients' social profile, based on the bio-psycho-social model of health and illness. Adverse economic data, low social cohesion and consequently social exclusion are often observed within the specific population. According to the quantitative research conducted, the population's nosological profile have serious effects on patients' everyday life and is directly related to social parameters. Fundamental social factors and access to health services affect people with a degree of disability of over 67% health in terms of disease prevention and treatment. At the same time, social determinants seem to affect negatively their health and treatment. There is a clear need for an information technology system that will monitor and produce appropriate key performance indicators to help design appropriate policies for the specific population.

Acknowledgment

This publication was supported by the project ENHANCE - ENabling Humanitarian Attributes for Nurturing Community-based Engineering (project No: 598502-EPP-1-2018-1-UK-EPPKA2-CBHE-JP (2018-2582 / 001-001), funded by the Erasmus + KA2 Cooperation for innovation and the exchange of good practices – Capacity building in the field of Higher Education. The publication reflects the views only of the authors, and the Commission cannot be held responsible for any use which may be made of the information contained therein.

Chapter 1
Development-Oriented Approaches to Humanitarian Responses

Yahya Gülseven
Başkent University, Turkey

ABSTRACT

As humanitarian crises have become increasingly protracted, strengthening the collaboration between development and humanitarian aid actors has been identified as a priority by the donor community, popularized in the concept of the humanitarian-development nexus (HDN). In recent years, aid practices based on the HDN approach have emerged increasingly in response to humanitarian crises. The remarkable thing about these aid practices is that donors try to address humanitarian crises by developing projects and programs aimed at equipping humanitarian victims with specific technical and vocational skills and integrating them into the labor market through value chains. Incorporating humanitarian crisis victims into labor markets according to the needs and demands of capital has recently been the main concern of the international donor community. This chapter aims to provide insights into the role and relevance of the development-oriented humanitarian responses in the attempts to incorporate humanitarian victims into the exploitative global capitalist relations at the global level.

INTRODUCTION

Humanitarian aid is designed to save lives and relieve suffering following emergencies and natural disasters whereas development aid specifically targets long-term, systematic issues by promoting economic and social development in developing countries. Although humanitarian and development aid systems have historically been based on different institutions and mechanisms, discussions about the two complementing each other or even being handled simultaneously have been on the agenda for a long time.

The need for and cost of humanitarian interventions has grown dramatically in recent years, mainly due to the protracted and recurrent nature of humanitarian crises. Every year, millions of people around the world are forced to leave their homes due to conflict, violence, and natural disasters. The number of

DOI: 10.4018/978-1-7998-9190-1.ch001

forcibly displaced people rose for the ninth year in a row to reach 82.4 million in 2020 (United Nations High Commissioner for Refugees, 2021). The massive increase in multiple and overlapping humanitarian and development needs has put enormous pressure on aid financing and forced donor countries to be more efficient and effective in addressing humanitarian crises. This trend has revitalized the long-standing discussion around better connectivity between humanitarian and development efforts. Especially in the aftermath of the recent global refugee crisis, the donor community has begun to prioritize the so-called development-oriented humanitarian interventions, allegedly aimed at empowering crisis victims and promoting self-reliance instead of being dependent on aid from outside. However, the empowerment and self-reliance of victims are narrowly and superficially defined in terms of their participation in the labor market as productive labor. This chapter critically examines these attempts to integrate humanitarian and development aid. The aim here is to provide insights into the role and relevance of the so-called development-oriented humanitarian aid strategies in efforts to deliver humanitarian victims into the hands of capital as cheap and flexible labor at the global level.

In the UN (United Nations) Secretary General's Report on the World Humanitarian Summit (WHS), which was held in İstanbul in 2016, the UN Secretary-General proposed a fundamental shift in donor community's approach to the issue of forced displacement: "one that goes from meeting immediate humanitarian needs to one that preserves the dignity and improves the lives and self-reliance of displaced persons" (United Nations General Assembly, 2016a). Strengthening the collaboration between development and humanitarian actors was identified as a top priority at the Summit. In line with this priority, the UN Secretary-General outlined a "New Way of Working," which calls on humanitarian and development actors to work together towards "collective outcomes" that reduce needs, risk, and vulnerability (United Nations General Assembly, 2016b)[1]. The Secretary-General, who views the displaced victims as "development opportunities" for countries of arrival, also urged national and community leaders to adopt social and labor policies to prevent long-term aid dependency and to treat the displaced people as "future assets" rather than burdens (United Nations General Assembly, 2016a).

Enabling forcibly displaced people to support themselves by acquiring skills to maintain their lives and eventually not being dependent on aid from outside is commonly referred to as a "development-oriented approach" to humanitarian crises (United Nations High Commissioner for Refugees, 2016). The so-called "development-oriented" responses to humanitarian crises, which have been popularized around the concept of Humanitarian Development Nexus (HDN), claim that development assistance can be used in responding to humanitarian crises and emergencies in a way that creates long-term "win-win" economic outcomes for both victims and donors (Betts *et al.*, 2017, p.14). It is argued that development-oriented approaches to humanitarian crises go beyond a focus on short-term temporary measures and provide sustainable solutions. In simplest terms, these approaches offer to empower the victims of humanitarian crises by building self-reliance, instead of being dependent on donors- an offer that sounds attractive at the first glance from a humanitarian and development point of view.

In the context of the recent global refugee crisis, for example, the leading donors have adopted development-based approaches, which considered the refugee crisis as an 'opportunity' for both refugees and host countries by promoting the skills and capabilities of the refugees and integrating them into labor markets. The discussions on addressing humanitarian crises by using long-term solutions rather than simply providing emergency relief are not new. We have been repeatedly reminded by the leading donors, especially the World Bank and the UN agencies, that long-term protracted humanitarian crises require development-based approaches that attack the roots causes of these crises rather than simply providing short-term emergency relief with limited humanitarian aid resources[2]. What is new is that aid strategies and interventions implemented within the framework of the HDN are increasingly focusing on employment generation and labor market accession for humanitarian victims. This chapter aims to explore how these development-oriented humanitarian aid practices are related to the attempts to incorporate humanitarian victims into labor markets. In this context, it also pays particular attention to the role of global value chains and how the so-called "value chain approaches" are used in development-oriented humanitarian interventions.

THE HUMANITARIAN-DEVELOPMENT NEXUS: IS IT OLD WINE IN A NEW BOTTLE?

Given the prominent role of humanitarian and development aid in the contemporary world, it is easy to forget that institutionalization of both is a recent phenomenon. Development aid, as we know it today, has its origins in the Cold War.[3] During the Cold War period[4], the reconstruction of wartorn countries and long-term development of newly independent countries were the main agendas of donors, while humanitarian action was largely left to non-governmental organizations (Osa & Hanatani, 2018, p. 4). It was only after the end of the Cold War that collective humanitarian support by the international community became fully functional through the UN. The institutionalization of today's humanitarian aid architecture can be traced to the adoption of the UN General Assembly Resolution 46/182 in 1991, which laid down the principles and framework for humanitarian assistance and continues to guide the work of the international humanitarian system (Gomez & Kawaguchi, 2018, p. 17). However, there were some important precedents. Countries have always engaged in the voluntary transfer of resources for developmental, humanitarian, military, and political purposes. Many of the ideas and practices that constitute humanitarian and development aid go back a lot further. Some scholars trace the origins of humanitarian and development aid in the 19th century, whereas some others stretch humanitarian and development aid back too far and provide examples of technical and humanitarian aid from ancient history, which are out of the scope of this chapter.[5]

The most comprehensive work undertaken to develop definitions of what constitutes foreign aid (what counts as aid and what does not) has been led by the Development Assistance Committee (DAC) of the Organization for Economic Cooperation and Development (OECD). The OECD-DAC does not provide aid but seeks to harmonize the development aid policies of its members. Its actions have focused on monitoring, assessing, and reporting the provision of official development assistance. Even though the specialized agencies of the the United Nations, the International Monetary Fund, and the World Bank are also engaged in this work, the DAC has been the primary forum in building consensus among donors about basic definitions and statistical norms of development assistance. Official Development Assistance (ODA) is defined by the OECD Development Assistance Committee (DAC) as "government aid that promotes and specifically targets the economic development and welfare of developing countries" (OECD, 2021). According to the OECD-DAC's reporting directives, humanitarian aid is a sector of ODA that aims specifically to "save lives, alleviate suffering and maintain and protect human dignity during and in the aftermath of emergencies" (OECD, 2012). It includes disaster prevention and preparedness, reconstruction relief, relief coordination, protection and support services, emergency food aid, and other emergency and distress relief. As can be understood from these definitions, OECD evaluates and reports humanitarian aid as a type of development assistance. OECD-DAC's definition of ODA includes resources provided for humanitarian aid, although these types of aid do not contribute directly to economic development.

In fact, both humanitarian and development systems have always been coordinated and managed separately (see Barnett, 2013). In addition to being built on different institutions, the sources from which both systems are funded have also been different. However, the purpose of this study is to critically examine the discourses and practices that attempt to bridge the divide between humanitarian and development aid, not to trace the origins of these two aid types. Following this purpose, it is sufficient to note that the need for collaboration between humanitarian and development actors was articulated at the outset of the present humanitarian system by the UN General Assembly in Resolution 46/182 of 1991 as a "continuum from relief to rehabilitation and development," and up until today, this remains at the top of the agenda of international aid actors (Kitaoka, 2018, p. xiv). Therefore, although the idea has come to the fore again with increasing humanitarian crises in recent years, attempts to bridge humanitarian aid and development aid are not new. Shusterman (2020) outlines four successive "waves" in the debate about integrating humanitarian and development action since the 1990s:

... The first wave was technical and assumed a clean transition from humanitarian to development in the early 1990s. The second, which did not take long to emerge, was more politically aware, probing not only 'how' to bridge the gap, but also "what for" and "why", with links to security, military intervention, and foreign policy added to the equation. A third wave materialized with the rise of "resilience" in the 2000s, and a fourth manifested with talk on the "humanitarian-development nexus", "new ways of working"... (p. 360)

During the 1990s, the debates about improving the humanitarian system revolved around the question of a disconnection between the actors and institutions of humanitarian aid and development cooperation. The humanitarian-development nexus is a continuation of long-standing efforts in the humanitarian and development fields, such as "Linking Relief Rehabilitation and Development" (LRRD), and the "gap approach" (Kaga & Nakache, 2019). The starting point of the discussions was the recognition that there were gaps between humanitarian and development systems in terms of funding, institutions, strategies, and coordination. Initially, many of the challenges on the so-called "relief–development continuum" were primarily seen as technical (Harmer & Macrae, 2004). The LRRD, which was originally proposed by the European Union (EU) in 1996, is based on the idea of a chronological transition from short-term relief by humanitarian actors to long-term development by donors. Underlying the conceptualization of LRRD was a linear one-way transition from a phase of relief to a phase of development (Levine & Mosel, 2014). Although its meaning was not clear, 'linking" meant providing relief in times of disaster and preparing the ground for the next phase of aid aimed at long-term development. Far from clarifying how relief can be linked to development, the focus of the LRRD approach shifted in the late 1990s, and particularly after the September 11 attacks in 2001 to linking aid with political and security objectives Far from clarifying how humanitarian aid will be linked to development, the focus of the LRRD approach shifted in the late 1990s, and particularly after the September 11 attacks, to linking aid closely with political and security objectives (Harmer & Macrae, 2004). From then on, besides the issue of linking humanitarian aid to development aid, efforts to link both to security and foreign policy came into the equation.

Another major multilateral effort to bridge the division between humanitarian and development sectors was the "gap approach" adopted by the World Bank and the United Nations High Commissioner for

Refugees (UNHCR) in 1999. Although the World Bank and the UNHCR had agreed to jointly develop innovative "gap-filling programs," the so-called "gap approach" has not been effectively implemented in practice (Gomez & Kawaguchi, 2018, p. 21). During the 1990s, there were also various bilateral attempts to link humanitarian and development aid. For example, the United States Agency for International Development (USAID) created the Bureau for Humanitarian Response in 1992 and, in the same year, the United Kingdom Overseas Development revised its organizational structure to cover relief assistance (Osa &Hatanani, 2018, p. 10).

While it was more or less clear that life-saving responses to emergencies implied humanitarian aid and that long-term initiatives to improve living standards constituted development aid, what kind of gap or disconnection existed between them and how to address it had not been put forward clearly in these approaches. Besides, discourses on linking humanitarian aid with development were far from being supported by strong institutional structures and solid cooperation mechanisms. The issue was further diluted after the events of September 11, when concerns about aligning both humanitarian and development aid with security and foreign policy objectives, rather than making them compatible with each other, came to the fore.

Moreover, as from the early 2000s, the resilience approach, which requires the simultaneous involvement of humanitarian and development actors, began to emerge. The European Union defines resilience as "the ability to withstand, cope, adapt, and quickly recover from stresses and shocks such as violence, conflict, drought, and other natural disasters without compromising long-term development" (European Commission, 2016). The resilience approach was suddenly embraced, at least on paper, by the leading donors. The official development cooperation agency of the United Kingdom, Department for International Development (DFID)[6], stated that the creation of resilience would be placed at the heart of its approach both to longer-term development and to humanitarian response (DFID, 2011a). Subsequently, the United States Agency for International Development, UN agencies, the EU institutions, and international Non-Governmental Organizations (NGOs) published policies, guidelines, and programs aimed at building resilience (Anholt & Sinatti, 2020, p. 312). However, the OECD points out that the resilience approach has proven difficult to translate into practice and implementation has been weak due to the lack of coordination and cooperation among the leading donors. (OECD, 2016).

In the last 10 years or so, an increasing number of protracted crises, such as the Syrian refugee crisis, have provided a further impetus for promoting coherence and complementarity among humanitarian and development actors and have been popularized around the concept of Humanitarian-Development Nexus (Zetter, 2019). From one point of view, the revival of the debates and efforts to harmonize humanitarian and development actors can be considered as a situation of having an 'old wine in a new bottle.' Compared to the previous approaches, however, the nexus approach has goals that are more concrete and is more ambitious in creating institutions and mechanisms in line with these goals. But this is not about the nexus approach being a more successful and sophisticated approach to harmonizing humanitarian and development assistance or promising to contribute to the more cohesive working of these two sectors.

The difference of the nexus approach is related to the specific meaning attributed to development in development-oriented humanitarian aid and the aid strategies and practices that have started to emerge in line with this definition. As will be discussed in what follows, the 'development' in the contemporary development-oriented humanitarian responses is limited to the inclusion of humanitarian victims in the labor market as cheap and flexible labor. From this point of view, having participated in the labor markets through value chains is presented as a durable solution for the victims of humanitarian crises, frequently without any attention being paid to extremely low wages and exploitative working conditions.

HUMANITARIAN CRISIS AS A 'DEVELOPMENT OPPORTUNITY': A JOURNEY FROM CRISIS ZONES TO SPECIAL ECONOMIC ZONES

Victims of humanitarian crises are no longer considered merely as objects of humanitarian care. Increasingly they are portrayed as a productive workforce whose formal integration into labor markets simultaneously can create self-sufficient actors and can contribute to development. A relevant example that illustrates this point is Betts and Collier's (2017) suggestion that the Syrian refugee crisis could be addressed through the establishment of Special Economic Zones (SEZs) near refugee camps in the countries neighboring the conflict areas, where forcefully displaced Syrian refugees could be provided with job opportunities. What these authors suggest is to provide multinational companies with tax incentives to encourage them to invest in these SEZs and make use of the Syrian refugees as a cheap source of labor. SEZs, according to the authors, would provide the Syrian refugees with "autonomy and opportunity," while also contributing to the development efforts of the refugee-hosting country, and thus creating a "win-win-win" situation for refugees, refugee-hosting countries, and multinational companies (Betts & Collier, 2017, pp. 156-181). For instance, policy recommendations of these authors to the Jordanian government on how to make use of the refugee crisis to industrialize and better integrate into the global economy is illustrative of their opportunistic approach to humanitarian crises:

At present ...Jordan cannot compete with low-income countries for cheap labor, nor can it compete with advanced economies on technology and innovation ...To industrialize, then, Jordan needs a small number of major businesses and a large number of skilled laborers to relocate to manufacturing clusters. The refugee crisis offers Jordan the chance to make this transition. Refugee camps and some urban areas could be reconceived as industrial incubator zones, where displaced Syrians could gain access to education, training, and the right to work (Betts & Collier, 2015, p. 86).

Similarly, SDZ Alliance, a non-profit/NGO joint venture of individuals and organizations, offers building the so-called 'Sustainable Development Zones' (SDZs) as a solution to the global refugee crisis. SDZs are defined as areas with distinct public and labor laws that are partially exempt from the legal regulations of the refugee-hosting countries in which they are located, and are claimed to be designed to help refugees stand on their own by providing jobs and protection (Nooke, 2019). The so-called Sustainable Development Zones are promoted as a win-win for both refugees and people of the hosting countries:

SDZs serve as islands of good governance. They offer protection as well as humanitarian assistance and social services for refugees and other migrants, but they also leverage mass displacement for the benefit of the local population by encouraging entrepreneurship, self-sufficiency, and new investment in jobs and infrastructure (Rücker, 2018).

These proposals are far from original and typical of the neoliberal worldview, which treats even catastrophic events and disasters as opportunities for entering new markets, making new investments, and exploiting the cheap labor of the vulnerable people. However, what makes it worth mentioning here is that their suggestion has been recently translated into action by some of the leading aid agencies. In March 2016, The World Bank, in partnership with the UK's Department for International Development launched a pilot project that involved the establishment of SEZs near Jordan's border with Syria (DFID, 2011b). The World Bank President asserted that the project aimed to provide the Syrian refugees with more jobs by attracting foreign investors to these special economic and industrial zones (World Bank 2017, March 27). The DFID administration, on the other hand, summarizes the purpose of the project as "turning refugee crisis into a development opportunity," and considers the project as a "paradigm shift" in the aid architecture, representing a transition from simply meeting immediate basic needs to a long-term "development-oriented approach" to emergencies (DFID, 2016). What this so-called development-based approach to humanitarian crisis offers is to get the forcefully displaced people out of refugee camps and put them in special economic zones as cheap labor, while encouraging the private sector to exploit the skills and productive labor of displaced people. Low-paid jobs without social rights and protection in these special economic zones are presented by DFID as an opportunity that "brings hope, dignity, and a more sustainable future" for refugees (see DFID, 2016).

This so-called development-based approach to humanitarian crises is also advocated by the OECD, in the context of the recent global refugee crisis. OECD suggests that refugees, who are often perceived as a burden for the host country, should be considered as an opportunity because they can "contribute to development by providing skills and resources, and spurring production capacity" (OECD, 2017a). The OECD considers forcibly displaced people as "potential assets for local growth and development" and encourages donors to invest in aid projects on vocational training to improve the skills and competencies of refugees (OECD, 2017b, p.13). In a like manner, the European Union, as an aid donor, has played a prominent role in the efforts to draw the forcibly displaced populations into labor markets as qualified labor. In response to the Syrian refugee crisis, the European Union established a regional trust fund to pool financial contributions from EU members, non-EU donor countries, the UN agencies, and some private aid organizations. The purpose of this trust fund, called the Madad Fund, is described as "provid[ing] coherent, comprehensive and joint aid response to the manifold and increasing needs arising from the Syrian crisis across the region" (The Agreement Establishing The EU Regional Trust Fund in Response to the Syrian Crisis, 2016). Madad Fund supports projects especially in the field of employment generation and vocational training for the forcibly displaced people in Egypt, Iraq, Jordan, Lebanon, Turkey, and the Western Balkans (OECD, 2017b, p. 77). By merging various financial instruments into one single mechanism under Madad Fund, the primary solution offered by the EU to the problems arising from the refugee crisis is to help forcefully displaced people develop new skills through training and find jobs in the host countries. Solutions to the complex and multi-dimensional issue of global forced displacement are simply reduced to employment generation strategies for drawing forcefully displaced refugees into labor markets.

The leading donors have been increasingly using aid to reduce the pressure for mass migration to the developed countries. In this regard, employment generation and facilitating accession to labor markets are gradually becoming an aid model for dealing with the refugee crisis. In recent years, there is an increase in the number of 'mutually beneficial' aid projects, where aid resources are provided to the governments of the refugee-hosting developing countries on the condition that they provide the refugees in their boundaries with work permits and employment opportunities. Following the example of the establishment of SEZs in Jordan, the World Bank, the European Union, and the UK's DFID decided to provide financial support for the construction of industrial parks in Ethiopia to provide job opportunities for the refugees in this country (BBC, 2016). Aid is provided to the Ethiopian government on the condition that jobs are provided to the refugees in these industrial parks.

GLOBAL VALUE CHAINS: ARE THEY A REMEDY FOR HUMANITARIAN VICTIMS?

The leading donors have not only been concerned with improving the productive labor of humanitarian victims, but they have been equally concerned with delivering them to the hands of capital, as evident from their efforts to complement their development-based humanitarian responses with value chain approaches. In the early postwar period, international trade was based on a classical division of labor, where the developed countries produced and exported finished products and imported raw materials, while the poor countries mostly exported primary goods and imported manufactures (Selwyn, 2018, p. 56). However, the organization of international production and trade has changed significantly in the last four decades and production has become globally integrated today. As of the 1970s, multinational companies began to move their production processes to low-wage developing countries to restore profit rates. The low labor cost was the main reason for this relocation.

This development was part of an effort to overcome the profitability crisis and to re-establish the conditions for capital accumulation. This change in global production has an obvious quantitative dimension, which can be seen in the considerable rise in trade and foreign direct investments in recent decades. However, the change in the structure and organization of global production is even more significant. Today, the entire production process in various industries is fragmented and geographically spread within and between countries, on a global scale. Value chains are organized and coordinated by a lead firm, usually a multinational company headquartered in a developed country. These so-called lead firms exercise power over their supplier firms by setting the rules for participation in the value chain. These functionally fragmented global value chains illustrate "the ultimate division of labor, in which every individual step in a manufacturing process can be assigned to the most appropriate workers, anywhere in the world" (Cammack, 2015, p. 65).

In addition to emphasizing the developmental role of labor productivity, almost all aid agencies have developed so-called "value-chain approaches to development" in recent years. From the World Bank to the OECD to the leading bilateral donor institutions, "value chain approaches to development" and humanitarian responses associated with the global value chains are becoming widespread. For example, the World Bank has recently established a team of experts with a specific mandate to improve the effectiveness of aid projects and programs based on value chain approaches (Gonzales, 2016). The World Bank considers global value chains as "a powerful driver of productivity growth, job creation, and increased living standards" (World Bank, 2014). Global value chain analysis has become a significant part of aid analysis and policy.

As a recent example of these value chain approaches to development-oriented humanitarian aid, The UN Refugee Agency, UNHCR, has developed the Approach to Inclusive Market System (AIMS) in collaboration with the International Labor Organization (ILO). The AIMS seeks to facilitate the inclusion of refugees and local host communities into local value chains through interventions that improve the target groups' skills and capacities as well as develop and improve the functioning of local value chains. UNHCR, an ardent supporter of the New Way of Working framework, defines traditional approaches to improving livelihoods for refugees as approaches that focus on improving the employability, skills, and know-how of refugees (Nutz, 2017). However, according to UNHCR, while the traditional approach provides livelihoods to victims of humanitarian crises in the short-term, it is not sustainable in the long term. According to UNHCR (2018), the AIMS is based on two sets of separate but interlinked interventions. On the one hand, it aims to improve the skills and abilities of the refugees, for instance through vocational training to integrate them into local labor markets. On the other hand, it focuses on matching labor with capital's needs and requirements by analyzing the labor demands and needs of companies that participate in local value chains.

These donor-supported value chain development projects are implemented under the assumption that integration into global value chains will improve living standards and reduce poverty. However, apart from this general assumption about the link between value chain development and poverty reduction, there are no convincing data to show that value chain development projects have been effective on the ground. There is little research on how donors implement these so-called global value chain approaches in practice. Mainstream global value chain research is recognizing that donor-led value chain projects show little or no concern for poverty reduction or broader development goals. For example, a study sponsored by the United Nations Industrial Development Organization (UNIDO) analyzes different donor-led value chain projects in different countries and concludes that donors mainly focus more on making participating firms in value chains more efficient with little concern for poverty reduction (Folke *et al.*, 2010, p. 8). Another study, commissioned by the Institute of Development Studies (IDS), was based on a review of thirty donor-led value chain development projects in different countries. Among other findings, the study emphasizes that the prime goal of these projects is private sector development, and there is not enough evidence on poverty reduction impacts from these projects to claim that they are effective in helping vulnerable people (Humphrey & Navas-Alemán, 2010). The study maintains that donors consider integrating local supplier firms into global value chains as being a remedy for the disadvantaged people in itself since it creates employment opportunities (Humphrey & Navas-Alemán, 2010, p. 39). In this regard, humanitarian efforts are carried out indirectly by promoting employment opportunities for vulnerable populations in small and medium enterprises in developing countries.

While the mainstream development aid discourse portrays the global value chain as a sphere of opportunity for the victims of humanitarian crises, critics argue that these value chains reproduce global poverty and inequality, rather than addressing it. One important contribution in this regard comes from Selwyn, who labels global value chains as "global poverty chains" (Selwyn, 2018, p. 27). Selwyn examines global textile, food, and high-tech industries and finds that a common feature of all these prominent value chains in the global economy is that they are based on a constantly growing and extremely exploited labor force (Selwyn, 2018, pp 48-56).

Exploitation in global value chains is not limited to exploitative working conditions and low salaries. There is evidence of slavery in various stages of global value chains from the production of raw materials to the manufacturing of basic commodities (Lebaron, 2015, pp. 32-36; Lebaron, 2016, pp. 381-396). While the legislation in developing countries is not sufficient to protect the workers, anti-slavery legislation passed by governments in developed countries allows multinational companies to overlook and turn a blind eye to slave labor used by supplier firms in their value chains:

Take the UK's 2015 Modern Slavery Act. While the act requires certain companies to report the voluntary efforts they are taking to prevent or address slavery in supply chains, it leaves open the possibility for companies to report that they are doing nothing, or to report that forced labor issues are too far down their supply chain for them to reach (Lebaron, 2015, p. 34).

Extensive research has shown the relationship between the proliferation of global value chains and increases in labor exploitation and abuses. Expansion of global value chains also coincided with the increase of Export Processing Zones (EPZs), which are special industrial zones offering incentives, such as tax and labor law exemptions, to attract foreign investment. EPZs now play a significant role in the global value chain as "spaces of legal exception where production takes place beyond the bounds of 'mainland' law" (Lebaron *et al.*, 2018). It is possible to find many academic studies and media reports about exploitative working conditions in global value chains, ranging from extremely low payment, long working hours, and health-damaging work to child labor, racial or gender discrimination, and forced labor at the extreme end (see Selwyn, 2018).

CONCLUSION

Attempts to integrate humanitarian aid and development aid have almost turned into efforts to create the necessary institutions and mechanisms to put victims of the humanitarian crisis at the service of capital as cheap and flexible labor. The donor community has been trying to create an image of the capitalist market as an area of potential opportunity for victims of humanitarian crises. The so-called Humanitarian-Development Nexus has portrayed the global value chain as a sphere of opportunity and a path to self-reliance. Enabling the displaced people to use their labor is presented as an opportunity as if these victims of humanitarian crises, who are divorced from the means of production, had any chance of survival other than using their labor.

As discussed above, there is plenty of evidence to suggest that global value chains function as spheres of labor exploitation and surplus extraction that further contribute to inequality and human rights abuses at the global level, rather than providing refugees with an opportunity to become self-reliant. Multinational companies have transferred their production processes to developing countries to take advantage of low wages. They were not only trying to escape higher wages in the developed countries, but they were also trying to avoid minimum health and safety requirements and take advantage of employing non-unionized workers. Multinational companies that govern the global value chains have achieved this through the support of bilateral and multilateral aid agencies and governments of the recipient countries.

These policies have been promoted with equal enthusiasm by the OECD-DAC members, and the multilateral humanitarian and development actors such as the UN's refugee agency (UNHCR) and the World Bank. However, the role and contribution of the aid practices related to the Humanitarian-Development Nexus in encouraging and sustaining the practices of forced labor and labor exploitation in the so-called global value chains have been under-investigated. An analysis of labor exploitation would be incomplete without considering the role of international humanitarian and development aid processes. Interventions resulting from attempts to harmonize humanitarian and development aid are playing an increasingly prominent role in submitting the poor and vulnerable populations (such as refugees) to capital for severe exploitation as cheap and flexible labor.

Poor living conditions resulting from humanitarian crises are regarded as solved by integrating victims into labor markets. The role of aid at this point is to provide the basic health and education to equip victims with qualifications and skills. Once they are thrown into the labor market as sufficiently healthy and educated labor force, it is considered as an opportunity for them to live a 'better life in dignity and self-reliance.' In this regard, the so-called global value chain approaches in donor interventions ignore the low salaries, lack of welfare benefits, and working conditions characterized by super-exploitation in global value chains or remain silent about it.

What is being challenged here is not that the victims of humanitarian crises have employment opportunities and live in a self-reliant way. What is challenged here is the reduction of development-oriented humanitarian aid to strategies for labor market accession, and the discourse, which portrays exploitative and poor working conditions in the context of value chains as the key to living a life in dignity and self-reliance. In this respect, it is considered that attempts to bridge the divide between short-term relief and longer-term development aid and so-called development-oriented humanitarian aid practices should be approached with caution and a critical perspective.

REFERENCES

Anholt, R., & Sinatti, G. (2020). Under the Guise of Resilience: The EU Approach to Migration and Forced Displacement in Jordan and Lebanon. *Contemporary Security Policy*, *41*(2), 311–335. doi:10.1080/13523260.2019.1698182

Barnett, M. N. (2013). *Empire of Humanity: A History of Humanitarianism*. Cornell University Press. doi:10.7591/9780801460616

BBC. (2016, September 21). Refugee crisis: Plan to create 100,000 jobs in Ethiopia. *BBC News*. Retrieved from https://www.bbc.com/news/world-africa-37433085

Betts, A., Bloom, L., Kaplan, J. D., & Omata, N. (2017). *Refugee Economies: Forced Displacement and Development*. Oxford University Press.

Betts, A., & Collier, P. (2015). Help Refugees Help Themselves: Let Displaced Syrians Join the Labor Market. *Foreign Affairs*, *94*(6), 84–92.

Betts, A., & Collier, P. (2017). *Refuge: Transforming a Broken Refugee System*. Penguin.

Bilzen, V. G. (2015). *The Development of Aid*. Cambridge Scholars Publishing.

Cammack, P. (2015). What International Organizations Do, and Why They Do It. *Spectrum: Journal of Global Studies*, *7*(1), 62–77.

Department for International Development. (2011a). *Humanitarian Emergency Response Review*. Retrieved from https://assets.publishing.service.gov.uk/government/uploads/system/uploads/attachment_data/file/67579/HERR.pdf

Department for International Development. (2011b). *Department for International Development Annual Report and Resource Accounts 2009-10: Third Report of Session 2010-11*. London: Stationery Office.

Department for International Development. (2016). *Jordan Compact Economic Opportunities Program: Summary Sheet*. Retrieved from http://iati.dfid.gov.uk/iati_documents/5519561.odt

Dreze, J. (1990). Famine Prevention in India. In J. Dreze & A. K. Sen (Eds.), *The Political Economy of Hunger: Famine Prevention* (pp. 13–122). Clarendon Press.

Folke, L., Riisgaard, L., Ponte, S., Hartwich, F., & Kormawa, P. (2010). Agro-Food Value Chain Interventions in Asia: A Review and Analysis of Case Studies. Vienna: United Nations Industrial Development Organization (UNIDO).

Gomez, O. A., & Kawaguchi, C. (2018). A Theory for the Continuum: Multiple Approaches to Humanitarian Crises Management. In Crisis Management Beyond the Humanitarian-Development Nexus (pp. 15–36). Routledge.

Gonzalez, A. (2016). Making Global Value-Chains Work for Economic Development and Shared Prosperity. In *Speech at Global Value Chain Development Report 2016 Background Paper Conference, Beijing*. Retrieved from https://www.worldbank.org/en/news/speech/2016/03/17/making-global-value-chains-work-for-economic-development-and-shared-prosperity

Gülseven, Y. (2020). Reconsidering Development Aid: A Systemic Analysis. *World Review of Political Economy*, *11*(2), 232–255. doi:10.13169/worlrevipoliecon.11.2.0232

Harmer, A., & Macrae, J. (Eds.). (2004). *Beyond the Continuum: An Overview of the Changing Role of Aid Policy in Protracted Crises*. Humanitarian Policy Group Research Briefing. Retrieved from https://cdn.odi.org/media/documents/341.pdf

Humphrey, J., & Navas-Alemán, L. (2010). Value Chains, Donor Interventions and Poverty Reduction: A Review of Donor Practice. *IDS Research Reports*, *2010*(63), 1–106. doi:10.1111/j.2040-0217.2010.00063_2.x

Kaga, M., & Nakache, D. (2019). *Protection and the Humanitarian-Development Nexus: A Literature Review* (Working paper No. 1). Retrieved from https://carleton.ca/lerrn/?s=Protection

Kitaoka, S. (2018). Foreword. In Crisis Management Beyond the Humanitarian-Development Nexus (pp. xiv-xv). Routledge.

LeBaron, G. (2015). It's Time to Get Serious about Forced Labour in Supply Chains. *Forced Labour in the Global Economy, Beyond Trafficking and Slavery Series*. https://cdn-prod.opendemocracy.net/media/documents/BTS-2-Global-Economy.pdf

LeBaron, G. (2016). Slavery, Human Trafficking, and Forced Labour: Implications for international development. In The Palgrave Handbook of International Development (pp. 381-398). Palgrave Macmillan.

LeBaron, G., Howard, N., Thibos, C., & Kyristis, P. (2018). *Confronting Root Causes: Forced Labour in Global Supply Chains*. https://eprints.whiterose.ac.uk/126167/1/Confronting_Root_Causes_Forced_Labour_In_Global_Supply_Chains.pdf

Levine, S., & Mosel, I. (2014). *Remaking the Case for Linking Relief, Rehabilitation, and Development: How LRRD can become a practically useful concept for assistance in difficult places*. Humanitarian Policy Group Overseas Development Institute Report. Retrieved from https://odi.org/en/publications/remaking-the-case-for-linking-relief-rehabilitation-and-development/

Nooke, G. (2019). *Migrant Cities instead of Refugee Camps*. SDZ Alliance. Retrieved from https://sdzalliance.org/migrant-cities-instead-of-refugee-camps

Nutz, N. (2017). *A Guide to Market-Based Livelihood Interventions for Refugees*. Project documentation. UNHCR. Retrieved from https://www.unhcr.org/594b7d7f7.pdf

OECD. (2012). *Towards Better Humanitarian Donorship: 12 Lessons from DAC Peer Reviews*. Retrieved from https://www.oecd.org/dac/peer-reviews/12lessons.pdf

OECD. (2016). *OECD Resilience Systems Analysis: Turning Strategy into Action.* Retrieved from https://www.oecd.org/dac/conflict-fragility-resilience/publications/Resiliance%20brief.pdf

OECD. (2017a). *Assessing the Contribution of Refugees to the Development of Their Host Countries.* OECD Development Center. Retrieved from https://www.oecd.org/officialdocuments/publicdisplaydocumentpdf/?cote=DEV/DOC(2017)1&docLanguage=En

OECD. (2017b). *Addressing Forced Displacement through Development Planning and Cooperation Guidance for Donor Policy Makers and Practitioners.* OECD Publishing.

OECD. (2021). *What is ODA?* Retrieved from https://www.oecd.org/dac/financing-sustainable-development/development-finance-standards/What-is-ODA.pdf

Osa, Y., & Hanatani, A. (2018). Introduction: Addressing the Humanitarian-Development Nexus since the Cold War. In Crisis Management Beyond the Humanitarian-Development Nexus (pp. 3–14). Routledge.

Rücker, J. (2018). *Sustainable Development Zones: Tools for Economic Development for Communities in Mass Displacement Scenarios.* Politas Consulting. Retrieved from https://refugeecities.files.wordpress.com/2018/07/sdzconcept-proposal-071118-final. pdf

Selwyn, B. (2018). *The Struggle for Development.* Polity Press.

Shusterman, J. (2020). Gap or Prehistoric monster? A history of the Humanitarian–Development Nexus at UNICEF. *Disasters, 45*(2), 355–377. doi:10.1111/disa.12427 PMID:31799696

The Agreement Establishing the European Union Regional Trust Fund in Response to the Syrian Crisis. (2016). http//ec.europa.eu/neighbourhood-enlargement/sites/near/files/neighbourhood/countries/syria/madad/20160526-revised-madad-fund-constitutive-agreement.pdf

UN High Commissioner for Refugees. (2016). *Better Protecting Refugees in the EU and Globally: UNHCR's Proposals to Rebuild Trust through Better Management, Partnership, and Solidarity.* Retrieved from https://www.refworld.org/docid/58385d4e4.html

UN High Commissioner for Refugees (UNHCR). (2018). *Refugee Livelihoods and Economic Inclusion - 2019-2023 Global Strategy Concept Note.* Retrieved from https://www.unhcr.org/publications/operations/5bc07ca94/refugee-livelihoods-economic-inclusion-2019-2023-global-strategy-concept.html

UN High Commissioner for Refugees (UNHCR). (2021). *Global Trends: Forced Displacement in 2020.* Retrieved from https://www.unhcr.org/statistics/unhcrstats/60b638e37/global-trends-forced-displacement-2020.html?query=global%20trends

United Nations General Assembly. (2016a). *One Humanity, Shared Responsibility: Report of the Secretary-General for the World Humanitarian Summit.* Retrieved from http://sgreport.worldhumanitariansummit.org/

United Nations General Assembly. (2016b). *The Outcome of the World Humanitarian Summit: Report of the Secretary-General.* Retrieved from https://digitallibrary.un.org/record/842411#record-files-collapse-header

World Bank. (2014). *Global Value Chains*. Issue Brief. Retrieved from https://www.worldbank.org/en/topic/global-value-chains

World Bank. (2017). *The Humanitarian-Development-Peace Initiative*. World Bank Brief. Retrieved from https://www.worldbank.org/en/topic/fragilityconflictviolence/brief/the-humanitarian-development-peace-initiative

World Bank. (2017, March 27). *World Bank Approves US$100 Million in Financing to Create 100,000 Jobs for Jordanians, Syrian Refugees* [Press release]. Retrieved from https://www.worldbank.org/en/news/press-release/2016/03/27/exceptional-financing-jordan-jobs-syrian-refugees

Zetter, R. (2019). Theorizing the Refugee Humanitarian-Development Nexus: A Political-Economy Analysis. *Journal of Refugee Studies*. Advance online publication. doi:10.1093/jrs/fez070

ADDITIONAL READING

Haslam, P. A., Schafer, J., & Beaudet, P. (2009). *Introduction to International Development: Approaches, Actors, and Issues*. Oxford University Press.

Lie, J. H. S. (2020). The Humanitarian-Development Nexus: Humanitarian Principles, Practice, and Pragmatics. *Journal of International Humanitarian Action*, *5*(1), 1–13. doi:10.118641018-020-00086-0

Neilson, J. (2014). Value Chains, Neoliberalism and Development Practice: The Indonesian Experience. *Review of International Political Economy*, *21*(1), 38–69. doi:10.1080/09692290.2013.809782

Neilson, J., Pritchard, B., & Yeung, H. W. C. (2014). Global Value Chains and Global Production Networks in the Changing International Political Economy: An introduction. *Review of International Political Economy*, *21*(1), 1–8. doi:10.1080/09692290.2013.873369

O'Brien, R., & Williams, M. (2013). *Global Political Economy: Evolution and Dynamics*. Macmillan. doi:10.1007/978-1-137-34114-3

OECD. (2019). *Lives in Crises What Do People Tell Us About the Humanitarian Aid They Receive?* OECD Publishing.

KEY TERMS AND DEFINITIONS

Global Value Chains: Global Value Chain refers to people and activities (design, marketing, distribution, and support to consumers, etc.) involved in the production of a product or service when the different stages of the production process are located across different countries.

Humanitarian Assistance: Aid that is provided to save lives, reduce suffering, and preserve human dignity during and in the aftermath of natural disasters and man-made crises.

OECD Development Assistance Committee: The committee of the Organization for Economic Cooperation and Development (OECD) that focuses on development cooperation matters. The OECD-

DAC sets international standards for development cooperation and reports the provision of official development assistance.

Official Development Assistance: In-kind and cash (soft loans and grants) transfers provided by official agencies, including states and local government, to increase development and welfare in developing countries.

Protracted Humanitarian Crisis: It refers to an environment where a significant proportion of the population is acutely exposed to disease, loss of life, and disruption of livelihoods over a prolonged period, with the state having limited or no capacity to help its people.

Special Economic Zone (SEZ): An area in a country that has its own specific administrative and legal rules and functions in line with specifically determined economic targets.

Value Chain Approach to Development: A development approach that aims to encourage the integration of developing countries into global value chains and improve the productivity of companies and workers along global value chains to reduce poverty and generate economic growth.

ENDNOTES

1 The UN Secretary-General and the heads of UNICEF, UNHCR, WHO, FAO, UN OCHA, WFP, UNFPA, UNDP together with World Bank and the International Organization for Migration, signed at the World Humanitarian Summit a Commitment to Action document, in which they agreed on a New Way of Working in humanitarian crises. The aim of the New Way of Working is described as not only meeting humanitarian needs, but also to decreasing needs, risks and vulnerability over time.

2 In 2017, the World Bank and the United Nations launched an initiative titled The Humanitarian-Development-Peace Initiative to increase collaboration between humanitarian and development actors. The stated aim of the initiative is to combine short-term emergency responses with longer-term, socio-economic solutions, humanitarian crises. For detailed information on The Humanitarian-Development-Peace Initiative, see World Bank (2017).

3 For a detailed description and analysis of the emergence of the international aid architecture in the Cold War, see Gülseven (2020).

4 The Cold War was a form of global social conflict between states and social forces associated with the rival social systems of capitalism and socialism. Although historians disagree on its starting and ending points, the period is generally considered to cover the period between the Truman Doctrine in 1947 and the Dissolution of the Soviet Union in 1991.

5 See, for example, Gerard Van Bilzen (2015, pp. 50-55), Dreze (1990, p. 19), Picard & Buss (2013, pp. 13-19). Bilzen (2015, p. 50) states that one of the earliest recorded humanitarian aid practices occurred in 226 BC when a huge earthquake devastated Rhodes and the nations around the Mediterranean sent food and equipment to the victims. He also notes that, in an early example of overseas technical aid in 300 AD, a number of Koreans were sent to China from Japan to teach the art of weaving and preparing silk for production. Dreze (1990) states that one of the oldest treatises on food aid by governments was written more than two thousand years ago in India. Picard & Buss (2013, p. 14) state that Alexander the Great provided Egypt with technical aid during the founding of the city of Alexandria.

6 The UK's Department for International Development (DFID) closed on 2 September 2020 and merged with the Foreign and Commonwealth Office (FCO) to create the Foreign, Commonwealth and Development Office.

Chapter 2
Adaptation Efforts and Policy Guidelines for Bangladesh at Global Warming of 1.5°C, 2°C, and 4°C:
Adaptation Efforts and Policy Guidelines for Bangladesh at Specific Warming Levels

Sujit Kumar Bala
Bangladesh University of Engineering and Technology, Bangladesh

GM Tarekul Islam
Bangladesh University of Engineering and Technology, Bangladesh

A.K.M. Saiful Islam
Bangladesh University of Engineering and Technology, Bangladesh

Motahar Hosen
Bangladesh University of Engineering and Technology, Bangladesh

ABSTRACT

Bangladesh is going to be the worst-hit country from the climate change. The impacts of climate change at specific warming levels (SWLs) of 1.5°C, 2°C, and 4°C upon the natural and social systems of Bangladesh are assessed. Eleven regional climate models' findings show that the average Bangladesh summer temperature is likely to rise from 3.24°C to 5.77°C while the mean annual precipitation is likely to increase up to 25% by the end of the century. The yield trends of Bangladesh in terms of Boro and Aman rice are projected to gradually decrease from 2-15% at a specific warming level of 1.5°C, 5-20% at a specific warming level of 2°C, and 5-25% at a specific warming level of 4°C. A few adaptation options are proposed for different SWLs and sea-level rise. Managing climate for SWLs could be for 'short term' up to 2020, 'medium term' up to 2050, and 'long term' up to 2080. The end goal is a sustainable, resilient, and transformed Bangladesh, where additional policy documents, strategies, and action plans to mainstream adaptation to combat climate change are necessary.

DOI: 10.4018/978-1-7998-9190-1.ch002

INTRODUCTION

Industrialisation, population growth, and depletion of natural resources as a result of development are threatening our ecosystem over the globe. Climate change is the result of these developmental activities. Climate change impacts every aspect of our life – food security, supply of fish, forest, and all-natural systems. Climate change is now considered one of the most burning global issues of our planet (El Sioufi, 2010). The US National Academy of Sciences, in a review (NAS, 2010), concludes that "the Earth system is warming and that much of this warming is very likely due to human activities" and mentioned it as a "settled fact." More new studies confirmed higher levels of GHG emissions for the last decade (Peters et al., 2011). The researches at the global and regional levels demonstrated higher impacts of climate change on key production systems, natural resources, and communities (Ravindranth et al., 2012). The IPCC has projected a global average temperature rise of between 3.5°C and 6°C relative to pre-industrial times, under a high emissions scenario by the end of the 21st century (IPCC, 2013). Approximately 1.1°C of global warming has already occurred (Arias et al., 2021). All these new and ongoing studies, observations, and models suggest more threats from climate change and are thought to happen much earlier and severer than it was thought to be. The latest IPCC agreement in Paris has committed to strengthening the global response to restricting warming levels to below 2°C and aiming for 1.5°C above pre-industrial levels (UNFCC, 2015). Therefore, there is a current need to assess the impacts between 1.5°C, 2°C, or even 4°C worlds and develop appropriate adaptation options and policy guidelines considering these specific warming levels.

Climate change imbalances a country and its natural ecosystems, i.e., forests, river basins, sea level, and socio-economic systems, through agriculture, fisheries, irrigation, and power projects. Such changes come through the changing temperature patterns, rising sea levels, and the intensification of natural disasters, affecting livelihood systems concerned with agriculture, water resources, and sanitation, forests, and biodiversity. It is a well-accepted fact that the poorest are the most affected and are the least able to adapt (Parry et al., 2007; Black et al., 2011). This is due to their direct reliance on natural resources and their livelihoods. A developing country like Bangladesh is to suffer more for severe vulnerability to its nature as well as a socio-economic system due to disruption of livelihoods as a result of climate change. Victims are supposed to be the poor of the country, as the country is not developed enough in all respects. Patnaik and Narayanan (2005) focused that socio-economic systems "typically are more vulnerable in developing countries where economic and institutional circumstances are less favourable."

Bangladesh is not an exception and is one of the most vulnerable countries to climate change. Due to physiographical location, this country is going to be the worst-hit place on the planet from extreme climate events such as droughts, floods, heavy rainfall, tropical cyclones, and storm surges (Rawlani & Sovacool, 2011). There is growing evidence and feel that Bangladesh is already undergoing the effects of climate change (Rahman et al., 2009). The densely populated delta of the Ganges-Brahmaputra-Meghna rivers is highly prone to hydroclimatic extremes such as river floods and storm surges. As per climatic predictions, climate change is very likely to affect the hydrology in the region (Gain et al., 2011; Ghosh and Dutta, 2011), and as the country embraces the confluence of the Ganges-Brahmaputra-Meghna (GBM) basin, so increase in flood hazards is also very likely to occur (Olsen et al., 2015). Climate change is generally expected to lead to a greater risk of fluvial (and pluvial) flooding in Bangladesh, but possibly also in more drought situations. This is a major societal concern as more than 40 million people live in flood-prone areas. The impacts of climate change are already evident worldwide, but the extent depends mostly on development factors like poverty, social stratification, political perplexity, etc. (Ayers et al.,

2014). We are feeling, touching, and tasting climate changes and are to taste them further in the future with the rising specific warming level like 2°C and above. The average temperature has risen by around 1°C over the last century. We are seeing fewer cold winters and hotter summers, rising sea levels on the coast by around 3 mm a year, and feeling the changing rainfall patterns. The frequency of heavy rainfall, flooding, and lightning over several parts of Bangladesh is on the increase and can be due to extreme climatic conditions. Bangladesh is a poster child of the potential impact of climate change.

Those changes are already affecting the agriculture sector of Bangladesh by flooding, drought, and wind storm. Food production and food security are most vulnerable to rising air temperatures (Wassmann, et al., 2009a). In terms of risks of increasing heat stress, current temperatures are already approaching critical levels during the susceptible stages of the rice plant in Bangladesh in March-June (Wassmann, et al., 2009a). Seawater inundation has become a major problem for traditional agriculture in Bangladesh (Ensor & Berger, 2009). The vulnerability to tropical cyclones in Bangladesh is also significant (Ali, 1999), and this vulnerability would increase due to climate change (Dasgupta, et al., 2010) (Karim & Mimura, 2008). This impact will be more with the addition of sea-level rise. It is predicted that 27 million people will be at risk for sea-level rise by 2050, and this has been more than double since 2008 (Wheeler, 2011). Cyclonic storm surges are the reason behind most extreme water level events along the coastal Bay of Bengal (Chiu & Small, 2016; Antony et al., 2016). Land subsidence - a process relatively unexplored in this region - enhances the natural complexity even further (Becker et al., 202). Future climate change is not certain. No doubt this country will face increased temperature, and the changes in temperature are not the same at the periods of 2030, 2050 and 2080, etc. As this climate change continues, so the levels of impacts and exposure of vulnerabilities would be different.

The frequency of heavy rainfall, flooding, and lightning over several parts of Bangladesh is on the increase and can be due to extreme climatic conditions. Bangladesh is a poster child for the potential impact of climate change. It is a least developed country, recognised by the UNFCCC as one of the most vulnerable countries to climate change impacts. Among other things, cyclones, floods, and saltwater inundation already threaten the livelihoods of some of the world's poorest people. 79% of Bangladeshis live in rural areas, yet the country has one of the world's highest population densities, 1015 people/km^2, as per the Population and Housing Census 2011 (BBS, 2011). The increased intensity and frequency of hazards under future climate change scenarios is a major challenge for development and a significant barrier to its vision of eliminating poverty and becoming a middle-income country by 2021. An integrated flood risk assessment that connects the physical flood hazard with socio-economic aspects is necessary (Roy et al., 2021).

DATA AND METHODS

Review of High-End Climate Impact on Bangladesh

Climate change impacts need to predict for design plans and programs to adapt to future conditions. The climate change impact systems must be modelled and simulated. For that, we need to venture into the future climatic world equipped with the best scientific tools, techniques, and knowledge to ascertain the changed climate evidence of the warming world and take risk-based decision approaches to meet and decide how to respond. Helix – High-end Climate Impacts and Extremes is a world-class independent research project funded by the European Union's Seventh Framework Program for research, technological development, and demonstration under grant agreement no 603864 to model and simulate future specific warming levels and their impacts upon the natural and social system of Bangladesh. Bangladesh University of Engineering and Technology (BUET) is one of the 16 international partners carrying out future climate change research considering regional contexts of Bangladesh climate for mentioned specific warming level of climate under the Helix project. This kind of methodological research for Bangladesh is almost the first to carry out for the country. Helix is to assist decision-makers and the research community in making adaptation to the changing climate more understandable and manageable by providing a set of credible, coherent, global, and regional views of different worlds at specific levels of global warming of 1.5°C, 2°C, 4°C and 6°C relatives to pre-industrial.

Bangladesh is highly proactive in challenging climate change together with the global community and joined the 195 countries and signed the historic Paris Agreement COP21 in December 2015. Bangladesh strongly supports the implementation of the Paris Agreement on climate change and believes that "Holding the increase in the global average temperature to well below 2°C above pre-industrial levels and to pursue efforts to limit the temperature increase to 1.5°C" will require the right policies and evidence on how they are working. The UN Sustainability Development Goals and the Bangladesh Government's full supports and efforts to fully comply with these global initiatives and commitments are to only benefit the climate efforts in Bangladesh, bringing wellbeing and equality of the society and the environment. It is essential to predict climate change impacts to design plans and programs to adapt for future changes to increase resilience and reduce climatic risk as much as possible.

Climate Change Projections at Specific Warming Levels

The Helix Climate Projections for Bangladesh are projections of future changes to the climate in Bangladesh to the end of this century. These are based on the strongest climate science to date. The projections are intended to give government and other organisations evidence to help them make informed, cost-effective, and timely decisions to prepare for the changing climate. They are created to be used by people working in all sorts of organisations in Bangladesh, and the information is to be available at (www.helixclimate.eu).

Temperature and Precipitation

Eleven Regional Climate Models show that the average Bangladesh summer temperature is likely to rise from 3.24°C to 5.77°C by the 2080s (Fahad et al., 2017). Though there is uncertainty, no doubt temperature will increase in the future, and greater warming is expected in Bangladesh. The southwest and the south-central part of Bangladesh are projected to experience a greater temperature rise in the future, and a 12% increase in mean rainfall is projected. Possible changes in rainfall are also exhibited both temporally and spatially. Bangladesh is projected to face an increase of up to ~12% in precipitation during the 2020s, 20% in precipitation by 2050s, and up to 25% during the 2080s. Brahmanbaria, Dinajpur, Nilphamari, and Rangpur districts are projected to be subject to increased precipitation during the 2020s and 2050s. Six districts in the northern region will likely face increased rainfall up to 25% during the 2080s. But rainfall will decrease by up to 8.22% over the southwest (SW) of Bangladesh (Fahad et al., 2017). Spatial distribution and percent change in cold nights (TN10p), warm nights (TN90p), cold days (TX10p), warm days (TX90p) is shown in Figure 1 (Khan et al., 2020). Spatial distribution and percent change in extremely wet days (R99p), total wet day precipitation (PRCPTOT), consecutive wet days (CWD), and consecutive dry days (CDD) (Khan et al., 2020) is shown in Figure 2.

Sea Level Rise

Projected sea-level rise and inundation patterns are projected considering existing polders with an average height of 4.53 m above mean sea level. Under the high-end climate scenarios, RCP 8.5, the mean sea level estimate for Bangladesh coast is to rise will be varying from 0.50 m to 1.00 m and extreme cases (by considering subsidence and rapid collapse of Antarctica ice sheets) up to 1.50 m where % of inundation of coastal area is to be 4.3%, 8.4%, and 11.30% respectively (Figure 3). Expected inundations of the Sundarbans for sea level rises are 12% for an increase of 0.50 m, 43% for rising 1.00 m, and 60% for a rise of 1.50 m (Figure 3). It is not possible to align particular magnitudes of sea-level rise with specific levels of global warming since the full response of sea level to warming is much slower, on the order of centuries.

Extreme Events - Floods

Extreme weather events are likely to become more common as our climate changes. Both high flows and low flows for the Brahmaputra, the Ganges, and the Meghna are projected to increase. Changes in % for variation of high flows and low flows for the Brahmaputra are projected to be 4% and 3% for 1.5°C, 5% and 10% for 2°C and 22% and 22% for 4°C as shown in Figure 4 (Mohammed et al., 2017). Frequencies and intensities of extreme floods are to increase, leading to the generation of high-water levels, inundation, and flooding all over the country, especially at a few hot spots and pockets of the country. Flash floods or early monsoon floods might increase in haor areas and in some border areas due to heavy non-time occurrence of rainfall. Monsoon, however, is to be wetter and flood-generating. However, we must also continue to be prepared for both hot summer and cold winter for natural variability in Bangladesh weather.

Agriculture

The yield trends of Bangladesh in terms of Boro and Aman are projected to gradually decrease from 2 to 15% at a specific warming level of 1.5°C, 5 to 20% at a specific warming level of 2°C, and 5 to 25% at a specific warming level of 4°C (Figure 5). This impact is more on Boro crop than Aman crop. The districts of Sylhet, Sunamganj, Moulvibazar, and Hobiganj are to experience crop loss of 2% at a specific warming level of 1.5ºC, 2% at a specific warming level of 2ºC and 7% at a specific warming level of 4ºC. The Southern Region of the country is projected to taste loss of crop yield of Boro and Aman rice 8-10% at specific warming level of 1.5ºC, 8-10% at specific warming level of 2ºC and 10-12% at specific warming level of 4ºC. The south-central region of the coastal zone of Bangladesh is projected to suffer yield loss of Boro and Aman rice 10-12% at a specific warming level of 1.5ºC, 18-21% at a specific warming level of 2°C and 21-28% at specific warming level of 4ºC. The northern region of Bangladesh is to suffer yield loss of Boro and Aman rice 4% at specific warming level of 1.5ºC, 4% at specific warming level of 2ºC, and 10% at specific warming level of 4ºC. The whole country is to suffer in food loss due to specific warming of the country.

Figure 1. Spatial distribution and percent change in cold nights (TN10p), warm nights (TN90p), cold days (TX10p), warm days (TX90p). (Khan et al., 2020).

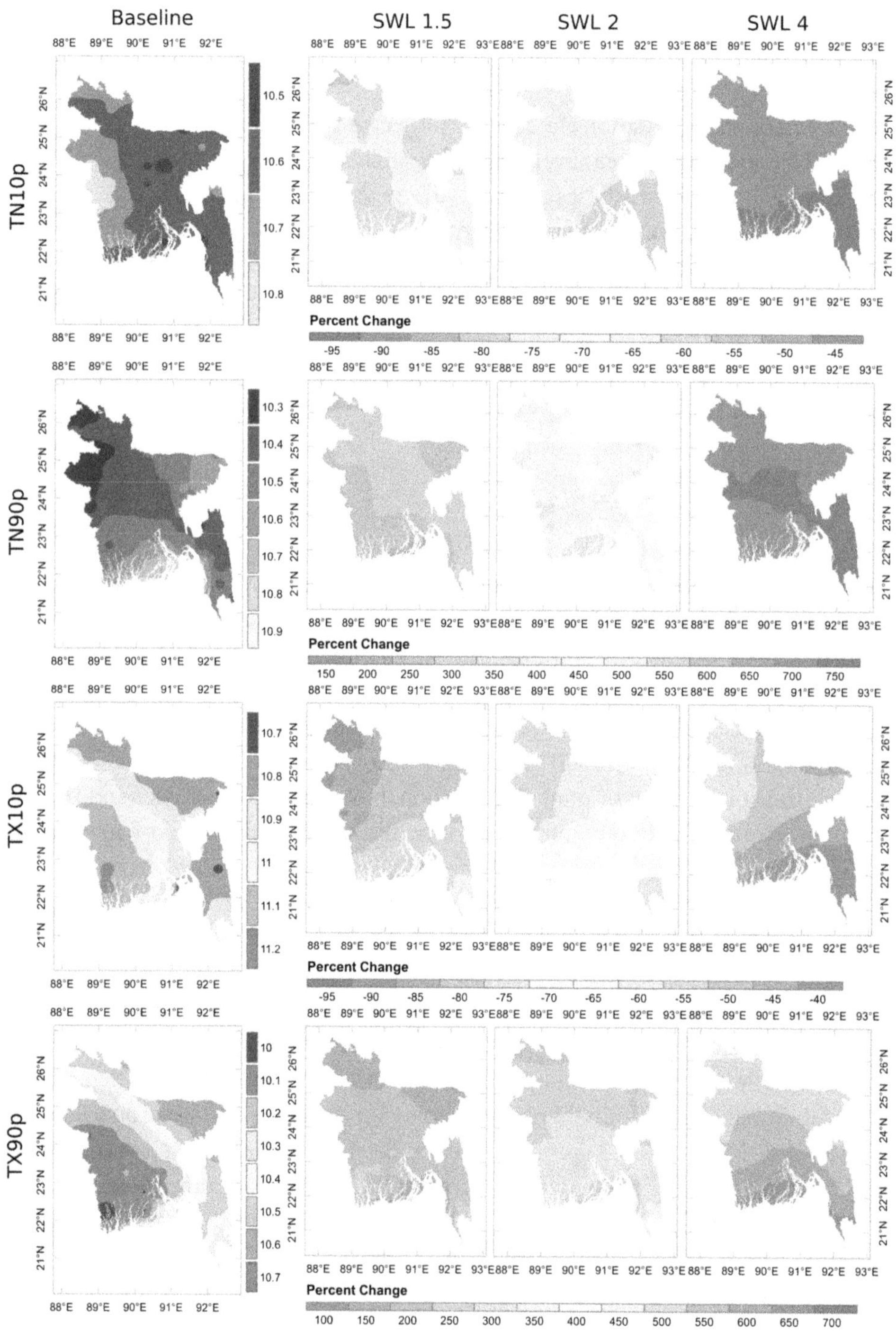

Figure 2. Spatial distribution and percent change in extremely wet days (R99p), total wet day precipitation (PRCPTOT), consecutive wet days (CWD), and consecutive dry days (CDD) (Khan et al., 2020).

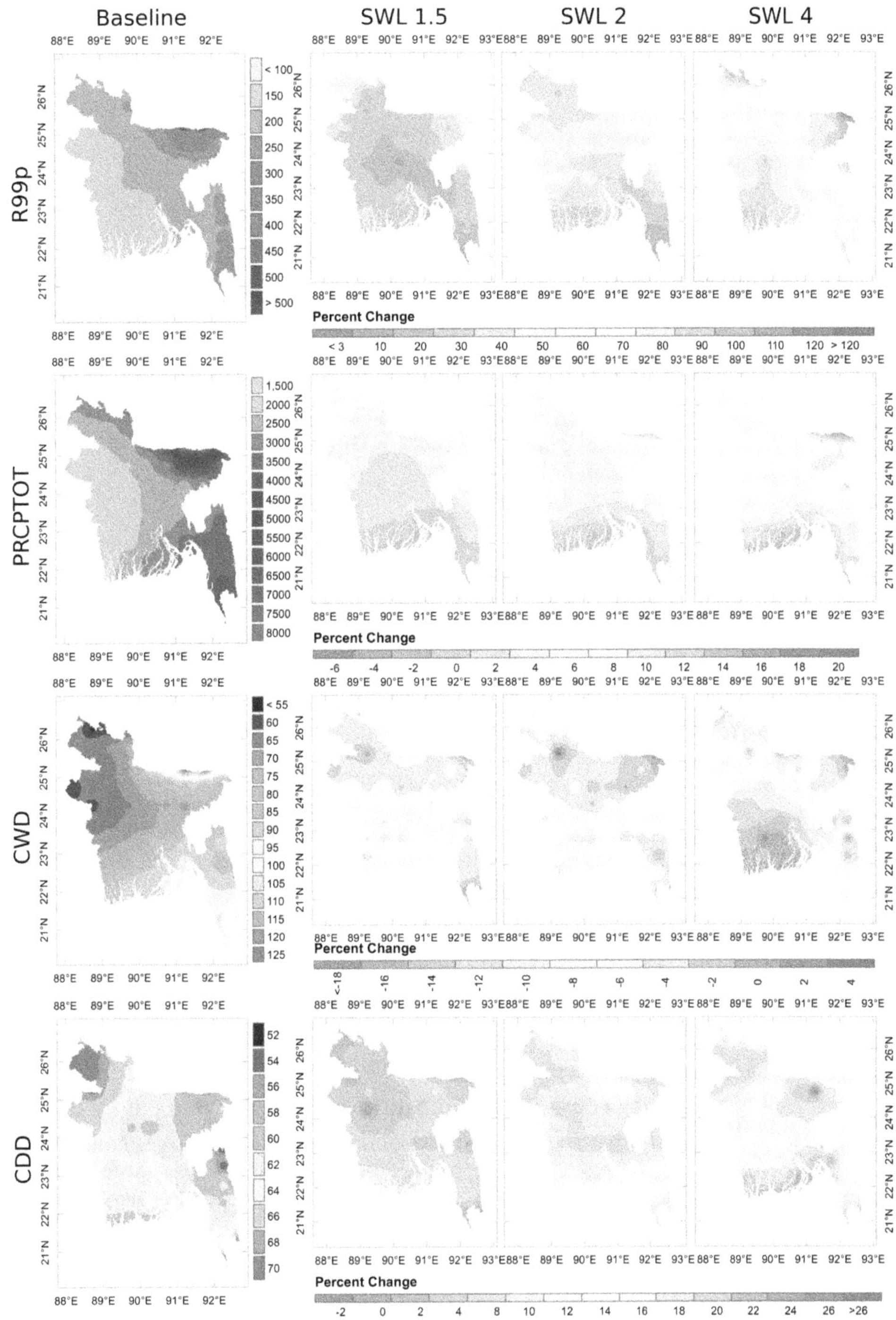

Figure 3. Inundation map for (a) 0.5 m, (b) 1.0 m and (c) 1.5 m mean SLR.

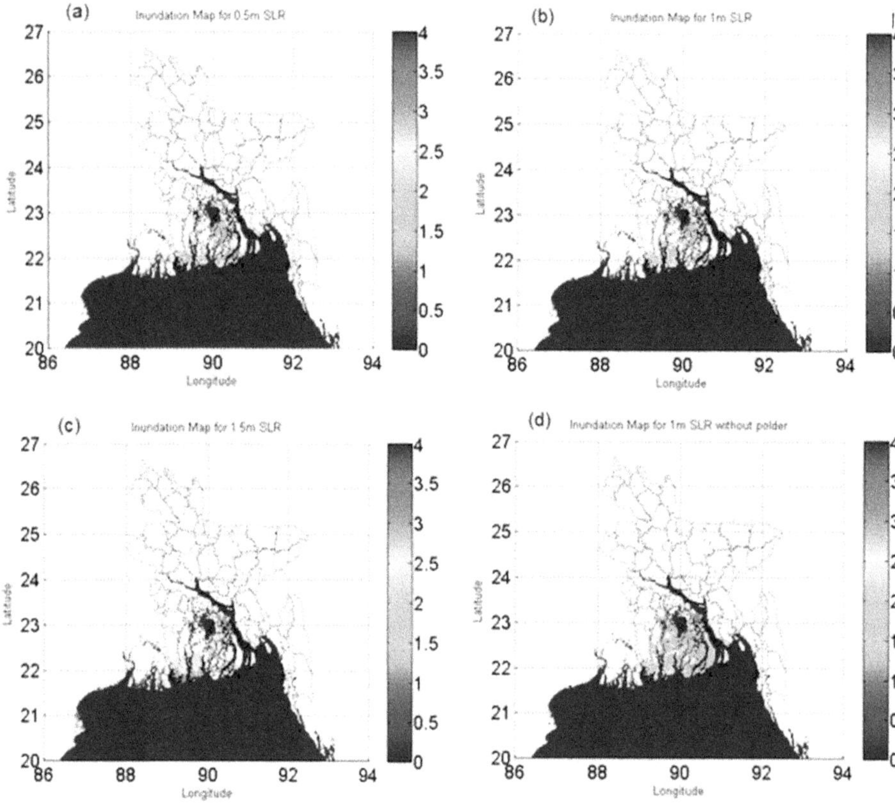

The weather and the effect on society vary across the country. The effect of climate change on our economy, infrastructure, society, and environment will vary from place to place. These will depend on how well we all plan for and adapt to climate changes.

Vulnerability for Coastal Bangladesh

Coastal Bangladesh is one of the hot spots of climatic variability, where already climate change impacts are being felt by the coastal community comprising mostly of poor layers of the society. An assessment of the vulnerability of coastal Bangladesh due to climate change was carried out by the Helix project considering future (2050) climatic change (Figure 6). The concepts of exposure, sensitivity and adaptive capacity were used for analysis of both present and future (2050) scenario vulnerability under the changing climate of the coastal belt. Composite vulnerability indexes through a translation of numerical values demonstrate the severity of future vulnerability of climate change. Severe future vulnerability (2050) of coastal Upazilas due to climate change has been increased by ten folds to 93 coastal Upazilas from 9 Upazilas. Coastal Bangladesh is comprised of 140 upzilas. So, the almost entire coastal belt would be hugely affected with high vulnerability. Similarly, the number of moderate and low vulnerable coastal Upazilas has been reduced hugely for the future scenario than the present one from 69 (sixty-nine) and 62 (sixty-two) Upazilas to 26 (twenty-six) and 21 (twenty-one) Upazilas respectively.

Figure 4. Return period curves of extreme precipitation (a), maximum discharge (b), and minimum discharge (c). The shaded regions show the range of values estimated by the 11 climate projections. The solid lines between the shades show the mean of the ensemble ranges. (Mohammed et al., 2017).

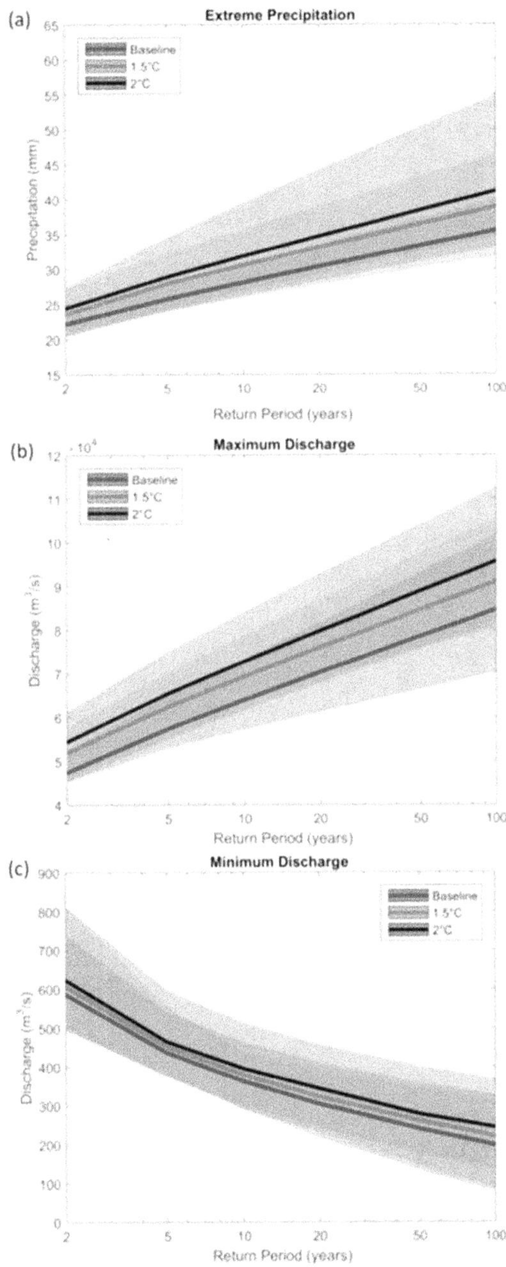

Figure 5. Spatial patterns of the changes in Boro yield at Specific Warming Levels (a) 1.5°C, (b) 2°C, and (c) 4°C.

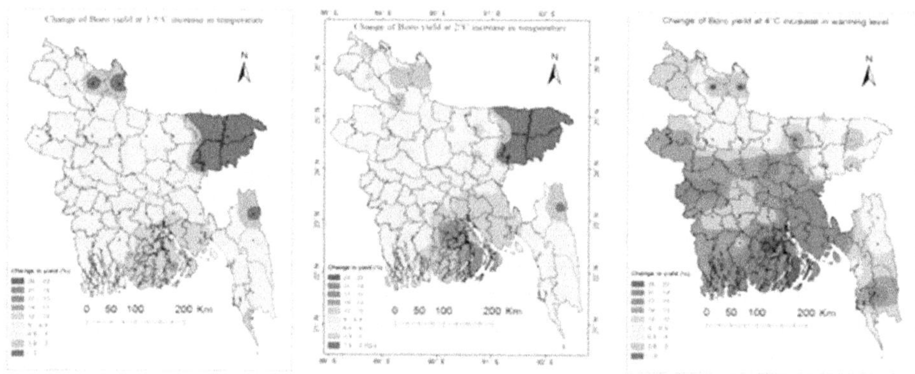

Figure 6. Climate vulnerability index (CVI) for (a) present and (b) future (2050) scenarios.

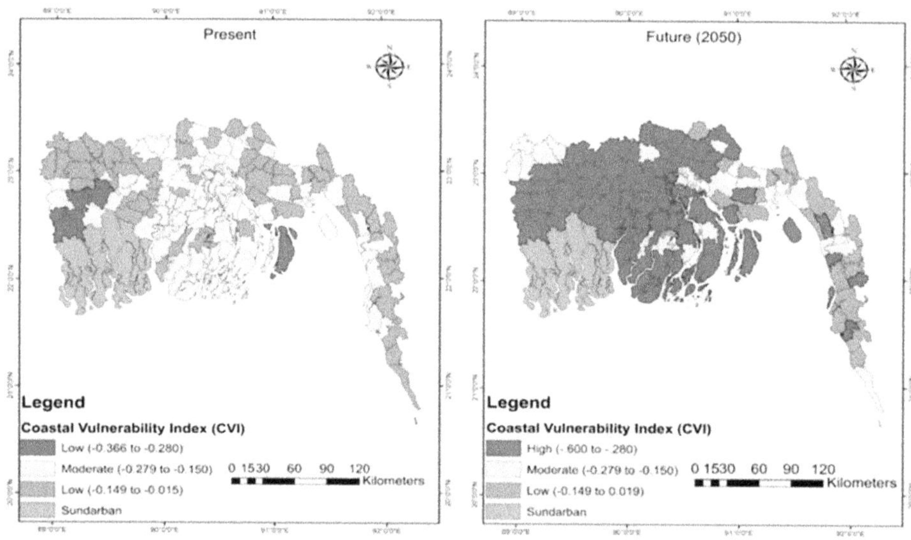

RESULTS AND DISCUSSIONS

Adaptation Options for Specific Warming Levels

The United Nations Framework Convention on Climate Change (UNFCCC) defines adaptation as "adjustment in natural or human systems in response to actual or expected climatic stimuli or their effects, which moderates harm or exploits beneficial opportunities. Various types of adaptation can be distinguished, including anticipatory and reactive adaptation, private and public adaptation, and autonomous and planned adaptation" (IPCC, 2014). Adaptation is an important tool necessary for enhancing resilience to get ready communities, regions, countries, and societies for meeting up the consequences of climate change. Adaptation is very much related to climate change bearing a characteristic of rapidly growing science embarking on almost all sectors and socio-economic setups.

The concept of adaptation has been gradually developed under the umbrella of the UNFCCC, and its importance was ascertained in the Bali Action Plan in 2008. At COP15 Copenhagen in 2009, it was agreed upon that "enhanced action and international cooperation on adaptation is urgently required to ensure the implementation of the Convention by enabling and supporting the implementation of adaptation actions aimed at reducing vulnerability and building resilience in developing countries, especially in those that are particularly vulnerable, especially least developed countries, small island developing states and Africa" (UNFCC, 2009).

The Paris Agreement emphasizes on global goal for an 'adequate adaptation response' in order to reach 'global temperature goal' through the submission of reports about progress as periodic global stocktakes (Chandni Singh et al., 2021). Chandni Singh et al. tried to define conceptual and methodological challenges of an adaptation goal and to show what effective adaptation looks like and how it can be enabled. In their review, they represent a way of exercising influence in adaptation decision-making through eleven principles as maximizing economic benefits; improved wellbeing; vulnerability reduction or adaptive capacity enhancement; enhanced resilience; sustainable adaptation; avoiding maladaptation; ecosystem-based adaptation; community-based adaptation; adaptive governance; ensuring equity and justice; transformation. They can serve as the Global Stock-take as well as national and sub-national exercises on tracking and monitoring adaptation.

Brayton Noll et al. (2021) showed that household adaptation is essential to effectively address increasing climate-induced risks. Household adaptations are in general Brayton Noll et al. carried out surveys in the United States, China, Indonesia, and the Netherlands, taking 3,789 samples and explored variations in factors shaping households' adaptations to flooding, the costliest hazard worldwide, and found that social influence, worry, climate change beliefs, self-efficacy, and perceived costs exhibit universal effects on household adaptations, despite countries' differences. Disparities also occurred in response efficacy, flood experience, beliefs in governmental actions, demographics and media, etc. So, climate adaptation policies should exercise caution because context-specific socio-behavioral drivers may discourage or even reverse household adaptation motivation.

Alexandre K. et al. (2020) discuss three scientific frontiers so that decision-makers and practitioners can take decisions and act on the design and implementation of concrete adaptation policies and actions. The frontiers are intended to go beyond i. incremental vs. transformational and ii. maladaptation vs. adaptation dichotomies and on iii. adaptation measures' effectiveness and roles in designing context-specific adaptation pathways. It was argued that the three abovementioned frontiers represent the most burning challenges to the Adaptation Science community to help address climate-related societal needs. It was also demonstrated that they are intertwined as moving one frontier forward will facilitate moving the others forward.

The Need for Adaptation

Bangladesh is the most populated country in the world and one of the poor in the third world. The poor are the most vulnerable exposed to climate change. The country is growing economically rapidly being, and economic, social, and ecological costs due to climate change are to be huge. As evident from the previous write-up, the country is set to undergo severe climate change impacts, which are to halt the developmental progress as well as enormous sufferings and loss.

So, adaptation is crucial for the survival of the country. The country also needs to keep up with the global community and its negotiations, strategies promoting adaptation to the projected effects of climate change as per Helix's findings with specific warming levels. Adaptations become more complex when warming up of globe is to surpass 1.5°C, 2°C, and 4°C bringing unimaginable consequences for the world and Bangladesh, where adaptation is the main option to take up.

The development of proper and adequate adaptation strategies in time and saving and protecting them from the effects of climate change is of utmost importance. Bangladesh bears growth-centric development strategies, and climate change issues also have followed a growth-centric pattern. The Bangladesh government has been actively working toward reducing GHG emissions and developing adaptive capacity with adequate policies and programs. Adaptation needs to be exemplified by more innovative strategies, incorporating climate change into all existing and future developmental programs of the country whilst also continuing separate policies on low carbon energy and adaptation. Here, the priority is to be given to the poor and vulnerable sectors keeping in mind the income-generating livelihoods and employment for the poor, promoting infrastructure growth, generating environmental services, conserving natural resources, and reducing vulnerability to climate risks.

The future climatic world needs to ascertain changed climate evidence due to global warming over the 21st century. Helix is to assist decision-makers and research community in making adaptation to the changing climate more understandable and manageable by providing a set of credible, coherent, global and regional views of different worlds at 1.5°C, 2°C, 4°C, and 6°C. They are to be used by people working in all sorts of organisations in Bangladesh. The future climatic world needs to ascertain changed climate evidence due to specific warming levels of 1.5°C, 2°C, 4°C, and 6°C for the 21st century.

Here some adaptation options are proposed in line with the specific warming level of future Bangladesh corresponding to 1.5°C, 2°C, and 4°C and sea level rise of 0.5m, 1m, and 1.5m. It is extremely difficult to foresee adaptation approaches for such specific warming levels for different sectors. Adaptation options are to be changed gradually before the onset of slow warming up of a specific level. Adaptation activities to be necessary for different specific warming levels are described in Table 1 to Table 5. Table 1 provides the adaptation options for the coastal region for inundation due to SLR of 0.5 m, 1.0 m, and 1.5 m. As can be seen from Table 1 that the main adaptation options are regular monitoring, evaluation, operation, and maintenance of the coastal polders; formulation of affordable insurance policies; land use planning, vulnerability zoning, introducing salt-tolerant crops; integrated agriculture-aquacultures. In the case of higher SLR, the coastal polders may need to relocate and retreat, devise enhanced and flexible insurance policies, and practice enhanced floating agriculture. Table 2 provides the adaptation options for the Sundarbans. The Sundarbans is a mangrove area in the delta formed by the confluence of the Ganges, Brahmaputra, and Meghna Rivers in the Bay of Bengal. The wetland is remarkable for protection from the tidal surge generated from the cyclonic depression in the Bay of Bengal. About one-third of the total area is used as a protected area for the conservation of biological diversity. The major adaptation options for the Sundarbans are enhanced monitoring & evaluation, operation & maintenance of coastal polders, relocation, retreat, formulation of affordable insurance policies, promoting intensive floating agriculture, creating alternative livelihoods opportunities, practicing intensive integrated agriculture-aquaculture, adoption of new technologies, and critical knowledge. Table 3 offers the adaptation options for storm surge-induced areas of Bangladesh. Bangladesh, with its repeated cycle of floods, cyclones, and storm surges, has proved to be one of the most disaster-prone areas of the world. Enhanced monitoring & evaluation, operation & maintenance of coastal polders; construction of new polders, new cyclone shelters, risk transfer through insurance—afforestation and CBA; vulnerability zoning as per new risks

Table 1. Adaptation options for inundation due to sea level rise (SLR).

Inundation	0.5 m SLR	1.0 m SLR	1.5 m SLR
Inundated coastal area except polder protected (km^2)	2000	3930	5300
% of coastal area	4.3%	8.4%	11.3%
Affected Population, million (based on based on current population)	2.5	6.0	8.0
Adaptation			
For inundation of coastal area and population	Intensive monitoring & evaluation, Operation & maintenance of coastal polders; Risk Transfer (Insurance), Land Use Planning Anew, Vulnerability Zoning, Introduction of salt-tolerant variety; Integrated agriculture aquaculture (IAA-crop-fishery-aquaculture).	Enhanced monitoring & evaluation, Operation & maintenance of coastal polders; Relocate, Retreat, Enhanced & Flexible Risk Transfer (Insurance) policy, Intensive Floating Agriculture, Alternative Livelihoods; Intensive integrated agriculture aquaculture (IAA-crop-fishery-aquaculture), Adoption of new techs and critical knowledge.	Relocate, Retreat, Enhanced & Flexible New Insurance policy, enhanced Floating Agriculture, Alternative Livelihoods via integrated agriculture aquaculture (IAA-crop-fishery-aquaculture), Adoption of new techs and critical knowledge.

zones, construction of earthen mounds, etc. can be embraced as adaptation options for the storm surged induced areas. Table 4 presents the adaptation options for agriculture. Agriculture is the mainstay of the economy of Bangladesh, which is projected to be adversely impacted by climate change. Major adaptation options that can be espoused for agriculture are plantation of local saline resistant seeds and species, bolstering the existing livelihood patterns, ensuring global funding, migration, relocation, coastal action plan and, enhanced insurance and other tools for risk transfer, aquaculture, salt-tolerant crop varieties, etc. Table 5 furnishes the adaptation options for flood-affected areas. Bangladesh is predominantly a floodplain country where about one-fourth of the country gets inundated during the normal monsoon season. The important options that can be adopted to tackle the adverse impact of climate change on the flood-affected area are raising crest levels of dykes, upgrading drainage system, raising the floor level of houses, greening roofs, floodwater retention like floodplain restoration, and land-use changes improved flood forecasting and warning system, floodplain zoning and, settlement relocation.

National Policy Efforts

Adaptation is the main step to prepare for the potential effects of climate change and will be better protected against consequences like flooding, cyclone, saltwater intrusion, etc. The earlier we plan for adaptation, the less it will cost and the better equipped we will be to cope with potential changes due to climatic hazards. Mitigation is to be done by those who are responsible for carbon emissions and to be focused upon some intended contributions as per UNFCCC norms and settings.

Specific warming level deals with the future, and it is to happen there. We are to go for the incorporation of adaptation into developmental programs so that it gets embedded in all levels of action. Adaptation is to be considered from two viewpoints - adaptation to current and future climate variability. As per Parry et al. (2008), there are three types of adaptation measures, which might be described as anticipa-

Table 2. Adaptations options for the Sundarbans.

Inundation	0.5 m SLR	1.0 m SLR	1.5 m SLR
Inundated coastal area except polder protected (km^2)	2000	3930	5300
% of coastal area	4.3%	8.4%	11.3%
Affected Population, million (based on current population)	2.5	6.0	8.0
Adaptation			
For inundation of coastal area and population	Intensive monitoring & evaluation, Operation & maintenance of coastal polders; Risk Transfer (Insurance), Land Use Planning Anew, Vulnerability Zoning, Introduction of salt-tolerant variety; Integrated agriculture aquaculture (IAA-crop-fishery-aquaculture).	Enhanced monitoring & evaluation, Operation & maintenance of coastal polders; Relocate, Retreat, Enhanced & Flexible Risk Transfer (Insurance) policy, Intensive Floating Agriculture, Alternative Livelihoods; Intensive integrated agriculture aquaculture (IAA-crop-fishery-aquaculture), Adoption of new techs and critical knowledge.	Relocate, Retreat, Enhanced & Flexible New Insurance policy, enhanced Floating Agriculture, Alternative Livelihoods via integrated agriculture aquaculture (IAA-crop-fishery-aquaculture), Adoption of new techs and critical knowledge.

tory adaptation or proactive adaptation, autonomous adaptation or spontaneous adaptation, and planned adaptation. Bangladesh is a highly vulnerable country that has been on autonomous adaptation since time immemorial. Planned adaptation takes place on a medium and long-term basis. It is a policy-based decision based on consensus to maintain or achieve a desired goal and state for a country and globe as a whole. Planned adaptation to climate change is being implemented at global, regional, national, local levels and through five-year plans by various policy programs and developmental activities on a short,

Table 3. Adaptation options for storm surges.

Inundation (Rahman et al., 2019)	Cyclone name	Only cyclone	Cyclone & 0.5 m SLR	Cyclone & 1.0 m SLR	Cyclone &1.5 m
Inundated coastal area except polder protected (km^2)	SIDR	1484	3380	5777	7588
	AILA	1999	4226	6216	7497
Affected population, (million)	SIDR	1.90	4.1	7.0	9.10
	AILA	2.3	5.1	7.5	9.0
Adaptation		Enhanced monitoring & evaluation, operation & maintenance of coastal polders; construction of new polders, new cyclone shelters, risk transfer through insurance—afforestation and CBA; vulnerability zoning as per new risks zones, construction of earthen mounds, etc.		Research-based simulation of future climate change and impacts, migration and relocation, increased polder heights, migration and relocations of human as well as habitats.	

Table 4. Adaptation options for agriculture for SWLs.

Regional Impact	Crop loss for 1.5°C	Crop loss for 2°C	Crop loss for 4°C
The districts of Sylhet, Sunamganj, Moulvibazar and Hobiganj	2%	2%	7%
Adaptation	Short-term flood-tolerant crop varieties, increased submersible embankment height, diversified agriculture, hard and soft approaches, built up high earthen spots in haor areas for cutting paddy, Improved flash flood warning & dissemination, community-based adaptation, livelihood diversifications, dredging of haors and drainage networks. Changing of sowing time, technology like rubber dam construction, floating agriculture, the introduction of tower type agriculture, community-based fishery, participatory management of submersible embankments; intensified monitoring, evaluation, operation, and maintenance of submersible embankments, risk transfer like insurance.		More focus on a technological solution like more rubber dam, new technology introduction, up to date flash flood warning and dissemination, use of effective agricultural inputs, Insurance, and other tools for risk transfer
Southern Region	8-12%	8-12%	10-12%
Adaptation	Changing cropping patterns like Boro, T. Aman/salt-tolerant jute, Wheat-Mug dal (Pulse), etc., Short duration Boro varieties, Short duration, and salt-tolerant crop varieties, low consumptive and short duration crop varieties, excavation of mini ponds, and surface water harvesting for supplemental irrigation, selection of suitable date for rice transplantation. Risk transfer like insurance, the introduction of salt-tolerant crops like jute, sunflower, etc.		More focus on the technological solution, new technology introduction, use of effective agricultural inputs, Insurance, and other tools for risk transfer
The coastal zone of south part of Bangladesh (SC)	10 – 12%	18-21%	21-28%
Adaptation	Flood resilient crop storage, food preservation, homestead gardening, salt and flood-tolerant crop varieties with short duration, changing cropping pattern, raising of homesteads, coastal vulnerability zoning, diversified livelihoods, conservation of coastal bio-diversities, insurance, and other tools for risk transfer, aquaculture, integrated agriculture, and aquaculture, etc.	Coastal vulnerability zoning, planning, and rehabilitation & relocation, plantation of local saline resistant seeds and species, bolstering existing livelihood patterns, global funding, migration, relocation, coastal action plan and, enhanced insurance and other tools for risk transfer, aquaculture, salt-tolerant crop varieties, etc.	
Northern region	4%	7%	More than 7% but not specific
Adaptation	Changing cropping patterns like Boro-T. Aman, Chilli-T. Aman, Maize T. Aman, etc., Short duration and cold-tolerant Boro rice varieties, Submerged tolerant rice varieties during Kharif season, Drought tolerant rice varieties during Kharif, Short duration pulse varieties-BARI Mug-6, Cultivation of wheat for water use efficiency, selection of suitable date for rice transplantation. Risk transfer like insurance.		More focus on the technological solution, new technology introduction, use of effective agricultural inputs, Insurance, and other tools for risk transfer.

medium, and long term basis. To go for adaptation considering specific warming levels, medium, and long-term planning is necessary.

Haq et al. (2008) grouped policy responses to climate change in Bangladesh into national and international categories. National policy responses were divided into four groups such as vulnerability reduction, disaster management, and climate risk management; mainstreaming climate change into development and national planning; and climate change policies, planning, and institutions. International policy response options are the United Nations Framework Convention on Climate Change (UNFCCC) and climate change activities in selected donor strategies.

Table 5. Adaptation options for flood for SWLs.

Impact (Mohammed et al., 2017)	Change in % for 1.5°C				Change in % for 2°C				Change in % for 4°C			
	High Flow Magnitude	**High Flow Duration**	**Low Flow Magnitude**	**Low Flow Duration**	**High Flow Magnitude**	**High Flow Duration**	**Low Flow Magnitude**	**Low Flow Duration**	**High Flow Magnitude**	**High Flow Duration**	**Low Flow Magnitude**	**Low Flow Duration**
Brahmaputra River	4	26	3	-8	5	36	10	-16	22	137	22	-50
Meghna River	9	22	8	-16	12	43	11	-21	42	107	32	-60
Adaptation	raising dykes, upgrading drainage, raising floor house, greening roofs, floodwater retention like floodplain restoration and land-use changes, improved flood forecasting and warning system, floodplain zoning, settlement relocation		Conservation and groundwater recharge, wetland restoration, Use of surface water.		raising dykes, enhancing drainage network, dredging of the river, increasing floodwater retention, capacity enhancement of flood forecasting		Conservation and groundwater recharge, wetland restoration, Use of surface water.				Conservation and groundwater recharge, wetland restoration, Use of surface water.	

Bangladesh has a good number of policy response options concerned with climate change. They indirectly address the impacts of climate change through specific actions and programs like poverty alleviation, employment generation, crop diversification that reduces vulnerability and directly through disaster risk reductions activities related to climate change and extreme events; and also, targeting climate change by mainstreaming climate change into sectoral plans and national policies. Bangladesh has and had projects addressing food insecurity and food production shortfalls by crop diversification and generation of alternative employment opportunities at community development, agricultural development, credit facilities, and infrastructure improvement. Social safety nets, fish and shrimp production for domestic consumption and exports, food for work, gender-sensitive job generations, participatory management of resources employing community, etc., are promoted for rural poverty alleviation and employment generation. All these developmental programs and activities play an important positive role in enhancing the resilience of the poor at the bottom layers of society.

Bangladesh has a Participatory Disaster Management Program (PDMP) with a focus on disaster management and prevention and also adaptation to climate change. This is a 'soft' approach to reduce the impacts of disasters, based on preparedness, like awareness raising to reduce disaster risks and losses, to strengthening capacity for disaster management; enhancing knowledge and skills in handling disasters; establishing disaster action plans in the most disaster-prone areas; promoting local level risk reduction measures; and improving early warning systems (Kelkar et al., 2007).

In 2003, Bangladesh established a Comprehensive Disaster Management Programme (CDMP), refocusing the government towards greater emphasis on disaster preparedness and risk reduction. CDMP has several disaster management components, where an integrated approach is taken to climate change and disaster management. CDMP has three main areas of focus - capacity building of the Ministry of Environment and the Department of Environment to coordinate and mainstream climate change into their existing activities; to strengthening existing knowledge and information accessibility on impact prediction and adaptation and awareness-raising, advocacy, and coordination to promote climate change adaptation into development activities.

Department of Environment (DOE) has a 'climate change cell' for government capacity building for coordination and leadership on climate change issues. The cell also coordinates awareness-raising, advocacy, and mechanisms to promote climate change adaptation and risk reduction in development activities, as well as to strengthen existing knowledge and information accessibility on impacts and adaptation to climate change. The climate change cell is assisted by the CDMP component - Local Disaster Risk Reduction Facility (LDRRF) and improves coordination at the local level. Thus, both disaster management and climate risk management (CRM) are fairly met up in Bangladesh. There might be an Integrated National Framework for CRM and DRR proposed by UNDP – a broader understanding of climate change risks and impacts at all levels, as well as capacity building for assessing risks and analysing them with sectoral and cross-sectoral perspectives and implications (UNDP, 2007).

Climate Change Policies, Planning, and Institutions

Bangladesh is a signatory to the United Nations Framework Convention on Climate Change (UNFCCC). In 1992, the Government of Bangladesh signed the UNFCCCC and ratified it in 1994. The Ministry of Environment and Forest (MOEF) is responsible for coordinating the UNFCCC process in Bangladesh.

A National Climate Change Committee, comprised of members from all relevant government and non-government organisations, was constituted in 1994 for policy and guidance and to oversee the implementation of obligations under the UNFCCC process. In addition to the Climate Change Cell, other government institutions that are relevant for climate change include an inter-ministerial committee on climate change, headed by the Minister for Environment and Forests and with representation from relevant government ministries as well as NGOs and research institutions; and a National Environment Committee to determine environmental policies chaired by the Prime Minister with representation from MPs as well as government and civil society. There are also Academic Institutes and Research Organizations that have been carrying out research studies on impacts, adaptation, and vulnerability to climate change and participated in a range of national efforts that seek to address climate change directly.

The GoB prepared the National Adaptation Program of Action (NAPA) in 2005 (MoEF, 2005). NAPA identified vulnerable areas and fifteen projects as future adaptation strategies. NAPA was updated in 2009 and identified thirty-eight adaptation measures. The NAPA process has been advanced through the adoption of the Bangladesh Climate Change Strategy and Action Plan (BCCSAP) of 2009 (MoEF, 2009), which provides the climatic contexts, analysis of socio-economic realities, and outlines policies for promoting the well-being of vulnerable groups and elaborates a set of programs based upon broad areas of interventions. BCCSAP identified forty-five adaptation and mitigation measures based on six pillars: food security, social security, and health; comprehensive disaster management; infrastructure; research and knowledge management; mitigation and low carbon development and capacity development and institutional. After approval of NAPA, UNDP received funding from Global Environment Facility (GEF) to implement the climate projects set out in NAPA.

To gain an overview of existing activities and roles in the NAP process, the MoEF, with support of the German Development Cooperation (i.e., the Gesellschaftfür Internationale Zusammenarbeit, GIZ) and the United Nations Development Program (UNDP), conducted a stock-taking and mapping exercise to inform a National Stakeholder Dialogue on Stocktaking for Bangladesh's national adaptation plan (NAP) process on Thursday, 30 March 2017 in Dhaka. Around 80 representatives from government and non-government institutions, including representatives of academia and civil society, attended the one-day event.

The GoB furthermore adopted legislation called the Climate Change Trust Fund Act in 2010 with the view to establishing the Bangladesh Climate Change Trust Fund (BCCTF) to fund activities to address the adverse impacts of climate change. The Fund is financed from the national budget of Bangladesh. The GoB also created Bangladesh Climate Change Resilience Fund (BCCRF), originally called the Multi-Donor Trust Fund, in 2009, to implement the six pillars identified in the BCCSAP. This fund became operational in 2010. Moreover, Bangladesh is also taking necessary initiatives for accessing the fund from Green Climate Fund (GCF), and the Economic Relations Division (ERD) of the Ministry of Finance is acting as the National Designated Authority (NDA) of Bangladesh to GCF. NDA has also initiated a process to get Bangladeshi institutions accredited as National Implementing Entity (NIE) to GCF and identified fourteen national entities having high potential to be eligible to fulfill the criteria set by the GCF board to gain accreditation.

The government is nurturing and boosting the public-private partnership (PPP) strategy aiming to enhance a sustainable environment as an investment sector. Bangladesh Green Development Plan (GDP) aims to develop new programs in environment, energy, and climate change that address the climate change adaptation and mitigation needs of the poor. It focuses on demand-side energy management by providing access to low-carbon fuels and energy, and it should also create green jobs with a strong focus on improving the management of the natural environment, biodiversity, and ecosystems. The government has expressed interest in using REDD+ (reducing emission from deforestation and forest degradation) as a strategy to fulfill some of these objectives, and the UN has produced a document investigating the readiness of the country to participate.

The recently developed Disaster Management Act, adopted in 2012, recognised the impacts of climate change and provided guidance for setting up an institutional mechanism for disaster management, reducing vulnerabilities, rehabilitation, and providing humanitarian assistance to the victims of both disasters and climate change impacts. In addition to these plans, strategies, and legislation, the Renewable Energy Policy 2008, the Energy Efficiency and Conservation Master Plan (EE&CC Master Plan), the Road Map of National Adaptation Plans (NAPs), the National Sustainable Development Strategy, the Perspective Plan (Vision 2021) and the Seventh Five Year Plan (FY2016 – FY2020), the National Disaster Management Plan provided some guidance for mitigation and adaption measures needed for Bangladesh.

The GoB developed the Intended Nationally Determined Contributions (INDC) and submitted them to UNFCCC well before the adoption of the Paris Agreement. The GoB developed the INDC of Bangladesh with a view to taking necessary mitigation actions to reduce its growing emissions of GHGs and to play its role in global efforts to limit temperature rise to 2°C or preferably 1.5°C above pre-industrial levels agreed in the Paris Agreement. The INDC of Bangladesh takes both unconditional and conditional emissions reduction goals for the power, transport, and industry sectors, alongside further mitigation actions in some other sectors, which Bangladesh intends to carry out with subject to required technical and financial supports. The INDC of Bangladesh also outlined the adaptation goals and measures through identification of existing measures already taken and also for future needs, including the long-term vision for adaptation, drawing synergies with mitigation measures. It also provides a qualitative description of support needs for taking actions for mitigation and adaptation efforts. Now, the challenge for GoB is to implement INDC in accordance with the Paris Agreement through enacting new legal and policy frameworks or/and amending existing frameworks. However, in order to amend the existing legislation and policies and/or adopt new policies and legislation, it is important to examine the scope, gaps, and constraints of existing provisions. The Country Investment Plan (CIP) and the Delta Plan 2100 are good national policy documents for addressing environment, forestry, and climate change. The Seventh Five Year Plan (FY2016-FY2020) has dedicated chapter 8 on sustainable development: environment and climate change, where sections 8.5.1, 8.5.2, and 8.5.3 are very well elucidated upon climate change adaptation, resilience, and mitigation activities during the fiscal period. The Plan's objective is to promote sustainable environmental management systems for mitigation and adaptation through active participation of women and the poorest members of society. Numerous policies and plans exist that directly relate to climate change or which refer to related environmental topics and issues. A new Environmental Policy (updating the previous 1992 version) was put out for public consultation in 2013 and has been released in draft form. A National Action Plan for Short-Lived Climate Pollutants was published in January 2014. This Plan, focusing on black carbon, methane, ozone, and some hydrofluorocarbons, identifies the Climate Change Trust Fund as a potential funding source for initiatives to reduce these pollutants. Also, in 2014, a Climate Fiscal Framework was published by the Ministry of Finance to help identify the supply and demand sides of climate fiscal funds and to ensure that they are sustainable and transparent in the long term.

In 2013, realising the need for coordinated efforts on gender issues and to mainstream gender issues in development initiatives to increase resilience, the government has facilitated the preparation of the Bangladesh Climate Change and Gender Action Plan (MoEF, 2013). The ccGAP integrates gender considerations into four of the six main pillars as identified in the BCCSAP: (i) food security, social protection, and health; (ii) comprehensive disaster management; (iii) infrastructure and (iv) mitigation and low carbon development.

The government's Vision 2021 document, which sets out to eliminate poverty and make Bangladesh a middle-income country by 2021, is a key adaptation planning document for the country. Vision 2021 has social, economic, and environmental dimensions, and it declares that all efforts will be made to protect Bangladesh from the adverse effects of climate change. It aims to reduce air pollution from industry and transportation, and improve waste management, to improve the state of the natural environment by conserving forests and water bodies and preventing river erosion. In order to achieve the goals of Vision 2021, the 'Outline Perspective Plan (OPP) of Bangladesh 2010-2021: Making Vision 2021 A Reality' includes a dedicated chapter on the environment where the vulnerabilities, responses, and climate change management strategies are analyzed. The Plan emphasises climate change adaptation through the active participation of local communities and the private sector, rather than a top-down strategy.

Mainstreaming Climate Change into Development and National Planning

Mainstreaming in the context of climate change is the incorporation of climate change considerations into established or ongoing development programs, policies, or management strategies, in addition to implementing adaptation initiatives separately. In order for climate change adaptation to be sustainable and applicable on a wide scale to enable the people, especially the poor, to cope with the possible risks of climate, it must be incorporated, integrated, or "mainstreamed" into the policy apparatus of governments (FAO, 2009).

The Bangladesh government is integrating climate change into sectoral plans and national policies. Infrastructure development is being planned, designed, and implemented considering climate change impacts. Climate-Smart Agriculture is a global slogan now, and Bangladesh is also adopting new technology inventing salt-tolerant, water submergence tolerant crop verities to cope with climate change impacts. Research programs are being taken in light of climate change information. Bangladesh's interim poverty reduction strategy paper (I-PSRP) recognises the direct link between poverty and vulnerability to natural hazards and notes that the incidence of disasters is likely to increase rather than decrease as a result of global warming. In November 2007, the Government announced an initiative to formally incorporate the impacts of climate change into all development plans, including PSRP revisions. Many of the NWP and NWMP priorities are synergistic with climate change adaptation, such as the recommendation in the NWP for early warming and floodproofing systems. It is a fact that many other policies, including the National Environmental Management Plan (NEMAP), the National Land Use Policy, and the National Forest Policy, do not make specific references to climate change.

The UNFCCC, in its documentation, defined adaptation into the planning process. It consists of four key components: assessment of climate impacts and vulnerability; planning for adaptation; implementation of adaptation measures; and monitoring and evaluation of adaptation actions. The four components might be good guidelines for mainstreaming climate change in a national context. Each component needs to be elaborated in Bangladesh context and mainstream climate change step by step.

SUMMARY AND RECOMMENDATIONS

This study has taken a novel initiative to summarise the possible impact of climate change at specific warming levels for Bangladesh. This study has also developed appropriate adaptation options for different adverse impacts Bangladesh will likely face. Considering these adaption options, the following policy guidelines for Bangladesh are formulated to combat the adverse impact of climate change-

- Bangladesh is a disaster-prone country, and its people have survived through meeting challenges like floods - riverine and coastal, cyclones, river erosion, waterlogging, droughts, cholera, diarrhea, malnutrition, saline intrusion, etc. Floods are every year's event in our life. Autonomous adaptation is our natural characteristic. Like deeply flooded rice, our people adapt and continue to live with natural disasters defying all odds and events. In the course of time, the country has bagged in her pocket a lot of policies and legislations to protect her people from severe floods, cyclones, and all other disasters, which are on the increase in time. The country also has earned a good reputation and image before the international community through effective management of disasters. Still, through a review of the number of policy documents, especially climate concerned, it comes to the realisation that we need much more to do in mainstreaming them in terms of smart climate for the country.
- There have been major improvements in Bangladesh's response to climate change adaptation. Still, a substantial adaptation deficit exists under the current climate. We need a deeper understanding of the risk and vulnerability of climate change for the future climatic world of Bangladesh.
- We are to go for managing the climate for specific warming levels. It could be for Short Term for a period of 10 years, Medium Term for a period up to 2050, and Long Term up to 2080 and needs to be understood gradually in the course of time. The end goal is sustainable, resilient, and transformed Bangladesh, where we need additional policy documents, strategies, and action plan to mainstream adaptation to combat Climate Change.
- Green Bangladesh is to be made Greener with Green Philosophy – Green Design, Green Approach, Green Technology, Green/Eco Dish or "Ecotarian Plate," etc., and Carbon Sinking through Afforestation and Soil Treatment.
- Bangladesh has no data on the real costs of adaptive measures and needs to take into consideration the existence of appropriate institutions, availability of resources, and access to technology.
- Adaptation strategies and priorities are needed to be identified for each five-year Plan. Climate change is to hamper the daily livelihoods of millions of poor, especially farmers, through increased temperature. So, improvement of food grain productivity potential and water use efficiency of agricultural crops is a priority for achieving food security. Furthermore, the Plan needs to make space for adaptation responses to be incorporated in relevant programmes, including those relating to watershed management, coastal zone planning and regulation, forestry management, agricultural technologies and practices, and health.
- In line with developing adaptation strategies, there needs to be a strong focus on implementing the adaptation strategies through climate mainstreaming of developmental activities. Developmental programmes are to be structured so to address the vulnerable populations.
- The existing gap in adaptation funding is huge. This gap should be reduced by taking proper actions through a five-year plan.

- Adaptation to climate change is a relatively new science. More studies are required for a better understanding of the impacts of climate change in Bangladesh. Efforts are being made at the international stage that can provide an understanding for implementing relevant policies at national levels. Integrating, coordinating, and implementing adaptation to climate change is to be done at all national levels, which is still far away.
- Bangladesh does not have any Climate Change Act, which is essential for smart climate management and to bring all sectors under legislative obligations. The Climate Change Act can be a legally binding long-term framework for building the ability of Bangladesh to adapt to a changing climate.
- Economics of the National Adaptation Programme - Bangladesh lags behind science-based research documents on costs and benefits analysis of adaptation programmes. Adaptation needs to be justified based on estimates of costs and benefits. Planning, management, and policy formulation on appropriate adaptation require knowledge of the economics of adaptation. Adaptation is embedded in a wide range of existing policies and should reflect a series of voluntary agreements across government and society more widely.
- Government should have Long Term Environment Plan in line with SDGs to transform Bangladesh.
- Organic Bangladesh is the main short-cut way for achieving transformed sustainable Bangladesh through ensuring Organic Dish or "Ecotraian Plate," i.e., eco-friendly organic food for the whole country. It means switching our agriculture to an organic one where a food-producing system would be eco and nature-based. It demands resolutions of more outside issues than inner plate items. That oust side plate is very complex.
- Intervention for successful adaptation will be effective if it is bespoken, balancing financial, economic, societal, equitable, governance, and legislative and environmental interests.
- GoB needs soft engineering, and nature-based solutions (NBS) as everywhere can be protected. Climate Change Operational Framework in the GoB organogram is to be defined clearly.
- Flexibility is required in adaptation, with multiple choices to achieve a management goal (e.g., defined risk levels).
- Adaptation pathways provide one structured way to achieve this. These can be challenging to generate due to barriers in planning for adaptation.

REFERENCES

Alexandre, K. (2020). Frontiers in Climate Change Adaptation Science: Advancing Guidelines to Design Adaptation Pathways. *Current Climate Change Reports*, *6*(4), 166–177. doi:10.100740641-020-00166-8

Ali, A. (1999). Climate change impacts and adaptation assessment in Bangladesh. *Climate Research*, *12*, 109–116. doi:10.3354/cr012109

Antony, C., Unnikrishnan, A. S., & Woodworth, P. L. (2016). Evolution of extreme high waters along the east coast of India and at the head of the Bay of Bengal. *Global and Planetary Change*, *140*, 59–67. doi:10.1016/j.gloplacha.2016.03.008

Arias, Bellouin, Coppola, Jones, Krinner, Marotzke, & Naik. (2021). *Climate Change 2021: The Physical Science Basis.* Contribution of Working Group14 I to the Sixth Assessment Report of the Intergovernmental Panel on Climate Change Technical Summary.

Becker, M., Papa, F., Karpytchev, M., Delebecque, C., Krien, Y., Khan, J. U., Ballu, V., Durand, F., Le Cozannet, G., Islam, A. K. M. S., Calmant, S., & Shum, C. K. (2020). Water level changes, subsidence, and sea level rise in the Ganges–Brahmaputra–Meghna delta. *Proceedings of the National Academy of Sciences of the United States of America, 117*(4), 1867–1876. doi:10.1073/pnas.1912921117 PMID:31907308

Chiu, S., & Small, C. (2016). Observations of cyclone-induced storm surge in coastal Bangladesh. *Journal of Coastal Research, 32*(5), 1149–1161. doi:10.2112/JCOASTRES-D-15-00030.1

Dasgupta, S., Huq, M., Khan, Z. H., Ahmed, M. M., Mukherjee, N., Khan, M. F., & Pandey, K. (2010). *Vulnerability of Bangladesh to Cyclones in a Changing Climate:Potential Damages and Adaptation Cost.* World Bank.

Fahad, M. G., Islam, A. S., Nazari, R., Hasan, M. A., Islam, G. M. T., & Bala, S. K. (2018). Regional changes of precipitation and temperature over Bangladesh using bias-corrected multi-model ensemble projections considering high emission pathways. *International Journal of Climatology, 38*(4), 1634–1648. doi:10.1002/joc.5284

IPCC. (2014). *Impacts, adaptation, and vulnerability: Summary for policy maker.* Author.

Karim, M. F., & Mimura, N. (2008). Impacts of climate change and sea-level rise on cyclonic storm surge floods in Bangladesh. *Global Environmental Change, 18*(3), 490–500. doi:10.1016/j.gloenvcha.2008.05.002

Khan, M.J.U., Islam, A.S., Bala, S.K., & Islam, G.M.T (2020). Changes of climate extremes over Bangladesh at 1.5°C, 2°C and 4°C of global warming. *Theoretical and Applied Climatology.*

Major, D. C. (2010). *Adaptation assessment guidebook.* Academic Press.

Mohammed, K., Islam, A. S., Islam, G. T., Alfieri, L., Bala, S. K., & Khan, M. J. (2018). Extreme flows and water availability of the Brahmaputra River under 1.5 and 2 °C global warming scenarios. *Climatic Change.*

Noll, B., Filatova, T., Need, A., & Taberna, A. (2021). *Contextualizing cross-national patterns in household climate change adaptation.* Nature Climate Change Articles. doi:10.103841558-021-01222-3

Rahman, K., Ensor, J., & Berger, R. (2009). River erosion and flooding in northern Bangladesh. *Understanding Climate Change Adaptation: Lessons from Community-Based Approaches*, 39-54.

Rahman, S., Islam, A. S., Tazkia, A. R., Shaha, P. K., Krien, Y., Durand, F., Testut, L., Islam, G. M. T., & Bala, S. K. (2019). Projected changes of inundation of cyclonic storms in the Ganges–Brahmaputra–Meghna delta of Bangladesh due to SLR by 2100. *Journal of Earth System Science, 128*(6), 145. doi:10.100712040-019-1184-8

Rawlani, A. K., & Sovacool, B. K. (2011). Building responsiveness to climate change through community-based adaptation in Bangladesh. *Mitigation and Adaptation Strategies for Global Change, 16*(8), 845–863. doi:10.100711027-011-9298-6

Rosenzweig, C., Solecki, W. D., Blake, R., Bowman, M., Faris, C., Gornitz, V., Horton, R., Jacob, K., LeBlanc, A., Leichenko, R., Linkin, M., Major, D., O'Grady, M., Patrick, L., Sussman, E., Yohe, G., & Zimmerman, R. (2011). Developing Coastal Adaptation to Climate Change in the New York City Infrastructure-Shed: Process, Approach, Tools, and Strategies. *Climatic Change*, *106*(1), 93–127. doi:10.100710584-010-0002-8

Roy, B., Khan, M., Mostafa, S., Saiful Islam, A. K. M., Khan, M., Uddin, J., & Mohammed, K. (2021). Integrated flood risk assessment of the Arial Khan River under changing climate using IPCC AR5 risk framework. *Journal of Water and Climate Change*, *12*(7), 3421–3447. doi:10.2166/wcc.2021.341

Singh, Iyer, New, Few, Kuchimanchi, Segnon, & Morchain. (2021). Interrogating 'effectiveness' in climate change adaptation: 11 guiding principles for adaptation research and practice. *Climate and Development*. doi:10.1080/17565529.2021.1964937

Wassmann, R., Jagadish, S., Heuer, S., Ismail, A., Redona, E., Serraj, R., Singh, R. K., Howell, G., Pathak, H., & Sumfleth, K. (2009a). Climate change affecting rice production: The physiological and agronomic basis for possible adaptation strategies. *Advances in Agronomy*, *101*, 59–122. doi:10.1016/S0065-2113(08)00802-X

Wassmann, R., Jagadish, S., Sumfleth, K., Pathak, H., Howell, G., Ismail, A., Serraj, R., Redona, E., Singh, R. K., & Heuer, S. (2009b). Regional vulnerability of climate change impacts on Asian rice production and scope for adaptation. *Advances in Agronomy*, *102*, 91–133. doi:10.1016/S0065-2113(09)01003-7

Wheeler, D. (2011). *Quantifying Vulnerability to Climate Change: Implications for Adaptation Assistance*. Academic Press.

Wise, R. M., Fazey, I., Smith, M. S., Park, S. E., Eakin, H. C., Van Garderen, E. A., & Campbell, B. (2013). Reconceptualising adaptation to climate change as part of pathways of change and response. *Global Environmental Change*, 10.

Chapter 3
Sustainable Management of Humanitarian Water Supply, Sanitation, and Hygiene (WASH) Interventions

Modupe Olufunmilayo Jimoh
University of Warwick, UK

Samuel Oluwatosin Jacob-Oricha
Independent Researcher, Nigeria

ABSTRACT

WASH interventions have not always been successful, despite the significant efforts to improve coordination and delivery of quality interventions. Assessments have shown that sustainable management of interventions must transcend the immediate intervention period to allow continuous infrastructural service delivery, social and behavioural change, civil participation, improved governance, and more robust community and public sector systems. This chapter highlights the challenges of sustainable management, approaches, and lessons learned from the humanitarian response, especially in providing WASH facilities in conflict zones and fragile communities. It would also focus on a case study of conflict-affected Pulka Town, Gwoza Local Government Area, Borno State, Nigeria.

INTRODUCTION

The Global Sustainable Development Goal 6 (SDG6) aims to 'ensure availability and sustainable management of water and sanitation for all (United Nations, 2015). It includes eight targets that address drinking water, sanitation, and hygiene (WASH) services, wastewater treatment, water quality, water use, water management, transboundary cooperation, water-related ecosystems, official development assistance and participation of local communities. WASH services include water supply, hygiene promotion, excreta management, solid waste management, vector and disease control. Provision WASH services are

DOI: 10.4018/978-1-7998-9190-1.ch003

fundamentals for the human population for survival and dignity (Hutton & Chase, 2017). According to UNICEF/WHO Joint Monitoring Programme (JMP) for Water, Supply, Sanitation and Hygiene (WHO/ UNICEF, 2021), 1-in-4 people lacked safely managed drinking water in 2020, and 3.6 billion people lacked safely managed sanitation services. At the present rate of progress, by 2030, 1.6 billion people will still lack safe drinking water at home, and 2.8 billion people will still lack safe sanitation at home. The report recommended that to close the global gap, the rate of progress must be four times faster.

Unfortunately, the situation is worse in conflict areas, with people three times more likely to lack safe drinking water (Figure 1). To close the global gap in fragile communities, the rate of progress for the provision of drinking water, sanitation and essential hygiene services would have to move 23 times, 9 times and 5 times faster, respectively (WHO/UNICEF, 2021).

Figure 1. Access to WASH services in refugee camps by country in 2021 (%)
Source:WHO/UNICEF, 2021;UNHCR, 2021)

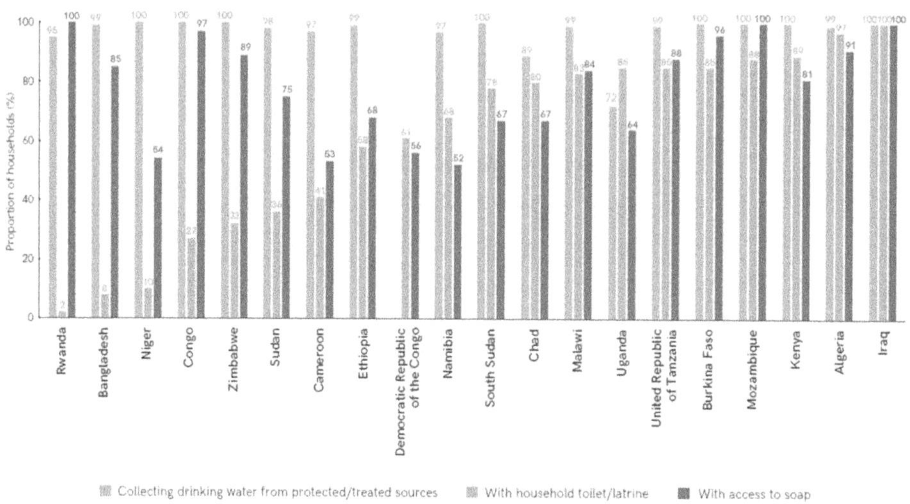

The importance of WASH services is much more emphasised in conflict areas where there is active or incidental destruction of existing facilities, higher density settlements or refugee camps (Truelove et al., 2020). Restrictions on the movement of victims living within or around the conflict zones also hinder access to WASH services. In addition, the influx of displaced persons to secured areas results in pressure on existing WASH facilities. Therefore, people in conflict or refugee camps are at a higher risk of contracting diseases related to unsafe water and poor hygiene (Hutton & Chase, 2017). Records have shown that children are also 20 times more likely to die from the effects of inadequate WASH services than from the violence of the conflict (UNICEF, 2019, 2021). A study of sixteen (16) countries that experienced conflicts revealed that more children die from inadequate water, sanitation, and hygiene than violence. Yearly diarrhoea disease (linked to poor WASH) killed 72,000 children under five between 2014 and 2016 compared to 3,400 children killed by direct acts of violence (Friedrich, 2019; UNICEF, 2019).

This chapter evaluates humanitarian WASH response frameworks and highlight sustainability indicators, successes, failures, and lessons learned from case study humanitarian projects.

SUSTAINABILITY INDICATORS IN WASH INTERVENTIONS

The sustainability of planned interventions is critical and multifaceted. For many humanitarian interventions, project failures set in after the hand-over or exit of the development agency or donor. Gensch et al. (2018) defined an exit strategy in an emergency context as "a planned approach of why what, when and how implementing organisations will end their humanitarian engagement". It considers that an exit strategy should include the steps towards transition, handing-over, decommissioning of infrastructure, and exiting or disengaging from activities, projects, programme areas or countries (Gensch et al., 2018). At the conceptual stage of any intervention, actors must factor in resilience, robustness, durability, preparedness, disaster risk reduction and prevention, hand over, and exit strategy into decisions and designs. It would mean that impacts could be significantly minimised and managed in the event of setbacks.

The first in the series is technical sustainability - locally sourcing technologies or technical capabilities. In cases where technical actors are from outside the community, community members must be carried along and trained on the expertise required for operation and maintenance. The ability of the community to facilitate the smooth running and maintenance of infrastructures after the exit of the actors is key to technical sustainability. Communities frequently abandon borehole projects because of faulty pumping systems that the community cannot or is unwilling to repair.

Secondly, the intervention must be sustainable financially. The community must have the capacity to financially support the operation and maintenance of the interventions to avoid its collapse. Financial viability depends on the economic power of community members and processes that allow projects to be self-sufficient. Recent designs have incorporated tokens or cards such that the money generated from the venture goes towards its operation and maintenance. Internally displaced Persons (IDP)'s solid waste management process can be designed around an integrated and circular economy system to facilitate sustainability.

Thirdly, the intervention must be socio-culturally sustainable. Actors must consider the cultural framework of communities when developing an intervention. It is vital for sanitation projects. The programs and projects should be convenient to use and acceptable to the beneficiaries, such that the willingness to use transcends the exit of the donors. Gender and religious concerns also fall into this category. Sanitation systems, for example, must provide privacy for girls and women and must be safe to access. Cultural taboos relating to engaging with sanitary products, such as faeces, should be addressed in preparedness for decommissioning or emptying latrines or toilets.

Fourth is environmental sustainability. Projects should factor in the impact of interventions on the environment and water bodies. Future occurrences, such as drying up of wells due to low water tables, offensive smells, the closeness of sanitation facilities to groundwater sources, the ecological effect of wastewater disposal, seasonal differences in the intensity of sunlight for solar panels, among others, should not be neglected. Beneficiaries should also be trained on safe excreta disposal, handling of chlorine for disinfection, and the use of personal protective equipment.

The fifth key to sustainability is inclusive and equitable design. WASH interventions should be such that the design of facilities allows access by any age, gender, or people with disabilities. Access should also be safe for children or the visually impaired and must consider the ease of use during the rainy season. Motorised water pumping infrastructures, raised platforms, and ramps for wheelchair access facilitate inclusiveness. Others are the provision of handrails, hand washing facility at reachable height and non-slip floors. Markings, labelling, WASH promotion messages and use of signals would be advantageous. Labels should be pictorial for users who cannot read easily. In addition, communications should use the local language to aid readability and understanding. For awareness programs, vocal campaigns and prints are most inclusive and effective.

Therefore, there is a need to approach interventions systemically in the quest for sustainability.

SUSTAINABLE APPROACHES, PRACTICES AND FRAMEWORKS FOR HUMANITARIAN INTERVENTIONS

Assessment

Intentional or accidental destruction of WASH services could happen via ground, aerial or sea attacks. Before planning WASH services, it is essential to assess the level of existing facilities and services before the conflict and at the time of planning. The initial WASH assessment could cover water, sanitation and hygiene conditions, existing infrastructures, environmental factors and topography. Other areas are stakeholders, available technical capacity, cultural and legal constraints, vulnerabilities, and security issues to provide base-level information necessary for a comprehensive plan (Gensch et al., 2018). A combination of multiple sources of information has proven helpful in emergencies. Assessment data sources can be through Primary or Secondary data sources. Primary data sources include observations, interviews and surveys. Others are focus group discussions and participatory/community mapping. Secondary data sources include local authorities, statistics and relevant urban development ministries. Non-governmental Organisations (NGOs), development agencies with previous knowledge of the area, United Nations (UN) agencies and satellite images can also provide secondary data. In addition, WASH needs assessment requires a multi-professional team for its success. They should be in water engineering, hydrogeology, ecology, sanitation, hygiene, data collection, management, community engagement, and translation. Interestingly, some regions only get attention to the condition of their WASH services after there has been a conflict in their area. An in-depth assessment is the bedrock of a successful intervention.

In-country Governance and Policy

In humanitarian emergencies, the national laws, regulations, standards, and codes provide the framework for emergency response. It is because states have the primary responsibility to protect the people within their territory. Local governments and government departments responsible for sanitation and other WASH interventions have the responsibility to coordinate the actions of other agencies. However, this does not always happen because of a lack of capacity and resources (Gensch et al., 2018). Before any intervention occurs, actors must fully understand all local laws governing water and sanitation provision and use, wastewater collection, treatment, discharge and reuse, and environmental standards. The more the local government buys into an intervention, the more chances of success and sustainability. It is encouraged that in places where national guidelines are not specific or existent, the alternative would be to use the Sphere Humanitarian Charter and Minimum Standards in Humanitarian Response. The Sphere is a handbook written collectively by many humanitarian agencies. It is a handbook that combines the ethical and moral backdrop of protection principles with the core humanitarian standards enshrined in the humanitarian charter. It further provides guidelines on the criteria to be adhered to (at the minimum) for WASH promotion, health, food security and nutrition, shelter and settlement (Sphere Association, 2018).

Design and Scale of Intervention

The design and scale of interventions should factor in sustainability. Interventions can either be short term or long term. In emergencies, humanitarian actors often provide a short-term intervention (e.g. water trucking) as an immediate response. A follow-on can then be a more pronounced, all-encompassing long-term project focusing on reaching the targets of SDG6. Actors for short-term interventions must clearly define duration because most short-term interventions might not be a cost-effective solution if run longer than necessary. The operation and maintenance of such schemes are often high. Whichever is planned and implemented by actors, the goal should be a continuous and sustainable impact, with reduced risk to abandonment.

The scale of interventions also varies from place to place. For example, interventions at the Cox's Bazar refugee camp, Bangladesh, were complex and broad. The vast number of refugees arriving at the refugee camp meant congestions in camps with limited space available (Truelove et al., 2020). The available 107 small and mid-sized sanitation management facilities were overwhelmed, leading to water contamination and disease outbreak concerns. The humanitarian actors had to set up rapid response schemes to reduce the risk of contamination and stave off disease outbreaks (UNICEF, 2019).

WASH intervention programs can be new initiatives, existing initiatives tested in other locations, improved initiatives in the recipient community, or a combination of styles. Examples can be the provision of new wells and boreholes, repairing existing wells and boreholes, redesigning latrines, and using local structures and technology for solid waste management. In Yemen, UNICEF and the World Bank interventions included the distribution of over 2,900 basic hygiene kits and the rehabilitation of WASH facilities in 29 rural health centres by installing water tanks and upgrading toilets, washbasins and sewage systems (UNICEF, 2019). Stakeholders can use simple or more intricate technologies for designs within interventions. During the COVID outbreak in 2020, many low-cost handwashing systems were designed within communities piloted. Many designs relied on pedals and sensors to activate the system to reduce disease transmission by hands touching the taps.

Finance is a crucial component of sustainable intervention. Between 2008 and 2017, humanitarian funding requirements increased from US$ 8 billion to US$ 26 billion, and in 2018, 10 countries used 60% of all humanitarian funds (Global WASH Cluster, 2020). Despite its significance, the WASH sector has been one of the least funded sectors. It has impacted the level and quality of delivery and its subsequent continuity. The financial planning for any intervention must therefore incorporate cost-effective measures.

Factors to consider when designing interventions:

- Intervention should be in the right place at the right time.
- The design team should be inclusive, diverse, and multidisciplinary.
- Personnel must have the skills and competence to deliver standard projects.
- Local capacity development strategies must be part of the project design.
- Project designers should ensure interventions are targeted and evidence-based.
- Interventions should be all-encompassing of all WASH services and cross-cutting with considerations for other sectors, environmental impacts, and human rights.
- Emerging threats, such as the effect of climate change on people, infrastructure and disaster risks, should be factored in.
- Design should incorporate preparedness and emergency surge mechanisms that are risk-informed.

- The financial team should use WASH Return on Investment tools for cost-benefit analysis at the planning stage.
- Public-private financing can be the key to sustainability beyond the exit of the actors.

Collaboration and Teamwork

People are at the centre of any intervention. People make it work from needs assessment to design, implementation, monitoring, reporting, and evaluation. Partners from the humanitarian and development sectors, local authorities, and governments also play a vital role. Therefore collaboration and teamwork would ensure that projects are not duplicated, and services are complimentary. It would also mean that information and knowledge-sharing occur towards a holistic intervention. The presence of many stakeholders in the provision of WASH services and other dependent services (such as food, economic empowerment, security etc.) has required collaborative working. It has led to a strategic partnership coordinating water and sanitation responses in emergencies. The strategic partnership is called the WASH cluster, and UNICEF leads it.

An example of a successful multi-agency collaboration was the establishment of rapid response mechanisms to control the spread of disease in Yemen. UNICEF, partners, and Yemen's General Authority for Rural Water Supply Projects raised 750 rapid response teams that provided critical hygiene supplies and information to 8.4 million people in more than 1.2 million households. The partnership also provided safe water to nearly 6 million people within 2018 by providing funding, power sources (fuel, electricity), critical supplies (spare parts, chlorination materials and services) and operational support towards the continuity of WASH services (UNICEF, 2019). The collaborative intervention resulted in short-term and long-term impacts.

Another initiative promoting collaboration and teamwork is the Agenda for Change movement. The "Agenda for Change" movement is a collaboration of partners within the WASH sector to advocate and support governments at local and national levels and other stakeholders for lifelong services by adopting a set of common principles (Agenda for Change, 2020). The actors' approach at the district, city, national and global levels are to be guided by the principles. It highlights the following behaviours as key to effective, equitable and sustainable synergies in the WASH sector:

- Enhance government leadership of sector planning processes.
- Strengthen and use country systems.
- Use one information and mutual accountability platform.
- Build sustainable water and sanitation sector financing strategies.

For a sustainable intervention, the team must be an appropriate mix of local and international staff and trainee and experienced staff. The combination must also be multidisciplinary to cater to WASH interventions' multifaceted (technical, social, and financial) requirements. Cluster leadership must account for all agencies and partners, their expertise, contributions, and goals. It should also clearly define the hierarchy of engagement, including communication strategies and conflict resolution.

Community Engagement

UNICEF (2020b) defines community engagement as "the active participation of people and communities in ways that mean their voices are heard, and their active contribution to decision-making is safe, equitable and effective – doing with, not doing to". Community engagement is crucial to the successful implementation and sustainability of any WASH intervention. The type and location of the crisis, the status of WASH services, security risks, and the actors involved in the intervention would influence the kind of community engagement strategy.

An assessment of community engagement with the Rohingya refugees in Coxs Bazar, Bangladesh, reported by Strategic Executive Group (2020), highlighted weaknesses in the level of access to information, who can access information, what type of information can be assessed, sense of belonging, complaints, and feedback mechanism. It highlighted that women, people with disabilities, and older people were more impacted and underserved by the current communication and community engagement strategies. In addition, it reported that about 40% of Rohingya refugees were unaware of the complaints and feedback mechanism or did not know how to use it. Despite three years of intervention, the assessment report from Bangladesh highlights the need for coordinated and inclusive community outreach.

Keys to effective community engagement:

- It should start with the needs assessment, design of the intervention and run through the intervention course.
- It should be inclusive, embracing existing leadership structures, power dynamics, influencers, gender dynamics, cultural practices, and religion.
- It should recognise the community's context and make it a baseline for change.
- Actors should use power responsibly and encourage mutual respect between all parties.
- It should foster collaboration between all actors and stakeholders.
- The process should incorporate information sharing and communication at the appropriate level, using multiple means of communication.
- It should lead to a partnership in the delivery of the intervention, such that local capacity building occurs.
- It should be evidence-based to motivate self-realisation for positive behavioural change, particularly for sanitation and hygiene services.
- Feedback, complaints, and resolution mechanisms should be clearly defined and used to encourage trust.
- In instances of resistance to intervention, it should be appropriately investigated, addressed, and adaptations made where necessary.
- For accountability, data and learning from monitoring and evaluation should be shared with the community.
- Community engagement strategy should consider risks and the security of persons, especially in conflict areas.

Effective community engagement results in increased community ownership of interventions during the design and implementation stage and beyond the exit of the humanitarian actors.

Table 1. Documented attacks on WASH services in conflict zones

Country	The Syrian Arab Republic	Yemen	Ukraine	State of Palestine
Need	15.5 million people needed WASH services.	24 million people needed humanitarian assistance.	2.8 million people in Donetsk and Luhansk Oblasts needed WASH services.	0.8 million children in need of adequate access to WASH services.
WASH Attacks	46 attacks on water facilities were documented by the UN in 2019.	122 incidents of airstrikes on water and sanitation infrastructure occurred between March 25 2015, and February 28 2020.	380 attacks on water facilities occurred between 2017 and 2021.	95 attacks on 142 WASH structures occurred between 2019 and 2021.

Source: (UNICEF, 2021)

Security Concerns

In times of conflict, random and targeted attacks destroy infrastructure, injure personnel, and cut off the power sources that keep WASH systems running. Humanitarian actors should factor in security concerns into the implementation of interventions. It should include considerations for the security of lives for staff of development agencies, NGOs, water supply companies and community members, and WASH infrastructures' security. Table 1 shows documented attacks on wash services in conflict zones worldwide. Most of these places have had years of war that have impacted the already poor infrastructures in some areas or existing urban services.

UNICEF (2019, 2021) expressed concern about the effect of insecurity and attacks on the sustainability of activities and infrastructures. The report enumerated responsibilities for various stakeholders, including humanitarian actors, states, the United Nations security council and donors. According to the report, the summary of the call for action regarding security is:

- Humanitarian aid must be allowed to reach children and families in need.
- Improved methods are needed to monitor attacks on WASH systems.
- Policies and legal measures are required to protect vital civilian infrastructure and personnel.
- There must be respect for the rule of war as stipulated in the international humanitarian law (IHL) and international human rights law (IHRL).

Monitoring, Evaluation, Accountability and Learning (MEAL)

The MEAL process helps identify what has been done, how it has been done, who the recipient was, the level of reach and how much more needs to be done. Monitoring and evaluating WASH infrastructures and interventions are essential in ensuring that the universal right to water and sanitation services is met even in conflict zones or IDP camps. Evaluation of interventions leads to the identification and articulation of learnings from the project to support subsequent decisions and actions. It is also necessary to assess the human and financial investment with success recorded in the intervention. MEAL processes should be incorporated from the design and carried out throughout the intervention. It should include monitoring the access to and utilisation of WASH services at household and community levels. The Sphere Humanitarian Charter and Minimum Standards in Humanitarian Response contains guidelines for monitoring and evaluating projects. In addition, the UNHCR developed a WASH monitoring system that incorporates primary and secondary data collection with a geographic information system (GIS) to create a dashboard that provides information on WASH-related indicators in emergency and non-emergency locations (UNHCR, 2020). This system offers a platform to measure and record the rate of successful impacts, failures, and the articulation of lessons learnt. Regular MEAL processes would ensure that humanitarian actors achieve the project goals, the technical integrity of WASH interventions is maintained, and WASH inclined behavioural change is carried on by the community, leading to increased sustainability of the intervention.

CASE STUDY OF PULKA, GWOZA LOCAL GOVERNMENT AREA (LGA), BORNU STATE, NIGERIA

Background to Case Study

Some empirical reviews and evidence showed that terrorism and insurgency in Nigeria started with an Islamist fundamentalist, Boko Haram, who regarded western education as forbidden. The first significant attack of Boko Haram came in September 2009 in the Northern part of the country, where many lives were lost, and properties worth millions were damaged (Walker, 2012). The insurgency affected the local communities, particularly those proximal to Sambisa Forest, such as Banki (Bama LGA), Gwoza and Pulka (Gwoza LGA). Boko Haram invaded, attacked, and took over Gwoza LGA, precisely Gwoza and Pulka towns, from August 2012 until October 2013, when the Nigerian Army liberated the communities from the holds of the insurgents (OCHA, 2020).

Pulka has an estimated population of 64,000 people, most of whom were displaced by the conflict in the northeast of Nigeria. Thousands have been relocated to Pulka from Banki by the state government, irrespective of warnings from UN Agencies and INGOs about the limited water availability (ACAPS, 2017). Pulka town is generally divided into 2 parts: the host community and IDP camps. The host community is further subdivided into sectors 1 – 9, while the IDP camps are 6 subdivisions, namely Camp A (Umbazah rock side), Camp B (Wege), Camp C, Camp D, and Camp D E and the Transit IDP Camp.

Water Sources for Pulka

People in Pulka town, Borno State, spend several hours under the scorching sun searching for water, often in unsafe and unprotected hand-dug wells. The violent activities in Gwoza and Pulka town led to the destruction of the significant sources of drinking water, such as Ajari borehole, Emir palace borehole, several hand pumps and wells in Gavisga, Bulabulin, Wege, Warabe and View centre in Gwoza and Pulka towns (Humanitarian Practice Network, 2017). The destruction of these water facilities by the Boko Haram Insurgency led to a significant decrease in drinking water accessibility in Gwoza LGA, especially Pulka town. Also, displacement of people from neighbouring villages and farm settlements into LGA headquarters and townships such as Gwoza and Pulka, respectively, further puts pressure on the few available drinking water sources as water demand increases (ACAPS, 2017). Access to water sources is a challenge due to the topography and geologic formation of the region. Pulka lies in a basement complex zone which has a characteristic of erratic occurrence of groundwater (Popoola et al., 2020). The layers in Pulka are rocky, and the water levels are said to be about 305 m deep (IOM, 2019).

Figure 2. Timeline of activities in Pulka, Gwoza local government area
*Source: OCHA (2020) *OCHA is the United Nations Office for the Coordination of Humanitarian Affairs*

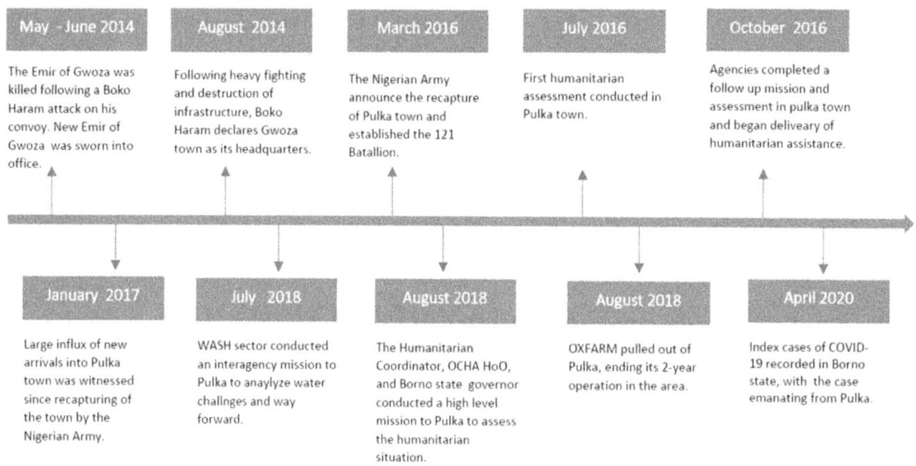

It, therefore, makes groundwater development extremely difficult. It has led to several failed boreholes, dried-up wells and hand pumps and explains why the river in Pulka is perennial (stop flowing in the dry season or dry years). The river source is also unprotected, making it unsafe for drinking purposes. Instances of use for drinking has led to an increase in diseases; therefore, it is used chiefly for laundry (MSF, 2021). There is an earth dam on the river to create an artificial reservoir. In 2018, the UNDP invited bids for the rehabilitation of the earth dam, reconstruction of the spillway, inlet and silt trap and the construction of a subsurface dam water treatment plant (Nechita et al., 2018; UNDP, 2018). The work is still ongoing (OCHA, 2020). Therefore, the principal water sources in Pulka are boreholes, wells and hand pumps, which are not widely functioning across Pulka town.

Oxfam Humanitarian Response in Pulka

Oxfam was the first major NGO to intervene and provide humanitarian services in Pulka (OCHA, 2020). However, after two years, it pulled out of Pulka amid rising security and operational challenges. Figure 2 shows a timeline of activities in Pulka, highlighting the start, type and end of intervention provided by Oxfam. Pulka had a history of water scarcity, especially in dry seasons. It meant community members travel a long distance to neighbouring communities to access water. International standards for humanitarian aid recommend that people in emergency conditions should have access to between 15 and 20 litres of water every day (Reed & Reed, 2013). It was, however, not the case in Pulka. It is a distant dream for the displaced people living in various IDP camps and Host communities in Pulka, as people struggle to access basic amenities like safe water.

The population of Pulka tripled from less than 20,000 before the conflict in April 2014 to an estimated 65,000 residents (MSF, 2018). With the collapse of public infrastructure and increasing population, water became one of the foremost needs of residents and IDPs in Pulka. Inadequate sanitation facilities resulted in solid waste dumping, unkept adults, open defecation, and diarrhoeal diseases. The multiplying effect was also evident in increased medical issues, like skin diseases and outbreaks of water-borne diseases in Pulka (MSF, 2018).

Nechita et al. (2018) reported that Oxfam's initial assessment identified 40 functional water sources, including motorised boreholes, hand pumps and open wells, with a cumulative yield of 368,200 litres/day. Of these sources, 38 were low yield, and only 2 had potentials for an acceptable yield. Oxfam's intervention started with trucking water from neighbouring communities into Pulka. It then moved on to the SHO (Dutch Public Appeal) humanitarian response project, which took place between March 6, 2017, and February 28, 2018. In response to the humanitarian emergency, Oxfam/SHO project activities involved the provision of water, sanitation, and hygiene (WASH) and emergency food security and livelihood (EFSL). Table 2 provides an evaluation of Oxfam's activities in Pulka.

Interventions after the Exit of Oxfam

After the exit of Oxfam, a team of humanitarian officers from Action Contre la Faim (Action against Hunger) and the rapid response mechanism (RRM) unit of the United Nations Population Fund Agency (UNPFA) visited Pulka in December 2018. Extracts from the field visit reports (Nechita et al., 2018) includes:

- Inadequate water supply persists due to the continuous influx of refugees and the stoppage of water trucking in July 2018.
- No partner was financing the fuelling of generators to power boreholes with generators.
- Solar-powered boreholes had limited performance due to the seasonally low intensity of the sun.
- Water availability at less than 8 litres per day per person.
- Water supply output has dropped to less than 50% of Oxfam's initial assessment of potential water yield.
- The only locations with constant water supply were the hospital, the MSF office, and the military zone due to the funding of the military borehole by MSF.
- Security concerns limit water trucking from Gwoza to Pulka and restrict the possibility of exploring other areas for a water source.

The information indicated that the gains made by Oxfam's were not sustained after their exit. The report made recommendations to address the challenges identified, focusing on the contributions required by different humanitarian actors present in the region and considerations for rehabilitating the earth dam.

Present Conditions in Pulka

A 2021 survey of 500 residents of Pulka and its IDPs by the author gave the following results:

- At peak raining seasons, 35% of drinking water is sourced from motorised boreholes, 22% from well-water, and 13% from the river and other sources.
- Due to increasing water demand, which is far higher than the water supply and rationing, 54% of responders could only access water once or twice a day.
- More than 48% of the populace have access to less than 15 Litres per day, while only 15% access up to 20 Liters of water per day.
- 60% of the populace agreed that drinking water was more accessible in Pulka before the insurgency.

Table 2. Evaluation of Oxfam's humanitarian intervention in Pulka

Key Items	Description
Project Activities (Water Supply)	Supply trucked water from other Gwoza Local Government Area (LGA) parts to Pulka as an emergency intervention for IDPs.
	Three boreholes were upgraded in Pulka to serve 3200 people. The upgrade included fitting higher-capacity water pumps and fitting the boreholes with diesel and solar power generators to increase the volume and duration of water supply.
	OXFAM supplied fuel to power all the boreholes with generators for both day and night water supply.
	Drilling of new boreholes funded by UNICEF.
	Direct supply to water at the IDP camps.
Project Activities (Sanitation and Hygiene)	Improved access to sanitation facilities (latrine and bathroom)
	Constructed five emergency latrine blocks for 250 IDPs in Camp B.
	Constructed five latrine blocks for 1250 IDPs in Camp C.
	Distributed soap and detergents, water jerry cans, and "aquatab" for water purification.
Project Activities (Empowerment)	Recruited and trained community members as WASH volunteers
	Created awareness about water, sanitation, and hygiene issues within the host community and the IDP camps.
	Cash for work schemes for jobs such as clearing refuse, rehabilitation of water, reservation pond, and restoration of small access roads.
	Unconditional cash transfers to 1300 households in Pulka to reduce malnutrition and increase nutritional intakes by community members.
Project Implementation Strategies	Efficient use of resources by repairing existing facilities when possible.
	Synergy and complementarity of services with other stakeholders.
	Coordination meeting and synergy with other humanitarian actors such as UNHCR, UNICEF, MSF, and IOM.
	Community engagement at planning and implementation stages.
	The project allowed women representation in most of the activities implemented.
	Establish and use complaint and response mechanisms and monthly coordination meetings to address complaints.
	All planned targets were accomplished within the project lifespan.
Challenges	The volatility of community and Environs
	Need for security clearance and military escorts.
	Cumbersome logistics and travel processes lead to delays.
	Sense of seclusion by some groups.
	Limitations in available resources.
Success	88.6% of beneficiaries ranked the project performance as excellent.
	Direct access to water by IDPs.
	Cleaner communities with no open defection and cleaner occupants. Improved sanitation situations.
	Reduction in disease incidences.
	Community empowerment and improvement of livelihood.
	Uplifted status of women.
	Prompt response to the project's complaint and response mechanism-built trust.
	Project activities around WASH provision and awareness have sustainable impacts.
Shortcomings	The project design did not include a monitoring and evaluation plan, logical framework, or baseline data to benchmark achievements that may have affected efficiency.
	Dissatisfaction by host communities on water supply. Despite the water supply interventions, there was still scarcity due to the increasing population of the host community (Pulka), water rationing/sharing with neighbouring IDP camps and security challenges restricting the citing of additional water sources.
	Only 2 of the newly drilled boreholes remained functionary because the water dried up due to the low water table.
	Motorised pumps were not sustainable because no other agency was paying for fuel after Oxfam left. The intensity of the sun limited the performance of solar-powered pumps.
	Cash related schemes are unsustainable.
Lessons learned	Incorporating Monitoring, Evaluation, Accountability and Learning (MEAL) assessment into the design is vital.
	Integrative programming of project activities was effective.
	Collaboration between implementing partners fostered knowledge sharing, skills exchange and reduced duplication of efforts.
	The use of the existing community structure for community engagement was effective.
	Project staff are critical for successful implementation.
	Localised decision making on operational issues by the base manager was timely and effective.
	Actors must consider project sustainability beyond their departure.

Adapted from Oladosun and Njobdi (2018)

- Although Pulka is not directly in the line of fire, the Boko Haram insurgency has led to an exponential increase in population, exacerbating water stress.

A coordinated WASH cluster has been activated to sustain the WASH efforts in Pulka and its environs (Gwoza LGA). They include a National NGO called LINDII Peace Foundation and international NGOs such as Danish Refugee Council (DRC), International Rescue Committee (IRC), Malteser International (MI), Mercy Corps (MC), Norwegian Church Aid (NCA), Norwegian Refugee Council (NRC). Other actors are UN agencies such as International Organization for Migration (IOM) and United Nations Children's Emergency Fund (UNICEF) (OCHA, 2021). According to OCHA (2021), although 869 humanitarian airlifts were carried out in and out of Pulka, Gwoza in the 6 months between January and June 2021, it is a reduction as humanitarian agencies have reduced staff due to target on humanitarian personnel and assets. The resultant effect is the suspension of some activities and the reduction in size and quality of humanitarian response.

A joint gap assessment is done monthly by actors across all locations to monitor and evaluate the progress of interventions and the needs met. Records of the gap analysis are accessible via an online 'WASH sector response monitoring dashboard'. The dashboard showed that in Gwoza LGA (comprising Pulka, Camps and Environs), between January and October 2021, there were: 139,030 operations and maintenance; 119,170 Hygiene Promotion; 64,100 Hygiene kits distributed; 79170 Waste Management Activities; 50,990 Desludging activities, 45,380 emergency and VIP latrines provided; 2,140 household latrines provided; 34,340 water trucking; and 12,000 borehole construction and rehabilitation. It has been through a financial investment of 7.8 million dollars, with the impact reaching only 139 034 people out of 211,629 million in need (WASH Sector Nigeria, 2021a).

Capital investment has continuously been the challenge for Pulka, in particular, to support water trucking and sustainable technological designs. There is a gap of 250,000 litres/day due to lack of sustainable water facilities, high needs, low aquifer yields between March and June, leading to low borehole yields (WASH Sector Nigeria, 2021b). Unfortunately, a significant funder was reported to have pulled out in July 2021. Therefore, it is imminent that sustainable solutions must be designed and implemented for Pulka. Sustainable solutions must also be considered to address the continuous influx of people into the Pulka area.

WAY FORWARD TOWARDS SUSTAINABLE WASH SERVICES DELIVERY

Improved strategy and collaboration is needed to achieve sustainability in the WASH sector, particularly in conflict-affected areas. It would further accelerate the progress towards achieving the Sustainable Development Goals. One of the initiatives towards this is the Roadmap 2020-2025. The Roadmap 2020-2025 has the vision that "By 2025, the WASH sector will have the capacity and resources to deliver in emergencies at scale, anywhere and at any time". The proposers believe that the vision would be achieved through a combination of strategic axes and pillars, reinforcing the importance of the availability and synergies of human and non-human resources. The axes focus on quality delivery, accountability and sustainability anchored on the pillars of capacity building, coordination and partnership, and financing (Global WASH Cluster, 2020). Figure 3 summarises considerations that lead to success and sustainability in humanitarian interventions.

Figure 3. Steps to sustainable intervention.
Source: Sphere Association (2018)

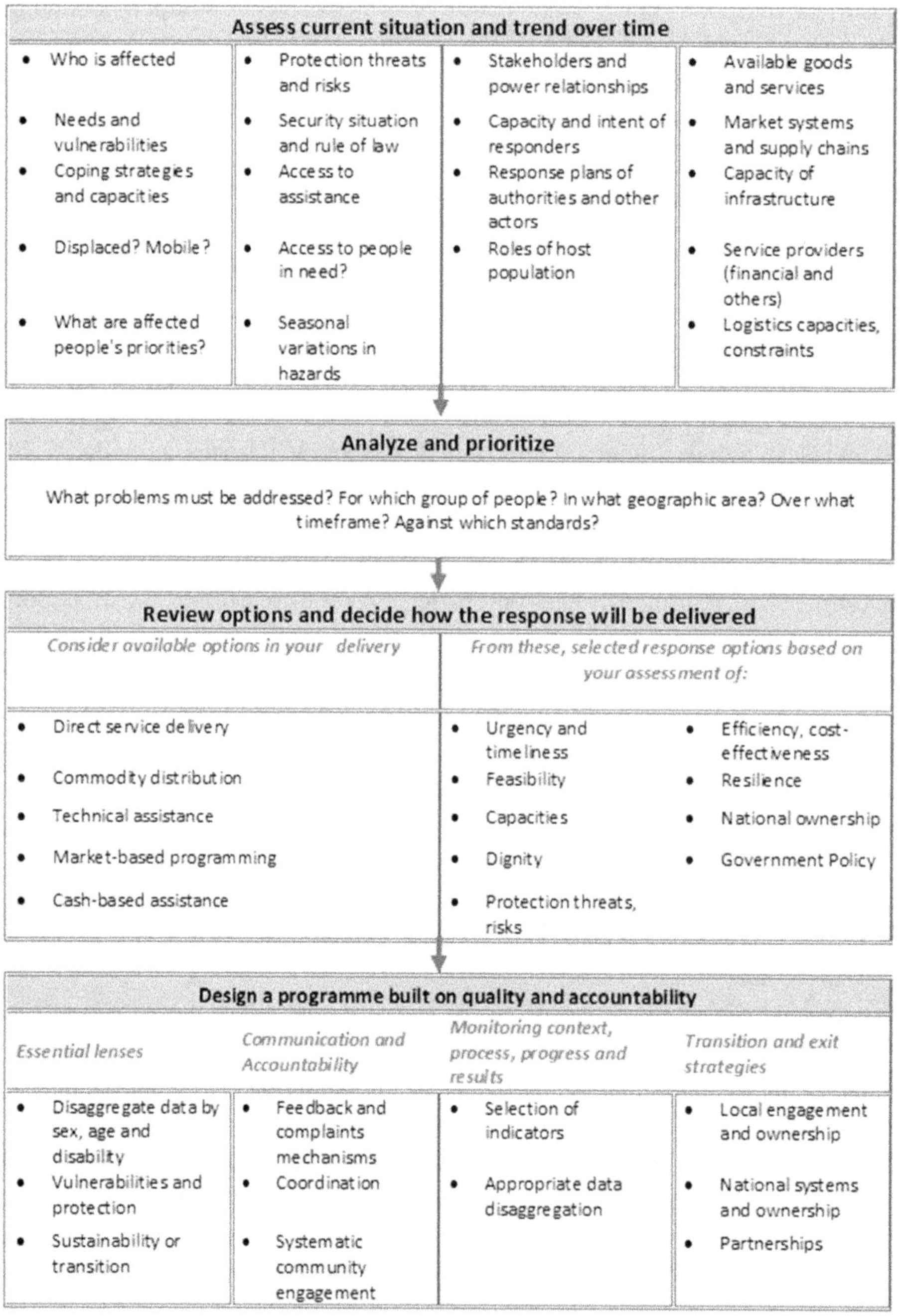

CONCLUSION

The core of sustainable WASH intervention is moving from delivering aid to ending needs. Maximising gains and transitioning the WASH sector from emergency life-saving interventions to achieving the targets set by the SDG6 requires strategic, operational models and mechanisms. In addition, there is no one-size-fits-all model for community engagement. WASH interventions should enhance long-term community goals and minimise environmental impact. It should manage the entire water, sanitation and hygiene chain and be built on a sound understanding of the diversity and vulnerabilities within the affected communities. Furthermore, adequate finance is key to establishing resilient services that operate during and after crises. It can be in the form of donors or public-private partnerships.

In conclusion, sustainability must be embedded in all sections of the programme cycle for the effectiveness of interventions. It can be achieved by incorporating water resource management with financial planning, community engagement, institutional coordination and MEAL.

FUNDING

This research received no specific grant from any funding agency in the public, commercial, or not-for-profit sectors.

REFERENCES

ACAPS. (2017). *Nigeria: Humanitarian Needs of Returnees from Cameroon, Chad & Niger* (ACAPS Thematic report: Returning refugees in Borno State, Nigeria, Issue). ACAPS. https://www.acaps.org/sites/acaps/files/products/files/20170731_acaps_thematic_report_returnign_refugees_in_borno_state-nigeria.pdf

Agenda for Change. (2020). *Strengthening WASH Systems: Concepts, Examples and Experiences*. Agenda for Change. https://washagendaforchange.org/wp-content/uploads/2020/04/20200227_agenda_for_change_systems_strengthening_experiences_final.pdf

Friedrich, M. J. (2019, May 21). Water and Sanitation Deficits Take a Toll in Armed Conflict Regions. *Journal of the American Medical Association*, *321*(19), 1861. doi:10.1001/jama.2019.6186 PMID:31112265

Gensch, R., Jennings, A., Renggli, S., & Reymond, P. (2018). *Compendium of Sanitation Technologies in Emergencies*. https://www.washnet.de/wp-content/uploads/emergency-sanitation-compendium.pdf

Global WASH Cluster. (2020). *Delivering Humanitarian WASH at scale, Anywhere and Any Time: Road Map for 2020-2025*. U. Global WASH Cluster (GWC). https://washcluster.net/sites/default/files/inline-files/RoadMap_2020-2025.pdf

Humanitarian Practice Network. (2017). *Special feature- The Lake Chad Basin: An Overlooked Crisis?* https://odihpn.org/wp-content/uploads/2017/10/he-70-web.pdf

Hutton, G., & Chase, C. (2017). Water Supply, Sanitation, and Hygiene. In C. N. Mock, R. Nugent, O. Kobusingye, & K. R. Smith (Eds.), Injury Prevention and Environmental Health. The World Bank. doi:10.1596/978-1-4648-0522-6_ch9

IOM. (2019). In North-East Nigeria, Access to Water Paves the Way for a Brighter Future. International Organization of Migration (IOM).

MSF. (2018). *Parched Pulka Searches for Water*. Medicine San Frontier (MSF). Retrieved November from https://www.msf.org/nigeria-parched-pulka-searches-water

MSF. (2021). *Pulka, Where Water is the Source of Life… and Disease*. https://www.msf.org/no-clean-water-pulka-nigeria-leads-disease

Nechita, C., Alfayo, L., & Yakubu, D. (2018). *Trip Report- Field Visit to Pulka Community, Gwoza LGA*. https://www.humanitarianresponse.info/sites/www.humanitarianresponse.info/files/documents/files/acf_trip_report_-_pulka_12122018.pdf

OCHA. (2020). *Fact Sheet: Pulka, Gwoza Local Government Area Borno State, North-east Nigeria*. N. UN Office for the Coordination of Humanitarian Affairs (OCHA). https://www.humanitarianresponse.info/sites/www.humanitarianresponse.info/files/documents/files/ocha_nga_pulka_factsheet_01052020.pdf

OCHA. (2021). *Periodic Monitoring Report- Nigeria-January-June 2021*. N. UN Office for the Coordination of Humanitarian Affairs (OCHA). https://reliefweb.int/sites/reliefweb.int/files/resources/20210906_nga_par_january_june.pdf_0.pdf

Oladosun, M., & Njobdi, S. (2018). *Report on End of Project Evaluation Project Title: SHO Four Famine Appeal 2017 - Nigeria*. https://www.researchgate.net/publication/337828350_End_of_Project_Evaluation_Project_Title_SHO_Four_Famine_Appeal_2017_-_Nigeria

Popoola, K., Talabi, A. O., Afolagboye, L. O., Oyedele, A. A., & Ojo, A. A. (2020). 2020/03/01). Groundwater potential evaluation in the basement complex terrain of Ekiti East Local Government Area, Southwestern Nigeria. *International Journal of Energy and Water Resources*, *4*(1), 81–90. doi:10.100742108-019-00053-5

Reed, B., & Reed, B. (2013). *How Much Water is Needed in Emergencies* (Technical Notes on Drinking-Water, Sanitation and Hygiene in Emergencies). https://www.who.int/water_sanitation_health/publications/2011/WHO_TN_09_How_much_water_is_needed.pdf

Sphere Association. (2018). *The Sphere Handbook: Humanitarian Charter and Minimum Standards in Humanitarian Response*. S. Association. www.spherestandards.org/handbook

Strategic Executive Group. (2020). *Joint Response Plan Rohingya Humanitarian Crisis*. United Nations (UN). https://reporting.unhcr.org/sites/default/files/2021%20JRP.pdf

Truelove, S., Abrahim, O., Altare, C., Lauer, S. A., Grantz, K. H., Azman, A. S., & Spiegel, P. (2020). The potential impact of COVID-19 in refugee camps in Bangladesh and beyond: A modeling study. *PLoS Medicine*, *17*(6), e1003144–e1003144. doi:10.1371/journal.pmed.1003144 PMID:32544156

UNDP. (2018). *Rehabilitation of Pulka dam, Gwoza LGA, Borno State.* United Nations Development Programme Office. Retrieved December from https://procurement-notices.undp.org/view_notice.cfm?notice_id=50804

UNHCR. (2020). *Water, Sanitation, and Hygiene (WASH) Monitoring System Briefing Note 2.* https://wash.unhcr.org/download/wash-monitoring-system-briefing-note-v2/

UNICEF. (2019). *Water Under Fire: For Every Child, Water and Sanitation In Complex Emergencies.* https://www.unicef.org/reports/water-under-fire-2019

UNICEF. (2020b). *Accountability to Affected Populations: A Handbook for UNICEF and Partners.* G. United Nations Children Emergency Fund (UNICEF). https://aa9276f9-f487-45a2-a3e7-8f4a61a0745d.usrfiles.com/ugd/aa9276_0fa31f247b194585b15e98562a269e0d.pdf

UNICEF. (2021). *Water Under Fire Volume 3: Attacks on water and sanitation services in armed conflict and the impacts on children.* https://www.unicef.org/reports/water-under-fire-volume-3

United Nations. (2015). *Sustainable Development Goals.* United Nations. Retrieved December from https://www.un.org/sustainabledevelopment/water-and-sanitation/

Walker, A. (2012). *What is Boko Haram?* https://www.usip.org/publications/2012/05/what-boko-haram

WASH Sector Nigeria. (2021a). *WASH Sector North-East Nigeria Response Monitoring Dashboard 2021.* Retrieved October 31 from https://app.powerbi.com/view?r=eyJrIjoiNTlmZGVhZGUtY-Tc1ZC00OGY5LWI3ZDAtODc4N2FhYThiMjdmIiwidCI6ImY2ZjcwZjFiLTJhMmQtNGYzMC04NTJhLTY0YjhjZTBjMTlkNyIsImMiOjF9&pageName=ReportSection

WASH Sector Nigeria. (2021b). *Nigeria WASH Response Newsletter: The Time is Now.* https://drive.google.com/file/d/1W5O8vpLzP-3OVlmKmNmjDU6enM0oSzMg/view

WHO/UNICEF. (2021). *Progress on Household Drinking Water, Sanitation and Hygiene 2000-2020: Five years into the SDGs.* https://www.who.int/publications/i/item/9789240030848

KEY TERMS AND DEFINITIONS

Community: A group of people with a combined interest living in a particular location.

Conflict: A protracted act of disagreement that can lead to war.

Humanitarian Actors: People that seek to promote human welfare.

Insurgency: Rebellious acts or armed uprising typically against the recognised government.

Internally Displaced Persons: People who are forced to leave their homes due to conflict or natural disasters but still reside within their country borders.

Interventions: Acts carried out towards reducing harm or difficulty.

Standard: A measure of quality or attainment.

Sustainable: The ability for longevity or continuous existence.

WASH: Water Supply, Sanitation, and Hygiene.

Chapter 4
The Presence and Impacts of Microplastics in Drinking Water:
Their Occurrence, Detection, Removal, and Implications

Alice Liddell
Queen's University Belfast, UK

Marco Geron
Queen's University Belfast, UK

Eoin Cunningham
Queen's University Belfast, UK

Beatrice M. Smyth
https://orcid.org/0000-0002-2983-2404
Queen's University Belfast, UK

ABSTRACT

This chapter discusses the issues of plastic, primarily microplastic pollution in freshwater and drinking water, with a focus on developing nations. Microplastics, generally defined as plastic particles with a size less than 5 mm, are beginning to gain attention as an emerging contaminant of concern. Whilst testing has recently begun on the contamination of freshwater and treated drinking water by microplastics in a number of developed regions, literature regarding microplastic pollution in the water of less economically developed countries is lacking. Microplastics pose a threat to human health, and therefore, it is important that cost-effective methods for the testing, detection, and removal of these plastic items from drinking water globally is considered with a higher level of urgency. It is argued that by achieving the aims laid out by the Sustainable Development Goals, particularly Goals 6 and 12, the threats from microplastic pollution will subside.

DOI: 10.4018/978-1-7998-9190-1.ch004

Figure 1. Chemical structure of high demand plastics
PP: polypropylene, PE: polyethylene, PVC: polyvinylchloride, PUR: polyurethane, PET: polyethylene terephthalate, PS: polystyrene

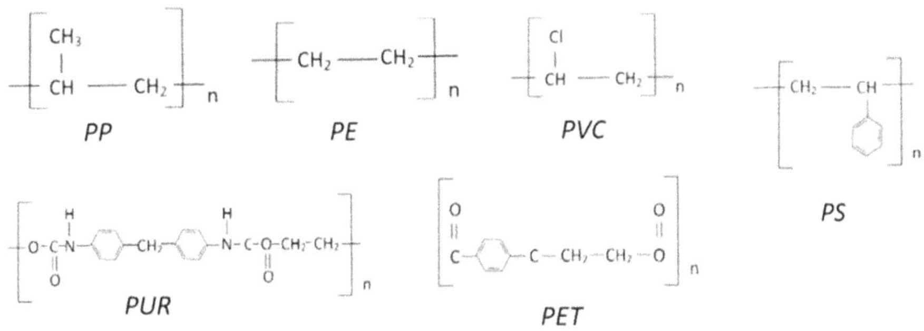

INTRODUCTION

The attractiveness of plastic, defined by its mechanical properties, inexpensiveness, and relative ease of production, has led to a continual rise in its manufacture, with an increase of 9 Mt plastic produced between 2018 and 2019 alone (Plastics Europe, 2020). The popularity of plastic items follows from the development of the first fossil fuel derived plastic, Bakelite, in 1907, encouraging the rapid development of similar plastics in the following years (polyester, nylon, polythene etc.) (Wagner & Lambert, 2018). The durability of plastic products, albeit seen as an attractive attribute to producers across all sectors, has recently led to concerns regarding plastic accumulation in the environment. Out of all the plastic ever produced, it is estimated that only 9% has been recycled, with 60% either being sent to landfill or released directly into the natural environment (Geyer et al., 2017).

Plastics are a branch of synthetic polymers, which are monomers that are specifically arranged in repeating units to deliver desirable properties. Polymers are long chains of molecules (monomers) and can be synthetic or naturally occurring. Polymers are derived from various hydrocarbon and petroleum materials, usually have a high molecular weight, and are connected by strong chemical bonds. Thermoplastics (which may be moulded at elevated temperatures) and thermosets (irreversibly hardened by curing) are the two main groups of plastics. Typical plastics (Figure 1) are made up of carbon and hydrogen with an extension to nitrogen, oxygen and chlorine groups.

The irresponsible disposal of plastic items is often evident adjacent to, or within, rivers and lakes. The result of this is the accumulation of plastic in freshwater waterways, either by direct dumping or their migration to water bodies through water run-off from the land. Currently when discussing the issue of plastic pollution in aquatic environments, the focus is on the quantity of plastic in the marine environment, with freshwater plastic pollution largely overlooked and understudied. A study published in 2021 (Meijer et al., 2021) showed that more than one thousand rivers were responsible for 80% of global riverine plastic emissions to the ocean, indicating the widespread contamination of rivers on a worldwide scale. Asian rivers were the most significant contributors to plastic pollution in the oceans, particularly those in areas with high population densities (Meijer et al., 2021). Although Asia has experienced rapid growth in terms of human development, the Asia and Pacific region accounts for almost half of the multidimensional poor worldwide, with South Asia home to more than 41% of the total (Conceição,

2019). This emphasises that the highest level of pollution occurs in regions that are the least economically developed, highlighting the need for improved sustainability in these regions.

In areas where plastic pollution of freshwater bodies is extensive, microplastic contamination will subsequently be significantly greater. Over time, larger plastic items (macroplastics) degrade, due to mechanisms such as mechanical fracture from water turbulence and photo-oxidation from UV light, causing weakening and eventual fragmentation into microscopic pieces (Enfrin et al., 2019). Microplastics are generally defined as any plastic item with a size < 5 mm (Wagner & Lambert, 2018). With no current methods for removing microplastics from environmental matrices, a large portion is accumulated in marine and freshwater environments. Small plastic fragments, which fit the current microplastic terminology, were first recorded in the Sargasso Sea, North Atlantic Ocean, in 1972 (Carpenter & Smith, 1972). Despite this, research regarding microplastic contamination remains in its infancy.

Following the consumption of microplastics, their fate within the human body is not yet well understood. Research has indicated the capabilities of small marine organisms to uptake microplastics (Van Cauwenberghe & Janssen, 2014), sparking concern regarding human health (Karbalaei et al., 2018). Several studies have been undertaken using mammalian cell lines to assess potential cytotoxic effects of micro/nano-plastics (Barbosa et al., 2020). The results of such studies have indicated: cytotoxic effects in mouse blood and DNA damage (Gopinath et al., 2019; Zheng et al., 2019), differential responses in expression patterns of genes in sea urchins (Bergami et al., 2019) and enhanced pharmaceutical induced cytotoxicity through the addition of nanoplastics to fish cell lines (Almeida et al., 2019). It is thought that smaller microplastics and nanoplastics are more capable of becoming internalised by cells, with predictions stating that microplastics with a size of approximately 20 μm will have the ability to penetrate organs, and those with a size of approximately 10 μm or less will have the ability to pass across membranes of cells and cross the blood-brain barrier (including within humans) (Campanale et al., 2020). Microplastics with a size of approximately 10 μm have recently been detected in the placentas of four women following childbirth, demonstrating their ability to pass into, and through, human organs (Ragusa et al., 2021).

Research has shown the presence of microplastics in both freshwater sources, such as rivers, and treated drinking water in several countries (Mintenig et al., 2019, Pivokonský et al., 2020). The pollution of the natural environment, stemming from the generation and subsequent mismanagement of plastic waste, is a global issue which can be recognised in every country, but it is generally the less economically developed nations which experience the brunt of the problems associated with plastic contamination (United Nations, 2021b). However, with the exception of China (Wang et al., 2020), there is scarce knowledge regarding microplastics in water sources and drinking water in developing countries. Whilst tackling microplastic contamination of water supplies is not currently the priority for many developing countries, it is an issue with sustainability at its core and straddles several of the United Nations Sustainable Development Goals (SDGs). The presence of microplastics in drinking water is an indication of failure to achieve SDG 12 Responsible Production and Consumption, with potentially serious implications for SDG 6 Clean Water and Sanitation and SDG 14 Life Below Water. The need for resilient infrastructure (SDG 9) to mitigate the impacts is clear, and both humanitarian and sustainable engineering have important roles to play in its delivery. The purpose of humanitarian engineering is to improve lives and livelihoods by delivering engineering solutions to disadvantaged people and communities, while sustainable engineering seeks to utilise resources carefully to protect the environment and the needs of future generations.

By considering the issues through the lens of humanitarian and sustainable engineering, the aim of this chapter is to provide an overview of microplastics, the current knowledge regarding microplastics in drinking water and the implications for developing nations. Given the potential direct impacts on human and environmental health, and the links between microplastics in drinking water and other environmental and humanitarian issues, this is a topic that should be considered with increased importance and urgency.

MICROPLASTICS IN DRINKING WATER

Microplastics in Freshwater and Regional Variations

As mentioned, plastic items eventually fragment, over time, into microscopic pieces within the natural environment. Subsequently, in areas where the mismanagement of plastic is high, the level of pollution from microplastics will be greater. Wealthier nations, including European countries and the United States of America, have, over many years, exported a considerable portion of their plastic waste to developing countries, which often have less stringent environmental laws (McCormick et al., 2016). One study showed that Malaysia, Indonesia and Vietnam mismanaged 55%, 81% and 86% of their own waste respectively, while also accepting plastic waste from the United States of America and other developed nations (McCormick et al., 2016). In the continent of Africa, over 500 shipping containers worth of plastic waste are imported each month, with 75% of this predicted to end up in landfill (Amankwa et al., 2021). Within this continent, 31/54 countries are considered to have low human development as of 2021 (considered as any country with a Human Development Index of 0 – 0.55) (World Population Review, 2021). The lack of infrastructure for handling the influx of plastic waste, and the delivery of plastic items which are of an insufficiently high standard to be recycled or re-used, alongside the increased demand for plastic items from growing local populations, has resulted in widespread plastic pollution in a number of developing nations (McCormick et al., 2016).

The main types of microplastics found in freshwater systems globally are PE, PET, PVC and PS, with these making up ~ 70% of the total microplastics identified (Li et al., 2020). Current knowledge suggests that fibres released from washing garments, micro-beads in personal care products, plastic water bottles and lids, tyre wear on roads and the outputs from wastewater treatment plants are some of the major contributors of microplastics into drinking water (Wagner & Lambert, 2018), however this is again expected to vary dependent on the location. Determining the sources of microplastics in water could provide valuable knowledge for subsequent remediation.

The wide variation of properties between individual plastic objects and the subsequent microplastics formed, makes it difficult to predict both their behaviour in water bodies and the likelihood of microplastics reaching drinking water systems from various sources. However, what is known is that the various densities of polymers (Table 1), alongside the type and shape of particles (fibre, pellet etc.), may alter their rising and settling velocities, and therefore their distributions through the water column. (Khatmullina & Isachenko, 2017). Furthermore, as reported in a modelling study on the fate of nano and micro-sized plastics in freshwater systems, it is thought that contaminants which adhere to the surfaces of microplastics significantly affect vertical distributions due to an increase in weight (Besseling et al., 2017). Therefore, the varying surface areas of microplastics will potentially play a more significant role in their distribution through the water column, compared to other properties such as density.

In addition to the mismanagement of plastic waste, there appears to be some correlation between the production of plastics and the level of microplastic pollution, however, this is a topic which requires additional research. In 2019, 368 Mt of plastic were produced globally, and China was responsible for the

Table 1. Highest demand polymers and their associated properties (LD: low density, MD: medium density, HD: high density, Mt: Megatonnes) (Plastics Europe, 2020; Lambert & Wagner, 2017).

Polymer	Abbreviation	Density (g cm^{-3})	Common uses	EU demand 2019 (Mt)
Polypropylene	PP	0.85-0.94	Food packaging, wrappers, microwave containers	~ 9.9
Polyethylene	PE	0.91–0.97	Bottles/caps (HDPE), water pipes (MDPE), personal care products, tyres (cords), plastic bags	LD: ~ 8.8 MD: ~ 6.4
Polyvinyl chloride	PVC	1.38	Clothing, containers, films	~ 5.0
Polyethylene terephthalate	PET	1.34–1.39	Bottles, strapping	~ 4.0
Polystyrene	PS	0.96–1.05	Containers, packaging, utensils	~ 1.5

Note: The EU data for 2019 has been included opposed to the latest 2020 data, due to Covid-19 resulting in considerable deviations from normal trends for 2020.

production of 31% of all plastic produced that year; Asia produced 51% of plastic in total, while Europe produced 16% (Statista, 2021). In one study the microplastic pollution of 37 freshwater bodies globally was analysed based on a regression model (Chen et al., 2020). It found that the level of contamination of the continents was in the following order: Asia > North America > Africa > Oceania > South America > Europe (no information is available for Antarctica).

The level of pollution was found to be significantly higher in developing countries than in developed countries, with China showing the greatest level of pollution overall (up to 3790 ± 800 microplastics/m^3), and Switzerland the lowest level of pollution (average of 0.04 ± 0.01 microplastics/m^3) (Chen et al., 2020). Out of the 37 freshwater bodies analysed, 20 were located within developing countries, with an average of 3030 ± 725 microplastics/m^3 reported within these locations. The remaining water bodies were located in developed countries, with an average of 189 ± 189 microplastics/m^3 reported. However, it should be noted that, of the 20 water bodies analysed in developing regions, 16 were located in China, further highlighting the lack of knowledge of microplastic pollution within developing countries, apart from China (Chen et al., 2020).

There is also a direct relationship between population and the level of microplastic pollution, with water bodies located in urban settings generally having a higher level of pollution than those in rural settings. The influence of the level of urbanization/population on the subsequent pollution from microplastics can be demonstrated even within much more localized settings. Two drinking water treatment plants (DWTPs), along the Úhlava River in the Czech Republic, were recently analysed to assess the quantity of microplastics entering and leaving each water treatment plant (Pivokonský et al., 2020). One water treatment plant extracted water from a dammed reservoir which was surrounded by protected land, and the other sourced water directly from the river, 90 km further downstream. In the water entering the upstream treatment plant, 23 ± 2 microplastics/L were recorded, whereas 1300 ± 35 microplastics/L were recorded in the water entering the treatment plant located further downstream.

Settlements and wastewater treatment plants were highlighted as potential sources responsible for the increase in microplastics in the raw water (Pivokonský et al., 2020), emphasising the influence of population density and urban development on the number of microplastics recovered. This has implications worldwide, as the global urbanisation rate has been rising rapidly (UN, 2018), but there are particular concerns for many developing regions, where the speed of urbanisation has been higher (UN, 2018).

Microplastics in Bottled Water

The demand for bottled water is continually rising on a global scale. It was stated in Euromonitor International's global packaging trends report that 480 bn plastic drinking bottles were sold globally in 2016, which was an increase of 300 bn within a decade (Euromonitor International, 2017). Additionally, a report from 2017 stated that a million plastic bottles were being consumed every minute globally and this was predicted to increase by another 20% by 2021 (Laville & Taylor, 2017). The increase in demand for bottled water is often due to population growth and bottled water being considered a more accessible and sanitary option for communities who do not have access to clean tap water. The World Health Organization and UNICEF stated that in 2020, around one in four people lacked access to safely managed drinking water within their homes (European Environment Agency, 2021), encouraging those who are without access to turn to packaged water. Whilst this is in some ways a successful method for providing communities with safe water who do not have access to it by other means, the use of plastic packaging for water consumption comes with its own set of problems.

When plastic water bottles are continually re-used the plastic degrades over time. The weakening and eventual fragmentation of the plastic bottles means that additional microplastics are released into the beverage within the container, especially if it is exposed to intense UV/sunlight or re-used several times. Consequently, the consumption of bottled water is an additional pathway for microplastic ingestion by humans. In one study, 259 bottles of water were tested for microplastic contamination, sourced from 19 locations globally (Mason et al., 2018). Of the bottles tested, 93% showed some sign of microplastic contamination. A range of 0 to 10,000 microplastics/L was reported, with an average of 325 microplastics/L found. It was predicted that the bottling process or the packaging itself was at least partially responsible for the contamination of the water (Mason et al., 2018). Polypropylene (PP) fragments were commonly detected in the bottled water; this is the material which is generally used to make the bottle caps (Mason et al., 2018). It is thought that the action of screwing and unscrewing the cap from the bottle contributes to the quantity of microplastics found in bottled water with plastic screw-top lids. Again, this will be more significant in cases where the bottles are used multiple times.

It is also predicted that, as with freshwater discussed above, the source of the water used for bottling (and the level of plastic pollution in it) plays a role in influencing the number and type of microplastics in the packaged water. Another influencing factor is the increased demand for bottled water, which adds to the problem of plastic pollution as a whole. Between 2007 and 2017, global bottled water consumption rose from an estimated 212 billion litres to 391 billion litres (Conway, 2020). Bottles which are not disposed of correctly are released into the natural environment, contaminating the land and waterways, and will eventually break down into smaller plastic pieces over time. To relieve the issues which are associated with an increased demand for bottled water, it is necessary that clean and safe drinking water from tap water sources is provided to communities globally. This will, in turn, reduce the pressures created from plastic pollution overall and subsequent microplastic contamination.

DETECTING MICROPLASTICS

Classification of Microplastics

The most frequently used definition of microplastics is any plastic particle < 5 mm in diameter. This definition was proposed in a meeting of experts hosted by the National Oceanic and Atmospheric Administration (NOAA) in 2008 (Arthur et al., 2009) and arose mainly from the suspected ease of organisms to uptake particles of such sizes. A lower cut-off value of 333 µm has also traditionally been accepted due to available mesh sizes of nets often used for microplastic sample collection, particularly within marine environments (Arthur et al., 2009) but, with recent advances, multiple studies have now successfully recovered considerably smaller microplastics. Recently, a lower size limit of 1 µm was proposed (with anything smaller being considered a nanoplastic). However, these size ranges are much more difficult to test for and results are limited in literature. Suggestions have also been made to reduce the upper size boundary from 5 mm to 1 mm. These revised size classifications have been considered in some recent studies (Meng et al., 2020).

Besides the upper and lower boundaries, size classifications within boundaries vary. For example, some studies classify the microplastics recovered into discrete size classes (e.g., 1-10, 10-50, 50-100 µm etc.), whereas other studies split their results into 'large' and 'small' fractions only. The lack of well-defined boundaries/classes makes it difficult to draw worthwhile inter-study comparisons and it is likely that standardisation will prove necessary if a sound basis for regulatory frameworks is to be provided (Hartmann et al., 2019). Other descriptive properties, such as origin, and physical and chemical properties, are not currently considered in the definition. In addition to size, microplastics are generally defined as either 'primary' or 'secondary' microplastics. Primary microplastics refer to those which are intentionally created to be the size that they are, for use within various industries. Examples of these include micro-beads which are used in various cosmetic products, fibres used within the clothing industry and plastic powders used for moulding. Secondary microplastics are microplastics which are created from the breakdown of larger plastic items via various mechanisms.

Methods for Detecting and Analysing Microplastics

To date, there are no standardised methods for the detection and analysis of microplastics. Various methods have been used in different studies, often employing expensive and specialised equipment within state-of-the-art laboratories. The methods which are applied depend generally on the sizes of the microplastics which are targeted in individual studies. For freshwater, primarily drinking (tap and bottled) water testing, it is preferable to analyse microplastics of the smallest sizes possible, as these are thought to be the most abundant and indicate the greatest potential risks to human health (Novotna et al., 2019). In areas where the drinking water is cleaned to high standards, few microplastics of the larger size categories (> ~ 100 µm) are generally reported, and therefore the need for specialised equipment which can analyse smaller sized particles (~ 20 µm) is required. It is also generally preferable for the chemical composition of the recovered microplastics to be analysed, allowing for insights into the potential sources of the microplastics.

Traditionally microplastics have been identified using the naked eye, providing a fast and simple method but one which is not applicable for smaller sized microplastics (< 2 mm) and is often linked with high error values (Wang and Wang, 2018; Löder and Gerdts, 2015; Shim et al., 2017). Alternatively, microscopy, for example a stereo-microscope, can be applied. Microscopic imaging techniques have higher reliability than analysis with the naked-eye, however, no compositional data is gained and smaller microplastics are still likely to be overlooked. Scanning electron microscopy (SEM) is another option, providing images with a higher resolution at a high magnification by firing an electron beam at the sample surface. This allows for an assessment of the surface characteristics of particles, for example measurements of their size, morphology analysis and identification of cracks due to degradation (Wang & Wang, 2018). When coupled with energy dispersive X-ray spectroscopy (SEM-EDS), information concerning the elemental composition of particles and inorganic materials or additives they contain may be gained (Wang et al., 2017), however this involves long processing times and is not considered feasible for large numbers of samples (Wang & Wang, 2018).

With the advancement of identification techniques, it is generally recommended that compositional analysis is applied to ensure that particles counted are in fact of plastic origin, and, if so, to determine what type of plastic they are. Vibrational spectroscopic techniques, such as Fourier Transform Infrared (FTIR) and Raman spectroscopy, are becoming increasingly popular for microplastic analysis. FTIR spectroscopy directs an IR beam at the sample, intensifying vibrational motions in the bonds between molecules. When the frequencies of such vibrational motions match the frequency of the incoming IR light, absorption occurs. This is shown as peaks on a spectrum, displayed as % transmission (y-axis) against wavenumber (a function of frequency) in cm^{-1} (x-axis), demonstrating which functional groups are present within a molecule or sample. Spectra can be compared to comprehensive polymer libraries, allowing determination of the composition of microplastics. Micro (μ) FTIR is an additional FTIR accessory which allows the operator to resolve the differential chemistry of a particle in a very small area and can analyse microplastics down to 10 μm. However, the additional equipment is expensive and not always accessible.

Raman spectroscopy is another form of vibrational spectroscopy which uses a monochromatic light source. The sample can absorb and re-emit/scatter a portion of the incoming light, with the backscattered light having varying frequencies dependent on the absorption, scattering or reflective properties of the polymers present. This again results in a unique spectrum. Micro (μ) Raman spectroscopy has the advantage of detecting and identifying particles ~ 1-10 μm, with one study stating that almost 80% of the microplastics between 5-20 μm found in their study would not have been detected with FTIR (Schymanski et al., 2018). However, long processing times are associated with μ-Raman spectroscopy and generally only a fraction of the filter is analysed and the results extrapolated.

Implications for Humanitarian Engineering and Developing Countries

Overall, microscopy may prove useful as a preliminary step for identification, but the high error rates mean that results may not be reliable and smaller microplastics cannot be assessed. SEM analysis is a promising method for the quantification of microplastics, with greater reliability than conventional microscopy, but again it is recommended that it be used in partnership with compositional analysis wherever possible. To allow for the analysis of the smallest sized microplastics (< ~ 20 μm), μ-Raman and μ-FTIR spectroscopy currently show the greatest potential for producing valuable results. However, spectroscopic techniques rely on expensive and specialised equipment, and so might be too resource intensive to enable microplastic testing/monitoring in some countries.

Reliable and accurate techniques for the analysis of microplastics within environmental samples are continuing to be developed and applied globally, however there is a clear lack of low-cost methods available. The majority of equipment used for microplastic analysis is specialized and expensive, creating a barrier for continual microplastic testing in developing countries. Microscopic imaging techniques (such as stereo-microscopes) can provide a basic analysis of larger microplastics within water systems, and can be employed at a relatively low cost, but they are unable to give an overview of smaller microplastics that are present nor any details of chemical composition, surface characteristics etc. Therefore, although the use of basic microscopes may provide useful preliminary results, it is recommended that they are applied in partnership with other techniques.

Perhaps, as testing continues, results from studies which employ spectroscopic techniques, or other methods which enable the analysis of microplastics on a smaller scale, can be scaled to predict the numbers of smaller microscopic plastic pieces within drinking water, in cases where more specialized equipment cannot yet be applied. It is not possible to remediate what we cannot measure, so low-cost and efficient methods for the detection and analysis of microplastics within water systems should be considered with a high level of importance.

MICROPLASTIC REMOVAL BY WATER TREATMENT

From Inlet to Outlet

Knowledge regarding the removal of microplastics via drinking water treatment remains limited, but some recent studies (Novotna et al., 2019; Pivokonský et al., 2020; Wang et al., 2020) have demonstrated the capabilities of drinking water treatment technologies to prevent microplastics reaching household drinking water. The removal of microplastics is dependent not only on the quality of the raw water drawn into the plant for treatment, but also on the specific treatment technologies applied. Some treatment stages have been shown to have a more significant impact on microplastic removal than others (Novotna et al., 2019). Dependent on location, varying technologies are used for water treatment. More economically developed countries tend to have more advanced technologies, but the construction and use of state-of-the-art drinking water treatment technologies is expensive and not feasible for application in many developing regions. This is an issue which is generally not discussed in the conversation regarding microplastics in drinking water, as the testing has to date been carried out primarily in developed nations.

Disparities in the overall removal of microplastics via water treatment, due to the level of advancement of DWTPs, can be highlighted in the results from the DWTP study discussed earlier, where sampling was carried out at two DWTPs in the Czech Republic. The treatment plant located upstream had an overall microplastic removal (from the raw water at the inlet to the treated water leaving the plant) of 40%, compared to 88% removal by the treatment plant located further downstream. The upstream plant had only basic treatment whereas the downstream plant involved a two-step removal of aggregates, followed by advanced treatment using granular activated carbon (GAC) filtration (Pivokonský et al., 2020). Similarly, in other studies which have tested the removal of microplastics by DWTPs which apply advanced levels of treatment, overall microplastic removals of 86%, 83% and 81% have been reported (Pivokonsky et al., 2018, Wang et al., 2020). Although there is variation in the overall removal rates reported, it is clear that DWTPs, particularly those with sophisticated technologies, act as a barrier to microplastics entering drinking water. Individual treatment technologies have differing impacts on microplastics, and these are discussed in detail in the next section to provide an understanding of the relative effectiveness of the technologies and their potential application in developing countries.

It is also worth noting that seasonal variations are likely to affect microplastic presence in the raw water entering DWTPs, and also overall removal of microplastics. One study, undertaken in Thailand, reported removal of microplastics via drinking water treatment of 67% in the dry season and 57% in the rainy season, with the average number of microplastics in the raw water entering the DWTP 1425 microplastics/L in the dry season and 1882 microplastics/L in the rainy season (Kankanige & Babel, 2021). In areas where flooding events or monsoon weather events are more prevalent, or where the infrastructure to deal with such weather events is not as developed, such weather conditions could result in a greater quantity of microplastics present within the water system. The increase in microplastics in the rainy season is likely due to greater volumes of surface run-off carrying microplastics into the water supply from overland contamination, the increased water turbulence disturbing and subsequently carrying microplastics previously within sediment to the DWTP, and a reduction in sedimentation overall (Kankanige & Babel, 2021).

Individual Treatment Technologies

Generally, the main stages of state-of-the-art drinking water treatment consist of clarification (coagulation/flocculation combined with sedimentation or dissolved air flotation (DAF)), filtration and disinfection. Some plants also include screening ('preliminary treatment') as a first step.

Clarification

Clarification depends on the processes of coagulation and flocculation, assisted by the addition of coagulant aids, generally in the form of alum (Al3+) or ferric salts (Fe3+), or a synthetic organic polymer, polyacrylamide (PAM), which cause negatively charged natural impurities to neutralise and clump together to form flocs. Some DWTPs allow flocs to settle to the bottom of a tank/clarifier for removal (sedimentation), whereas others pass air bubbles through the water (DAF), increasing the buoyancy of flocs, which are removed from the surface by a skimmer.

Studies which have tested for microplastics in DWTPs have shown promising results for the removal of microplastics through the application of this process. A removal of 41-55% of MPs > 1 μm, with preferable removal of fibrous microplastics, was reported in one study (Wang et al., 2020), with another reporting 62% removal of microplastics > 1 μm (Pivokonský et al., 2020). Due to the low density of microplastics, clarification using DAF may be a more efficient method for removal than sedimentation (Zhang et al., 2020). No studies were found that analysed coagulation/flocculation combined with DAF as an individual step, and this should be further explored.

Filtration

Either slow sand filtration or rapid gravity filtration (RGF) occur next, trapping particulate matter via adsorption to the filter media. With slow sand filtration, the top layer of sand is removed regularly for cleaning and re-used. With RGF, water and air are pushed backwards through the filter at high pressure every 24-48 hrs. This backwashed water is either released into nearby water courses or diverted back to the start of the treatment process, meaning that microplastics trapped within filter media can potentially be re-introduced into the natural environment or into the treatment system, presenting a second opportunity for smaller particles to pass through the filter.

Again, results have shown that filtration can be a successful method for removing microplastics from the water system. In a lab bench study, mimicking the drinking water filtration process, polyethylene microplastics were shown to be completely rejected by a 30 nm pore size polyvinylidene fluorine membrane, with little membrane fouling occurring (Ma et al., 2019). Zhang et al. (2020) reported that particles > 100 μm could be completely removed via filtration (cheese cloth and anthracite were used to mimic DWTP filtration), but particles 1-20 μm (especially 10-20 μm) were more likely to remain. However, these results show high removability compared to studies which sampled directly from DWTPs, where low removal of fibres was reported, likely due to their small diameter (Pivokonský et al., 2020). Fibres were not tested in the bench-scale studies, and this could account for higher removal rates reported.

Granular activated carbon (GAC) media is sometimes applied after RGF to remove additional organic compounds. When analysed as a separate step, GAC filtration has indicated a significant reduction in microplastic counts (Pivokonský et al., 2020). Overall, filtration shows promise in the removal of microplastics from DWTPs, and there are indications that the application of dual-media filtration (such as using additional GAC filtration) improves this step further.

Disinfection

Disinfection, via chlorination, UV treatment or ozonation, is often the final treatment stage, and is applied to kill pathogenic organisms. It has been reported than an increase of up to 16% the number of 1-5 μm microplastics was recorded following ozonation, indicating that this step assists in the breakdown of larger microplastics into smaller sizes (Wang et al., 2020). Therefore, instead of disinfection acting as a method for the removal of microplastics from the water supply, it instead breaks plastics into smaller pieces, which are of more concern to human health (Novotna et al., 2019). Additionally, it has been suggested that microplastic presence hinders the efficiency of disinfection processes. During ozonation, microplastics decrease the ozone molecules available, they can protect pathogenic microorganisms from damage during UV treatment, and they act as a protective substrate for bacteria during all disinfection processes (Enfrin et al., 2019). This will be particularly problematic in cases where pathogenic microorganisms and bacteria are present in the water in higher quantities and are associated with more severe health risks.

Solutions for Developing Countries

DWTP technologies have been proven to show success in the removal of microplastics from the water system. However, it has been predicted by the World Health Organization and UNICEF that, by 2030, there will still be 1.6 billion people in the world without access to basic clean water (European Environment Agency, 2021); these technologies are therefore currently only an advantage to those who live in developed countries, where state-of-the-art water treatment is accessible. In communities where access to water treatment facilities is not reachable, the potential for contracting illnesses from waterborne diseases, due to harmful pathogens being present in the water systems, is significantly higher. There is concern that microplastics will increase the potential for contracting illnesses further, due to microplastics acting as vectors for other harmful substances. Biofilms which develop on microplastics, resulting in harmful substances adhering to their surfaces, will be ingested with any microplastics consumed. For example, zinc was shown to desorb in the gut of an earthworm much more readily from microplastics than it does from soil (Hodson et al., 2017). This raises the concern that micro/nanoplastics can increase the bioavailability of contaminants, potentially posing threats to human health through leaching of adhered contaminants into the gut. Again, the threat will be greater for communities located in developing regions, where both the levels of microplastics in the water, and the harmful contaminants, will be greater.

To reduce the pressures that both microplastics and potential contaminants adhered to them have on human health, it is vital that safe drinking water is provided to all. One potential solution for developing countries, where clean water treated by state-of-the-art drinking water treatment technologies cannot yet be provided, is the wider distribution of point-of-use systems. These systems are a low-cost and low-energy alternative to DWTPs, employed on site, with the goal of reducing the level of harmful pathogens within the water. Based on the review of technologies above, and the evidence of different DWTP treatment methods successfully removing microplastics from the water system, it is predicted that these alternative, low-cost systems would also be successful in the removal of microplastics.

An example of a point-of use system is the use of sachets which contain ferric sulphates and calcium hypochlorite, providing an on-site alternative to coagulation and flocculation processes applied within DWTPs (Pooi & Ng, 2018). Bio-sand filtration is a promising alternative to conventional filtration processes applied within DWTPs and can make use of locally sourced materials. Similarly, biochar is another alternative to conventional filtration processes, providing an absorbent filter medium which can be applied at low-cost (Pooi & Ng, 2018). Working with individual communities to provide point-of-use systems presents an achievable method to provide safe water which is free of disease-causing pathogens and potentially also reduce the risk of harm to human health from microplastics within the water.

MICROPLASTICS AND THE SUSTAINABLE DEVELOPMENT GOALS

The Sustainable Development Goals address global challenges, with the aim of providing a better future for all. There are 17 goals in total, which have been adapted by several governments and corporations to improve sustainability on a global scale. Each goal is split into targets, which are measurable against specific indicators. Despite the widespread pollution from microplastics, there is currently only one goal which explicitly mentions (micro)plastic pollution of the aquatic environment. Under SDG 14 (Life Below Water), indicator 14.1.1b relates to reducing the pressure of plastics on the marine environment. The authors argue that action to manage and mitigate the impacts of microplastics on human health globally aligns with additional Sustainable Development Goals, in particular SDGs 6 and 12. This provides an opportunity for updating these goals, to present governments with data and recommendations for tackling this emerging pollutant.

The Sustainable Development Goal 6, Clean Water and Sanitation, is particularly relevant to the issues discussed. This goal aims to 'ensure availability and sustainable management of water and sanitation for all' by 2030 (United Nations, 2021a). Although the impacts on human health from microplastics are not yet fully understood, the provision of clean drinking water will reduce the risks associated with this emerging contaminant. It has been shown that water treatment removes a significant portion of microplastics from drinking water, alongside pathogens and harmful contaminants which may adhere to their surfaces. Reaching this goal will address concerns surrounding the potential additional harmful effects of microplastics on human health in developing countries, from a lack of treated water and an increased amount of plastic pollution, in comparison to those which are developed. Additionally, the provision of clean water will reduce the need for the use of bottled water. This will in turn reduce pressures of macroplastic pollution of freshwater bodies and subsequent microplastic pollution from the breakdown of mis-managed bottles in water systems. Furthermore, the potential detrimental impacts of bottled water consumption (due to more microplastics likely being present in bottled water than tap water, from the breakdown of the plastic bottle), will be reduced.

SDG 12, Sustainable Consumption and Production, again directly relates to the issues discussed in this chapter. This goal aims to 'substantially reduce waste generation through prevention, reduction, recycling and reuse,' by 2030. Achieving this goal will massively impact the generation and release of microplastics into freshwater systems, and subsequently into drinking water intended for human consumption. Tackling plastic pollution as a whole and reducing the quantity of plastic waste which reaches water bodies, will in-turn significantly contribute to tackling the issues of microplastic contamination and pollution in the environment as a whole. Although water treatment can remove microplastics from the water system, we should focus on tackling this issue at source by reducing our plastic production and consumption, avoiding mis-management of plastic waste, and preventing additional uncontrolled plastic pollution from entering the environment.

CONCLUSION

The pollution of freshwater from both macroplastics and microplastics is evident globally, but developing countries face the most significant consequences of plastic mis-management. The level of pollution of freshwater and drinking water by microplastics is only beginning to gain attention, and the contamination of water in developing countries is still far from understood. Evidence of contamination from already developed regions, which have state-of-the-art drinking water treatment facilities and high-end waste management facilities, allows us to predict that a much more severe level of pollution is occurring within not yet developed countries. Although microplastic pollution of drinking water is not currently the main priority for clean water provision in many regions, with water-borne diseases taking precedence, this emerging contaminant deserves attention before pollution levels become more extreme. Research has provided evidence of the detrimental impacts of microplastics on the health of marine and mammalian life, with impacts on human health still not fully understood. It is now important to understand the links between microplastic concentration, sizes and characteristics, and effects on human health. Low-cost testing techniques are required for the identification and characterisation of microplastics; this is currently a gap in the market and needs to be addressed. Point-of-use water treatment systems show promise for providing safe water in developing countries and should be further investigated for their effectiveness at microplastic removal. On a wider scale to ensure sustainability and protection of ecosystems, solutions to the problem of microplastic pollution need to be focused on the source, and sustainable production, consumption and waste management processes are essential. Tackling microplastic pollution requires an inter-disciplinary approach, from an understanding of human health impacts and the links with microplastic characteristics and relationships with contaminants/bio-films, to the development of methodologies for testing and monitoring pollution levels, and the techniques which can be applied for microplastic removal from the water system. Humanitarian engineering solutions which provide safe and clean drinking water to communities in all countries need to continue to be developed and provided as a matter of urgency.

ACKNOWLEDGMENT

This research was supported by the Engineering and Physical Sciences Research Council (EPSRC) (UK) [project reference: 2280886]. We thank Northern Ireland Water for their assistance in the conceptualisation of this work.

REFERENCES

Almeida, M., Martins, M. A., Soares, A. M. V., Cuesta, A., & Oliveira, M. (2019). Polystyrene nanoplastics alter the cytotoxicity of human pharmaceuticals on marine fish cell lines. *Environmental Toxicology and Pharmacology*, *169*, 56–65. doi:10.1016/j.etap.2019.03.019 PMID:30953935

Amankwa, M. O., Tetteh, E. K., Mohale, G. T., Dagba, G., & Opoku, P. (2021). The production of valuable products and fuel from plastic waste in Africa. *Discover Sustainability*, *2*(1), 31. Advance online publication. doi:10.100743621-021-00040-z

Arthur, C., Baker, J., & Bamford, H. (2009). *Proceedings of the International Research Workshop on the Occurrence, Effects, and Fate of Microplastic Marine Debris.* NOAA Technical Memorandum NOS-OR&R-30. https://marinedebris.noaa.gov/proceedings-international-research-workshop-microplastic-marine-debris

Barbosa, F., Adeyemi, J. A., Bocato, M. Z., Comas, A., & Campiglia, A. (2020). A critical viewpoint on current issues, limitations, and future research needs on micro- and nanoplastic studies: From the detection to the toxicological assessment. *Environmental Research*, *182*, 109089. Advance online publication. doi:10.1016/j.envres.2019.109089 PMID:32069751

Bergami, E., Krupinski Emerenciano, A., González-Aravena, M., Cárdenas, C. A., Hernández, P., Silva, J. R. M. C., & Corsi, I. (2019). Polystyrene nanoparticles affect the innate immune system of the Antarctic sea urchin Sterechinus neumayeri. *Polar Biology*, *42*(4), 743–757. doi:10.100700300-019-02468-6

Besseling, E., Quik, J. T. K., Sun, M., & Koelmans, A. A. (2017). Fate of nano- and microplastic in freshwater systems: A modeling study. *Environmental Pollution*, *220*, 540–548. doi:10.1016/j.envpol.2016.10.001 PMID:27743792

Campanale, C., Massarelli, C., Savino, I., Locaputo, V., & Uricchio, V. F. (2020). A Detailed Review Study on Potential Effects of Microplastics and Additives of Concern on Human Health. *International Journal of Environmental Research and Public Health*, *17*(4), 1212. Advance online publication. doi:10.3390/ijerph17041212 PMID:32069998

Carpenter, E. J., & Smith, K. L. Jr. (1972). Plastics on the Sargasso sea surface. *Science*, *175*(4027), 1240–1241. doi:10.1126cience.175.4027.1240 PMID:5061243

Chen, H., Qin, Y., Huang, H., & Xu, W. (2020). A regional difference analysis of microplastic pollution in global freshwater bodies based on a regression model. *Water (Basel)*, *12*(7), 1889. Advance online publication. doi:10.3390/w12071889

Conceição, P. (2019). *Human Development Report 2019: beyond income, beyond averages, beyond today: Inequalities in human development in the 21st century.* United Nations Development Program. http://hdr.undp.org/sites/default/files/hdr2019.pdf

Enfrin, M., Dumée, L. F., & Lee, J. (2019). Nano/microplastics in water and wastewater treatment processes – Origin, impact and potential solutions. *Water Research*, *161*, 621–638. doi:10.1016/j.watres.2019.06.049 PMID:31254888

Euromonitor International. (2017). *Global trends in food and drinks packaging: a report compiled for auspack 2017.* https://go.euromonitor.com/00-EV-APAC2017-AUSPACK_Thank-You.html?aliId=eyJpIjoidWZmM1dPUzBiblV3RDVRVSIsInQiOiJXdDBjRVN1NXR2SG5aRjRKa1plWEVnPT0ifQ%253D%253D

European Environment Agency. (2019). *Industrial wastewater treatment – pressures on Europe's environment (23/2018).* https://www.eea.europa.eu/publications/industrial-waste-water-treatment-pressures

Geyer, R., Jambeck, J. R., & Law, K. L. (2017). Production, use, and fate of all plastics ever made. *Science Advances*, *3*(7), 25–29. doi:10.1126ciadv.1700782 PMID:28776036

Gopinath, P. M., Saranya, V., Vijayakumar, S., Mythili Meera, M., Ruprekha, S., Kunal, R., Pranay, A., Thomas, J., Mukherjee, A., & Chandrasekaran, N. (2019). Assessment on interactive prospectives of nanoplastics with plasma proteins and the toxicological impacts of virgin, coronated and environmentally released-nanoplastics. *Scientific Reports*, *9*(1), 1–15. doi:10.103841598-019-45139-6 PMID:31222081

Hartmann, N. B., Hüffer, T., Thompson, R. C., Hassellöv, M., Verschoor, A., Daugaard, A. E., Rist, S., Karlsson, T., Brennholt, N., Cole, M., Herrling, M. P., Hess, M. C., Ivleva, N. P., Lusher, A. L., & Wagner, M. (2019). Are We Speaking the Same Language? Recommendations for a Definition and Categorization Framework for Plastic Debris. *Environmental Science & Technology*, *53*(3), 1039–1047. doi:10.1021/acs.est.8b05297 PMID:30608663

Hodson, M. E., Duffus-Hodson, C. A., Clark, A., Prendergast-Miller, M. T., & Thorpe, K. L. (2017). Plastic Bag Derived-Microplastics as a Vector for Metal Exposure in Terrestrial Invertebrates. *Environmental Science & Technology*, *51*(8), 4714–4721. doi:10.1021/acs.est.7b00635 PMID:28355064

Kankanige, D., & Babel, S. (2021). Contamination by ≥6.5 µm-sized microplastics and their removability in a conventional water treatment plant (WTP) in Thailand. *Journal of Water Process Engineering*, *40*, 101765. doi:10.1016/j.jwpe.2020.101765

Karbalaei, S., Hanachi, P., Walker, T. R., & Cole, M. (2018). Occurrence, sources, human health impacts and mitigation of microplastic pollution. *Environmental Science and Pollution Research International*, *25*(36), 36046–36063. doi:10.100711356-018-3508-7 PMID:30382517

Khatmullina, L., & Isachenko, I. (2017). Settling velocity of microplastic particles of regular shapes. *Marine Pollution Bulletin*, *114*(2), 871–880. doi:10.1016/j.marpolbul.2016.11.024 PMID:27863879

Laville, S., & Taylor, M. (2017, June 28). A million bottles a minute: world's plastic binge "as dangerous as climate change." *The Guardian*. https://www.theguardian.com/environment/2017/jun/28/a-million-a-minute-worlds-plastic-bottle-binge-as-dangerous-as-climate-change

Li, C., Busquets, R., & Campos, L. C. (2020). Assessment of microplastics in freshwater systems: A review. *The Science of the Total Environment*, *707*, 135578. doi:10.1016/j.scitotenv.2019.135578 PMID:31784176

Löder, M. G. J., & Gerdts, G. (2015). *Methodology Used for the Detection and Identification of Microplastics—A Critical Appraisal. In Marine Anthropogenic Litter*. Springer International Publishing., doi:10.1007/978-3-319-16510-3_8

Ma, B., Xue, W., Ding, Y., Hu, C., Liu, H., & Qu, J. (2019). Removal characteristics of microplastics by Fe-based coagulants during drinking water treatment. *Journal of Environmental Sciences (China)*, *78*, 267–275. doi:10.1016/j.jes.2018.10.006 PMID:30665645

Mason, S. A., Welch, V. G., & Neratko, J. (2018). Synthetic Polymer Contamination in Bottled Water. *Frontiers in Chemistry*, *6*, 407. doi:10.3389/fchem.2018.00407 PMID:30255015

McCormick, A. R., Hoellein, T. J., London, M. G., Hittie, J., Scott, J. W., & Kelly, J. J. (2016). Microplastic in surface waters of urban rivers: Concentration, sources, and associated bacterial assemblages. *Ecosphere*, *7*(11), e01556. Advance online publication. doi:10.1002/ecs2.1556

Meijer, L. J. J., van Emmerik, T., van der Ent, R., Schmidt, C., & Lebreton, L. (2021). More than 1000 rivers account for 80% of global riverine plastic emissions into the ocean. *Science Advances*, *7*(18), 1–14. doi:10.1126ciadv.aaz5803 PMID:33931460

Meng, Y., Kelly, F. J., & Wright, S. L. (2020). Advances and challenges of microplastic pollution in freshwater ecosystems: A UK perspective. *Environmental Pollution*, *256*, 113445. doi:10.1016/j.envpol.2019.113445 PMID:31733965

Mintenig, S. M., Löder, M. G. J., Primpke, S., & Gerdts, G. (2019). Low numbers of microplastics detected in drinking water from ground water sources. *The Science of the Total Environment*, *648*, 631–635. doi:10.1016/j.scitotenv.2018.08.178 PMID:30121540

Novotna, K., Cermakova, L., Pivokonska, L., Cajthaml, T., & Pivokonsky, M. (2019). Microplastics in drinking water treatment – Current knowledge and research needs. *The Science of the Total Environment*, *667*, 730–740. doi:10.1016/j.scitotenv.2019.02.431 PMID:30851606

Pivokonsky, M., Cermakova, L., Novotna, K., Peer, P., Cajthaml, T., & Janda, V. (2018). Occurrence of microplastics in raw and treated drinking water. *The Science of the Total Environment*, *643*, 1644–1651. doi:10.1016/j.scitotenv.2018.08.102 PMID:30104017

Pivokonský, M., Pivokonská, L., Novotná, K., Čermáková, L., & Klimtová, M. (2020). Occurrence and fate of microplastics at two different drinking water treatment plants within a river catchment. *The Science of the Total Environment*, *741*, 140236. doi:10.1016/j.scitotenv.2020.140236 PMID:32603938

Plastics Europe. (2020). *Plastics – the Facts 2020: An analysis of European plastics production, demand and waste data*. https://plasticseurope.org/knowledge-hub/plastics-the-facts-2020/

Pooi, C. K., & Ng, H. Y. (2018). Review of low-cost point-of-use water treatment systems for developing communities. *Npj Clean Water*, *1*(1), 11. Advance online publication. doi:10.103841545-018-0011-0

Ragusa, A., Svelato, A., Santacroce, C., Catalano, P., Notarstefano, V., Carnevali, O., Papa, F., Rongioletti, M. C. A., Baiocco, F., Draghi, S., D'Amore, E., Rinaldo, D., Matta, M., & Giorgini, E. (2021). Plasticenta: First evidence of microplastics in human placenta. *Environment International*, *146*, 106274. doi:10.1016/j.envint.2020.106274 PMID:33395930

Schymanski, D., Goldbeck, C., Humpf, H. U., & Fürst, P. (2018). Analysis of microplastics in water by micro-Raman spectroscopy: Release of plastic particles from different packaging into mineral water. *Water Research*, *129*, 154–162. doi:10.1016/j.watres.2017.11.011 PMID:29145085

Shim, W. J., Hong, S. H., & Eo, S. E. (2017). Identification methods in microplastic analysis: A review. *Analytical Methods*, *9*(9), 1384–1391. doi:10.1039/C6AY02558G

Statista. (2021). *Distribution of global plastic materials production in 2019, by region*. https://www.statista.com/statistics/281126/global-plastics-production-share-of-various-countries-and-regions/

United Nations. (2021a). *The Sustainable Development Goals Report*. https://unstats.un.org/sdgs/report/2021/The-Sustainable-Development-Goals-Report-2021.pdf

United Nations. (2021b). *Neglected: Environmental Justice Impacts of Marine Litter and Plastic Pollution*. https://wedocs.unep.org/bitstream/handle/20.500.11822/35417/EJIPP.pdf

Van Cauwenberghe, L., & Janssen, C. R. (2014). Microplastics in bivalves cultured for human consumption. *Environmental Pollution*, *193*, 65–70. doi:10.1016/j.envpol.2014.06.010 PMID:25005888

Wagner, M., & Lambert, S. (2018). *Freshwater Microplastics:Emerging environmental contaminants*. Springer. doi:10.1007/978-3-319-61615-5

Wang, W., & Wang, J. (2018). Investigation of microplastics in aquatic environments: An overview of the methods used, from field sampling to laboratory analysis. *Trends in Analytical Chemistry*, *108*, 195–202. doi:10.1016/j.trac.2018.08.026

Wang, Z., Lin, T., & Chen, W. (2020). Occurrence and removal of microplastics in an advanced drinking water treatment plant (ADWTP). *The Science of the Total Environment*, *700*, 134520. Advance online publication. doi:10.1016/j.scitotenv.2019.134520 PMID:31669914

Wang, Z. M., Wagner, J., Ghosal, S., Bedi, G., & Wall, S. (2017). SEM/EDS and optical microscopy analyses of microplastics in ocean trawl and fish guts. *The Science of the Total Environment*, *603–604*, 616–626. doi:10.1016/j.scitotenv.2017.06.047 PMID:28646780

World Population Review. (2021). *Most Developed Countries in Africa 2021*. https://worldpopulationreview.com/country-rankings/most-developed-countries-in-africa

Zhang, Y., Diehl, A., Lewandowski, A., Gopalakrishnan, K., & Baker, T. (2020). Removal efficiency of micro- and nanoplastics (180 nm–125 μm) during drinking water treatment. *The Science of the Total Environment*, *720*, 137383. Advance online publication. doi:10.1016/j.scitotenv.2020.137383 PMID:32325555

Zheng, T., Yuan, D., & Liu, C. (2019). Molecular toxicity of nanoplastics involving in oxidative stress and desoxyribonucleic acid damage. *Journal of Molecular Recognition*, *32*(11), 3–9. doi:10.1002/jmr.2804 PMID:31373076

ADDITIONAL READING

Campanale, C., Savino, I., Pojar, I., Massarelli, C., & Uricchio, V. F. (2020). A practical overview of methodologies for sampling and analysis of microplastics in riverine environments. *Sustainability (Switzerland), 12*(17), 6755. Advance online publication. doi:10.3390u12176755

Fahrenfeld, N. L., Arbuckle-Keil, G., Naderi Beni, N., & Bartelt-Hunt, S. L. (2019). Source tracking microplastics in the freshwater environment. *Trends in Analytical Chemistry, 112*, 248–254. doi:10.1016/j.trac.2018.11.030

Johnson, A. C., Ball, H., Cross, R., Horton, A. A., Jürgens, M. D., Read, D. S., Vollertsen, J., & Svendsen, C. (2020). Identification and Quantification of Microplastics in Potable Water and Their Sources within Water Treatment Works in England and Wales. *Environmental Science & Technology, 54*(19), 12326–12334. doi:10.1021/acs.est.0c03211 PMID:32852201

Koelmans, A. A., Mohamed Nor, N. H., Hermsen, E., Kooi, M., Mintenig, S. M., & De France, J. (2019). Microplastics in freshwaters and drinking water: Critical review and assessment of data quality. *Water Research, 155*, 410–422. doi:10.1016/j.watres.2019.02.054 PMID:30861380

Kosuth, M., Mason, S. A., & Wattenberg, E. V. (2018). Anthropogenic contamination of tap water, beer, and sea salt. *PLoS One, 13*(4), 1–18. doi:10.1371/journal.pone.0194970 PMID:29641556

Lares, M., Ncibi, M. C., Sillanpää, M., & Sillanpää, M. (2018). Occurrence, identification and removal of microplastic particles and fibers in conventional activated sludge process and advanced MBR technology. *Water Research, 133*, 236–246. doi:10.1016/j.watres.2018.01.049 PMID:29407704

Shen, M., Song, B., Zhu, Y., Zeng, G., Zhang, Y., Yang, Y., Wen, X., Chen, M., & Yi, H. (2020). Removal of microplastics via drinking water treatment: Current knowledge and future directions. *Chemosphere, 251*, 126612. Advance online publication. doi:10.1016/j.chemosphere.2020.126612 PMID:32443234

Stock, F., Kochleus, C., Bänsch-Baltruschat, B., Brennholt, N., & Reifferscheid, G. (2019). Sampling techniques and preparation methods for microplastic analyses in the aquatic environment – A review. *Trends in Analytical Chemistry, 113*, 84–92. doi:10.1016/j.trac.2019.01.014

KEY TERMS AND DEFINITIONS

Developed Country: A country which has a higher level of industrialization, quality of life and economy in comparison to countries which are less developed.

Developing Country: A country with a less developed industrial base, which is seeking to become more advanced both socially and economically.

Drinking Water Treatment Plant: A facility where a series of treatments are applied to water to remove pathogens and other harmful substances, providing the public with safe drinking water.

Freshwater: A naturally occurring body of water which does not contain high levels of dissolved salts.

Raw Water: Water which is delivered to drinking water (or wastewater) treatment plants which has not yet been treated.

Sustainable Development Goals: A set of 17 goals which have been laid out, each with specific aims, to achieve a better quality of life for all by 2030.

World Health Organization: A United Nations body which aims to improve health standards to the highest level globally.

Chapter 5
Coastal Hazards Management:
Hard Engineering Solutions along the Taiwanese and Vietnamese Coastline – Unintentional Consequences and Future Humanitarian Engineering Implications

Viola Marcia van Onselen
National Taiwan Normal University, Taiwan

Tsung-Yi Lin
https://orcid.org/0000-0003-3934-4428
National Taiwan Normal University, Taiwan

Phu Le Vo
Ho Chi Minh City University of Technology, Vietnam

Thao Danh Nguyen
Ho Chi Minh City University of Technology, Vietnam

ABSTRACT

Both natural and anthropogenic forces could play significant roles in coastal erosion in Western Taiwan and Central Vietnam's coastlines. Intensive socio-economic development, sea level rise, more frequent and strong intensity of disasters are predicted to occur in a future of global climate change, which poses an urgent need for coastal hazard management strategies. This chapter describes main causes and discusses the applied engineering interventions to reduce coastal erosion at these sites. Hard engineering structures are often constructed in these areas, but they seem to be merely short-term costly solutions and have a negative impact on the coastal environment and its residents. Nature-based solutions and soft engineering approaches are proposed, which seem to be sustainable and less expensive than hard engineering options. These possible future solutions can be applied in coastal settings to meet the principles of sustainable and humanitarian engineering with multiple benefits to reduce the risk and negative impacts on both humans and the environment.

DOI: 10.4018/978-1-7998-9190-1.ch005

INTRODUCTION

Although Vietnam and Taiwan are in different geographical zones in Asia, the future threat of coastal erosion hazards is similar. The coastal zones have rapidly evolved over the past few decades and these developments have resulted in numerous human interventions in the natural systems. Especially in Vietnam, the coastal zone has a high population density with people building very close to the shoreline. Coastal tourism, social-economic developments and industries all seek fortune and exploit resources in this area (Lin et al., 2017). Tourism numbers keep increasing, data from 2019 show that Vietnam received 16.3 million international tourists, which is an increase in 15.4% year-on-year (Ngyuen, 2019). As the tourism number keeps on growing, the number of constructions for tourism facilities along the coastline keeps rising. Simultaneously with this rapid economic growth, coastal sites have suffered severe impacts of global climate change resulting in an increasing number of coastal hazards like typhoons, more severe floods, saline intrusion and coastal erosion (Nguyen et al., 2019). Since the beginning of the last century, many places have been subject to erosion, causing coastline retreat. Countless beaches are already completely eroded, people become more vulnerable to hazards and an increasing number of people lose their houses. Local farmers are forced to exploit even more resources, or damage natural habitats like mangroves in search for adaptation strategies (Bruun, 2012). In response to these risks, diverse coastal engineering structures have been constructed, often focused on fast results, which is visible in the many 'hard solutions' along the coastline of Taiwan, such as tetrapods, cement seawalls and groins (Shih, 2016). Vietnam is now following the same path as Taiwan, by 'hardening' large parts of the coastline with artificial engineering structures. As the tourism and fishery industries grew, pressures of economic developments have also led to rapidly increasing artificial structures, which are causing an imbalance in the shape of the coast, ultimately leading to the erosion of local beaches (Hoang, 2020; Noshi et al., 2015). In many cases, these protective measures have been inadequate in dealing with strong waves, especially in case of higher storm surges. Even worse, such concrete structures negatively affect beach preservation in adjacent areas as well, due to effects of downdrift erosion (Saengsupavanich, 2020).

In the face of global climate change and increasing number of coastal hazards, coastal protection should adapt to these risks. What should be the strategy? And how to balance the development of environmental ecology and social economy and other issues? As an answer to these questions and challenges, humanitarian engineering methods can offer various co-benefits by stimulating integral and multidisciplinary approaches. An emerging theme in coastal management promotes working with nature, instead of working against it. These novel humanitarian strategies are underlined by the strong connections between human and nature, and restoration of coastal ecosystems, adaptive planning and soft engineering approaches are suggested as alternatives to hard structures (Gesing, 2019a; van Slobbe et al., 2013). These strategies require the inclusion and input of multiple stakeholders and bottom-up approaches are encouraged. Therefore, as people's awareness of coastal science and the consequences of climate change have risen, setting up traditional "hard" structures is no longer the only recognizable option. Restoration of coastal ecosystems will lead to a higher number of ecosystem services and more resilient settings. Moreover, more people are willing to adopt adaptive planning such as the setback zone approach, based on appropriate setback planning by the government. Adaptive "soft" engineering, such as re-building dunes is also welcomed, as landscape and ecology conservation are taken into consideration. Coastal communities in more developed areas, on the other hand, should be educated about the cause and effect of future coastal changes, and local residents or stakeholders should be allowed to have a chance to give active input for their own future (Seddon et al., 2021).

The purpose of this chapter is to visualize examples of individual cases from Vietnam and Taiwan, concerning the changing trends of the natural process, discuss various causes of coastal erosion and explore the implemented and possible future strategies as countermeasures to coastal erosion and other challenges in these coastal zones. Case studies from Central Vietnam and Western Taiwan are discussed, as these areas suffer from severe coastal erosion and face similar issues. The cases studies discuss the predominant engineering solutions to coastal erosion in these areas; hard engineering solutions. The changes in coastline dynamics after these interventions are brought forward in the individual case studies for both situation in Vietnam and in Taiwan, to clarify the impact on shoreline erosion. Furthermore, suggestions will be provided for humanitarian engineering options in line with the theme of this book: to provide vulnerable individuals and communities with engineering solutions that improve both lives, livelihoods and degraded environments.

BACKGROUND AND ISSUES

Vietnam and Taiwan both are highly dependent on their coastal resources and often socio-economic development is prioritized over nature conservation, leading to rapid urbanization and industrialization and rapidly rising population density in the coastal zones (Gowing et al., 2006; Lin, 1996; Shih, 2016). This has resulted in degradation of coastal ecosystems and increased vulnerability to natural hazards, since many ecosystem functions are lost and the resilience capacity decreases. Both countries are dealing with intense coastal erosion, an issue that poses major concerns for the future development and fate of the coastal zone. Several sites with intense coastal erosion were investigated both in Western Taiwan and the Central Vietnam coastlines, to identify the consequences and main causes of erosion at each particular site. Both natural forces and anthropogenic activities play a role in the intensifying shoreline erosion. This section will discuss the underlying factors and human interventions leading to increased coastal erosion along the shoreline, such as development of dams upstream, sand mining, land use changes, etc. (Nguyen et al., 2019; Lin et al., 2017). The different causes (both natural and anthropogenic), the severity and current status of coastal erosion in Western Taiwan and Central Vietnam will be discussed based on data collection and images taken during field trips from previous studies on coastal erosion in Central Vietnam and Western Taiwan.

Coastal Erosion in Central Coastline of Vietnam

Globally, people are highly dependent on coastal resources, but especially in Vietnam people rely on these resources due to its geographic position. Vietnam is an elongated country with a 3,260 km long coastline, housing most of the Vietnamese population, which is growing at a fast pace. Especially the low-lying delta areas which are sensitive to climate change and prone to flooding, are densely populated. The climate along the central coast is influenced by seasonal monsoons (NE and SW) resulting in frequent storm surges and tropical typhoons (Nguyen et al., 2019). Socio-economic development in the coastal areas has posed numerous challenges and puts pressure on the coastal system. Pollution, high smog levels in cities, floods, saltwater intrusion and coastal erosion are some of these challenges, which are related to human induced global climate change. Industries in the coastal zone are focused on (amongst others) petroleum, sand-mining and tourism (Long et al., 2006). Together with severe side-effects of the ongoing socio-economic developments in the narrowing coastal strip of Vietnam, the coastal zone is facing many threats. The long stretches of beaches in Central Vietnam attract many tourists and both small and large-scale construction projects can be found everywhere along the coastline. The population density is still increasing, since people want to move here for economic benefits. Simultaneously coastal erosion is threatening new residential areas and many habitats are lost; the width of some beaches is being reduced hundreds of meters to even kilometers over the past decades (Bui & Ly, 2016). Resorts and other businesses are being built on unstable grounds, while local authorities spent a lot of money to stabilize beach erosion. This mainly results in beaches being turned into concrete structures and massive sandbag walls and the newly built resorts and other buildings will quickly be threatened again by coastal erosion, causing them to collapse.

It is already evident that sea level rises faster over the last decades in Vietnam, the relative sea level has increased 5 cm over the last 30 years. Such a rise in sea level would eventually inundate wetlands and lowlands, accelerate coastal erosion, increase the salinity of estuaries and aquifers and have a huge impact on the coastal ecosystems (Furukawa, 2007). The population in coastal areas will become more vulnerable to flooding and storm events and this might trigger displacement of coastal communities. Natural hazards are increased by global climate change; typhoons, storm surges, tropical storms, strong waves, and the rise of sea levels, not only directly impact the coastline, but also have adverse effects, such as floods or saline intrusion. This will affect coastal infrastructure, deteriorate coastal (tourist) facilities and livelihoods and cause problems further inland in agricultural zones.

Additionally, human factors such as sand-mining (along with coastal dike systems) cause an increasing lack of sediment transport to the coastal areas. Upstream irrigation works affect the river flow regime and reduce the amount of sediment which is transported to the sea. This lack in sediment budget also induces coastal erosion and land loss problems (Bui & Ly, 2016). Coastal engineering structures to protect against coastal erosion can also have small- or large-scale consequences in the distribution of sediment along the coastline, leading to coastal erosion at certain sites (more on this later in later in this chapter). Other activities that contribute to coastal erosion are land reclamation, dredging channels and damaging mangrove forests. Coastal erosion combined with other human activities has considerably declined marine biodiversity and natural resources. In Central Vietnam, Cua Dai Beach in Hoi An is one of the examples of severely eroded beaches, where the coastline has retreated hundreds of meters over the last few decades (Hung et al., 2017). In the case of newly constructed hotels at Cua Dai Beach, the shielding embankment collapsed, and the hotel buildings in severely eroded areas were destroyed due to loss of sand and collapsing foundations (see Figure 1a). These buildings are quickly constructed, without paying careful attention to the causes and severity of coastal erosion. Seawalls are set up to protect the newly constructed tourism facilities, but the disappearance of the beaches in front of these seawalls result in the collapse of the cemented structures. The local government stacked woven cloth and sandbags into a defensive wall on the severely eroded coast, but this is only a temporary solution to this ongoing problem (Figure 1b).

Coastal Erosion in Taiwan

Taiwan is a mountainous island with a total shoreline length of around 1500 km, including offshore islands. Due to its geography (almost two-thirds of Taiwan is covered by mountains), most of the population tends to concentrate along the coastal regions, especially along the low-lying west coast. The west coast is built up of soft sediments that are deposited by the main rivers which form deltas along the shore and flow into the sea (Hsu et al., 2008). During the recent economic development in Taiwan, population

Figure 1. Cua Dai beach in Vietnam has experienced severe coastal erosion. (a) Damaged seawall and collapsed houses of Cua Dai Coastal Tourism and Leisure Resort; (b) Woven cloth and sandbags into as a defensive wall on the severely eroded coast. Source: Lin et al., (2017).

density among the West coast has drastically increased, due to the development of new industries, and available land has become scarce. Coastal land reclamation projects for the construction or expansion of harbors, industrial parks and recreation areas, have reshaped the coastline of Western Taiwan. Other human activities that impact the coastline are sand mining from riverbeds, for the purpose of construction materials, leading to decreased sediment transport to the coastline. The construction of reservoirs upstream is also a major cause of coastal erosion in Taiwan, many reservoirs have been built in the last few decades, disturbing the sediment transport to the coastline. Moreover, over extraction of groundwater for aquaculture activities may have direct or indirect effects on the changing coastline. The total number of aquaculture farms along the Taiwan coast has steadily increased since the 1970s (Hsu et al., 2007). A large amount of groundwater has been extracted, which lowers the water table and permits soil compaction of the land subsurface layer, resulting in land subsidence followed by seawater intrusion and eventually shoreline retreat (Chang et al., 2018). The annual rainfall in Taiwan is concentrated in the wet season (from May-October) and can reach 2.5 times the world average. Most rivers have a steep slope and during intense rainfall large amounts of sediments are discharged to the river mouths. Commonly, during the dry season these sediments are redistributed along the coastline by natural coastal processes like waves, tides, currents and winds (Chang et al., 2018). The East coast suffers more from typhoons, while the West coast is prone to seasonal changes and strong winds during the winter period. Natural causes of coastal erosion therefore include short-term beach dynamics, sea-level rise and geohazards.

As the shoreline is retreating and human activities keep affecting the system dynamics, natural buffer zones like coastal dunes and important coastal ecosystems disappear, ultimately leading to the destruction of homes and properties. Moreover, the lack of a consistent and integral coastal management system, improper land use is visible all over Taiwan (Shih, 2016). Based on a recent survey, more than 80 percent of Taiwan's sandy beaches have undergone erosion over the past 3 decades. Since coastal erosion has intensified at most of these sandy beaches at an alarming rate, it has become an island wide problem (Hsu et al., 2007). An example of such a sandy beach is the Golden Beach, a popular attraction in southwest Taiwan. This beach is at risk because of severe coastal erosion; after a viewing platform has collapsed, the beach is slowly disappearing and constructed barriers are collapsing, see figure 2a (Lin, 2016). The pace of coastal erosion has alerted local authorities and residents, since the base of the viewing platform is suffering from subsidence and the platform collapsed. Now people are afraid of the possible collapse

Figure 2. a) collapse of seawall at Golden Beach, Tainan. Image of T.-Y. Lin (2011); b) Dune erosion and shoreline retreat at Wangziliao barrier, Tainan County. Source: Hsu et al. (2007).

of the seawall, which might threaten their homes. Coastal sand dune erosion is also an issue in Taiwan (Figure 2b), mainly due to human interventions in upstream areas, like the construction of reservoirs or dams and sand mining, resulting in a decrease of sediment influx to the shoreline (Hsu et al., 2007).

It should be realized that a coast is a dynamic system; due to these dynamics shorelines are subject to constant change influenced by wind, tides, waves and currents, to realize a dynamic equilibrium. Human activities and modification of the coastline disrupt this equilibrium by disturbing the natural processes. In both countries the rapidly increasing artificial engineering structures that are constructed to prevent coastal erosion, also cause an imbalance in the shape of the coast, leading to the erosion of local beaches. Due to a lack of careful evaluation, or because many local authorities focus merely on short-term solutions, more and more artificial structures are introduced along the coastlines. In fact, this causes more interference in the natural dynamic system and the erosion situation could be worsening, as will be discussed in the next section.

LESSONS LEARNT FROM ARTIFICIAL ENGINEERING INTERVENTIONS

Due to global climate change, sea level rise, increasingly frequent storms and stronger waves are predicted to occur in the future, which poses an urgent need for appropriate coastal hazard management strategies (Alcántara-Ayala & Goudie, 2011). The next two sections describe solutions to coastal erosion and lessons learned from implemented structures and management strategies. This section will take a closer look at the most commonly seen coastal engineering structures that are applied to counteract coastal erosion in Taiwan and Vietnam. Case studies are discussed to explore what studies have found so far on these different types of engineering constructions.

Hard or so-called 'grey' engineering structures are reshaping coastlines worldwide. These structures have numerous functions; from stabilizing beaches to dissipating waves, but they can have negative impacts on the coastal ecosystems or even increase coastal erosion. In general, a beach with sufficient width and sediment supply can function as a natural buffer against waves and storm surges. However, the growing pressure of economic developments have led to reshaping the coastline by using hard engineering structures in response to coastal erosion or for navigation purposes. These engineered structures, such as seawalls, groins, and detached breakwaters have traditionally been used in Taiwan and Vietnam

Figure 3. Most common coastal defence structures in Vietnam and Taiwan. Source: Google Earth.

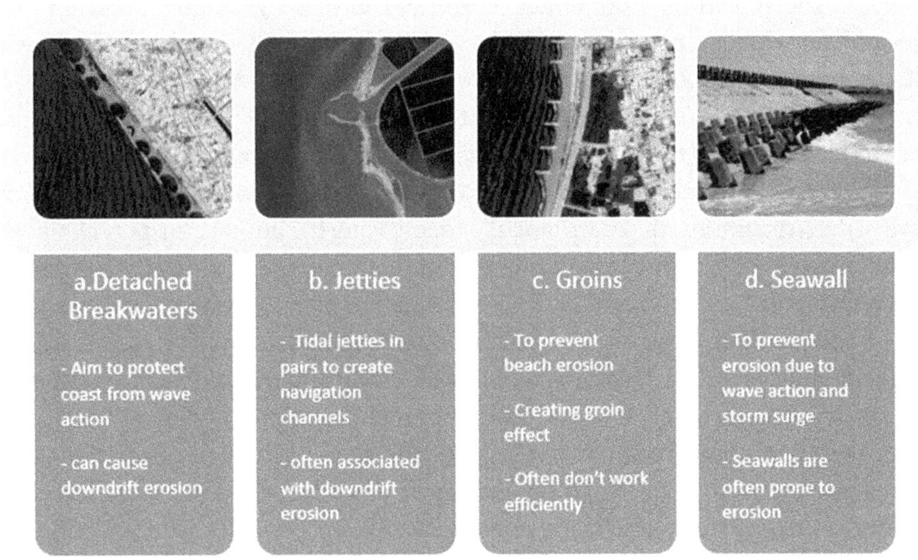

to protect the shoreline and prevent damage or loss of life in hazardous typhoon events (see Figure 3). Along the central Vietnam coastline gravel revetments or earth dikes have also been used in the past. However, these measures have not proved suitable in counteracting strong waves, especially during high storm surges (Hsu et al., 2008).

Introducing Several Hard Engineering Structures; Purposes and Side Effects

Inlet Jetties are constructed for navigation purposes, since tidal inlets and river mouths are often shallow and variable in location due to natural dynamics. Therefore, many tidal- or river inlets have been regulated by creating a channel from the harbour to the open sea by the construction of a pair of jetties (Bui & Ly, 2016). This can improve navigation conditions and could mitigate inland flooding. However, sometimes continuous dredging of sediments is needed to keep the channel functional, additionally the construction of jetties induces downdrift erosion (Hoang, 2020). Over time negative side effects will become visible because these structures interrupt the sediment transport along the coastline (Noshi et al., 2015). The natural sediment flux along a shoreline (littoral drift) will be intercepted by the jetties and the sediment input on the downdrift side decreases, leading to erosion in adjacent areas. Eventually downdrift erosion requires further measures, since the negative sediment budget will not be complemented naturally. When in a natural equilibrium with the bypassing littoral drift, normally shallow sandbars form across the inlet of tidal inlets and river mouths. However, when the inlet or river mouth is stabilized to accommodate navigation, this bar is typically cut off by the jetties

Groins (groynes) are often installed to trap moving sand at erosion sites. They can also be seen in combination with jetties in case groins are installed at the side of downdrift erosion. They are often constructed in parallel rows extending from the shoreline into the sea. For groins to function effectively, an adequate sediment supply from a longshore current is required. Studies have shown that during storm events, when waves approach the shoreline form a different angle, the protective role of groins is reduced and the shielded beach will erode nevertheless (Alves et al., 2020; De Vriend & Van Koningsveld, 2012; Hsu et al., 2008). The so-called 'groin effect' occurs when these structures interrupt the longshore sediment transport and sand accumulates along the updrift direction along the structures, while erosion occurs in the downdrift direction (Figure 3).

Another construction that is often seen to protect the shoreline or harbour activities along the coastline from wave action, are **detached breakwaters**. These structures are installed parallel to the coastline and they provide shelter from wave action. These structures also result in sediment accumulation in the lee zone (in between the structure and the coastline), thus decreasing the littoral transport along the shoreline. There are different types of breakwaters, but with offshore breakwaters typically a tombolo is created when sediment accumulates (see Figure 3). The construction of breakwaters results in changes in the wave field, nearshore current system (sometimes creating dangerous currents for recreational users) and longshore sediment transport. The reduction in longshore sediment transport can significantly increase erosion at nearby beaches, as seen with other structures interrupting the sediment flux (Lin, 2016).

Seawalls or revetments are structures that separate the shoreline directly from the water. They are designed to prevent coastal erosion and property damage from strong waves or storm surge and flooding. To resist the force of strong waves and storm surge, seawalls are designed as massive sloping concrete structures. In Taiwan and Vietnam, they are typically enforced by so called 'tetrapods' which are concrete tetrahedral structures designed to dissipate the force of incoming waves (Lin, 2016). The construction of seawalls, particularly seawalls with steep slopes, appear to have an adverse effect on the fronting beaches or result in downdrift erosion (Pearson et al., 2004). If seawalls are not properly designed, waves can top over the wall and erode the shoreline behind it. It is also commonly seen that seawalls are built up out of insufficient, cheap materials which quickly leads to erosion and loss of function of these structures. Moreover, erosion occurs immediately in the front of a seawall due to the increase in reflected wave energy (Hsu et al., 2008). According to the article of Hsu et al. (2008), failure of seawalls or revetments in Taiwan happens on the front side due to active wave energy, but also on the back sides, because of extreme pore pressure. This pressure leads to soil fluidization inside the seawall, resulting in the destruction of the seawall. After seawalls are collapsed huge funds are necessary to repair or rebuild the structure and in case that they are poorly constructed, future coastal developments or land reclamation projects are at risk.

Case Studies on Artificial Coastal Engineering Structures in Taiwan and Vietnam

- Groin effect and seawall collapse: a local case study on coastal engineering structures in **Duc Long Ward and Doi Duong beach at Phan Thiet city in Vietnam** observed coastline dynamics and changing beach morphologies over more than a decade (Lin et al., 2021). A single groin was installed in the river mouth of the Ca Ty estuary and a land reclamation project has been initiated at the downdrift side of the groin. Remote sensing (RS) and Geographical Information System (GIS) are effective tools to determine the shoreline change over time and for this study shoreline changes have been mapped in GIS to detect erosion patterns at Doi Duong beach (Figure 4). Erosion has been observed at Doi Duong beach, because sand that would have naturally spread along the shoreline has been blocked by the groin. Land reclamation projects were initiated when houses collapsed due to this downdrift erosion and the land reclamation project even further halted the sand supply to adjacent areas (Takagi et al., 2012). Over 40m of beach has been lost in a

Figure 4. The process of land reclamation at Duc Long ward after the construction of a groin and inlet channel, which led to shoreline recession at Doi Duong beach. Source: Takagi et al. (2012).

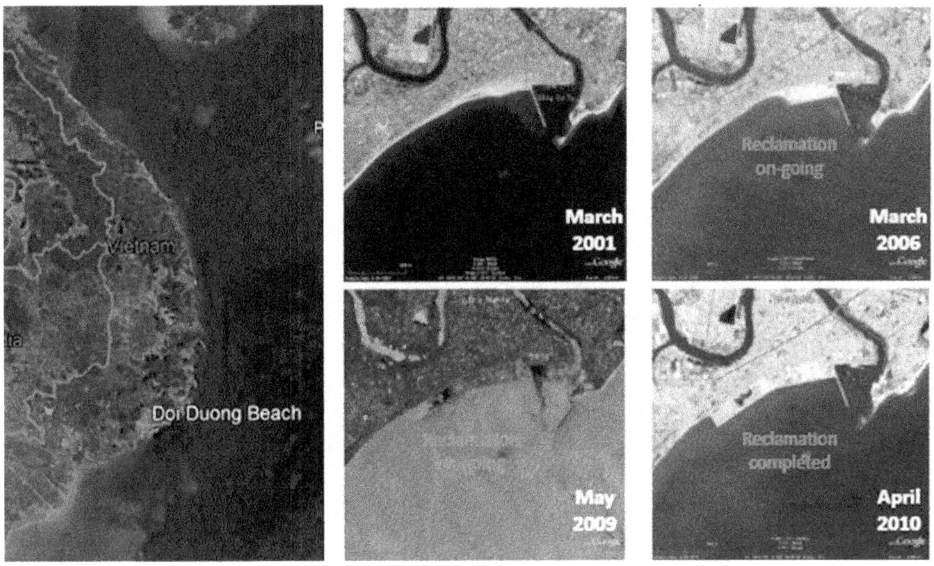

period of 9 years, related to the construction of new land and groins (see Figure 5). This has led to increased threats for residential areas along the shoreline and the low seawall that is protecting the newly reclaimed land is at high risk of flooding, and waves are constantly washing over the seawall (Takagi et al., 2012). This poorly constructed seawall could have negative impacts on the future developments on this new strip of land and appropriate management of the coastal areas is required to mitigate to the increase future threats of natural disasters such as, storm surge, typhoons, etc.

- Inlet Jetties and sandspit disturbance: at **Loc An beach in Vietnam**, a pair of jetties has been constructed on an active sandspit. To collect data on shoreline erosion detected shoreline positions were extracted from using long-term satellite images and historical maps studied. In the field, evidence of coastal erosion is uncovered and explored in the beach morphologies, such as formation of beach scarps, sharps edges, fallen trees, collapsed, or damaged houses and structures, etc. After construction of the inlet jetties for navigation purposes, the NE jetty blocked the south-westward longshore drift, resulting in sand deposition northeast of the jetty, resulting in the formation of a wide foreshore. (Figure 6). At the downdrift side, adjacent to the SW jetty, downdrift erosion will increase because the sediment flow is interrupted. After the construction of the jetties, the navigation channel frequently needs to be dredged, as it turned out that the channel in between the jetties traps a lot of sediment. Thus, by interrupting the natural dynamics, the sandspit formation is halted and there is no more equilibrium as the surrounding morphology gradually changes and more erosion will occur at fixed points (Lin et al., 2021; Hoang 2019).

Figure 5. The amount of shoreline erosion and collapsing houses in the erosion area at Doi Duong beach. Source: Takagi et al. (2012).

▪ Groin effect: for a local case study **Jinshawan, Taiwan** satellite images and numerical simulations were used to analyse morphological changes at a recreational beach Jinshawan (golden sand bay) (Hsu et al., 2007).). Breakwaters were installed at the Hommei fishing port in northern Taiwan and after 1984 when the northern breakwater was extended, the sediment transport has been further blocked. This has led to severe erosion of the neighbouring Jinshawan beach (Figure 7a). Although the breakwater helps to provide shelter for the harbour (by reducing wave heights), it also results in changes to the wave field and the nearshore current system and sediment transport pattern have been altered. Reversed currents are one of the side effects of the breakwater extension, and the sediment from the adjacent beach is transported towards the harbour entrance. Numerical simulations indicated the changes in the wave-induced nearshore current pattern before and after construction of the breakwater (see figure 7 b+c). Due to these changes in wave patterns, sand has been removed from the Jinshawan beach and the beach has been disappearing (Hsu et al., 2007).

In Taiwan already more than 50 percent of the coastline has been 'constructed', since hard engineering structures have been applied in coastal management strategies for the last few decades. Vietnam is now following the same fate (Shih, 2016). Above examples have shown that these engineering structures, which are designed to prevent erosion, actually often cause downdrift erosion over time. These hard

Figure 6. The situation in Loc An, where after jetties have been constructed sediment is trapped against the NE Jetty and downdrift erosion is seen on the westside of the jetties.

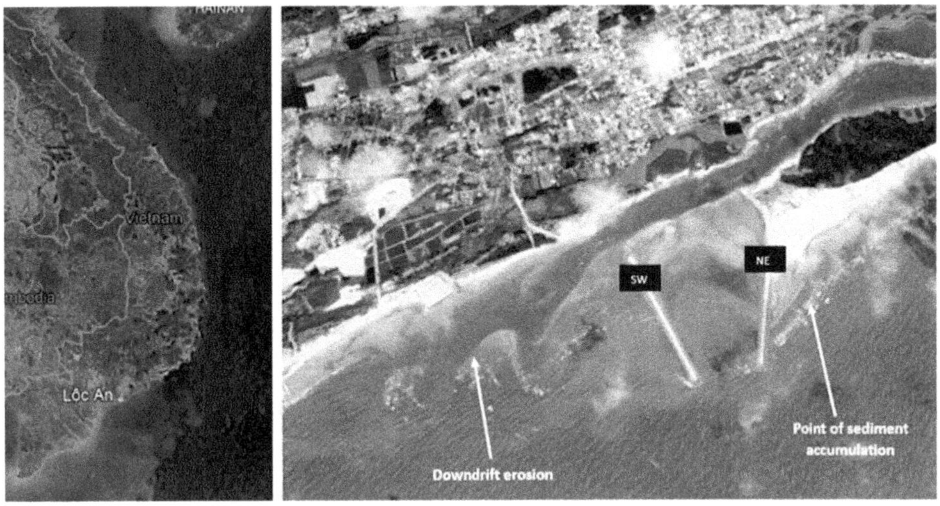

structures along the coastlines will disturb the local system dynamics and transform the geography of the area. Moreover, these engineering structures are merely short-term solutions, because they are often rapidly constructed at low cost, using unsustainable materials that are sensitive to erosion over a longer period of time. Due to the process of downdrift erosion, the construction of these structures becomes a recurring process, slowly transforming sandy shorelines into an artificial coastline (Uda & Yamamoto, 1994). An issue that leads to less sustainable approaches, is that in both countries different interests are represented by various sectoral authorities. Combined with the high number of stakeholders that are involved in coastal planning, fragmented approaches are implemented without long-term strategies. These different responses to similar issues limit the effectiveness of measures to prevent coastal erosion (Vo

Figure 7. a) Golden beach (Jinshawan) and north breakwater extension at Hommei fishing harbor;(b) wave-induced nearshore currents before breakwater extension; (c) wave-induced nearshore currents after the extension. Source: Hsu et al., (2007).

et al., 2018). Other underlying factors for poorly managed and constructed engineering structures are; limited exchange of information between different sectors, reluctance to bear costs for engineering projects, corruption, poor awareness and lack of data on dynamics, activities and threats in the coastal zone.

Different solutions are slowly being introduced to deal with coastal erosion, such as hybrid approaches (combining 'grey/artificial' and 'green/natural' methods) or green approaches such as Nature-based Solutions (Waryszak et al., 2021). In Vietnam and Taiwan there are not many references or data on the implementation of Nature based approaches along the coastline. However, these solutions are more sustainable and could last for a longer period of time, since they take into account the natural processes and system dynamics, using nature as part of the solution. The next section will highlight some of these approaches and explain more about strategies to create awareness amongst different stakeholders and support integral decision making.

SOFT ENGINEERING APPROACHES AND FUTURE OPPURTUNITIES

From a humanitarian engineering viewpoint, new alternatives are being developed and implemented, taking into account the important role and function of coastal ecosystems. Grey engineering structures usually alter the natural environment and rather work against nature than with nature, while nature preservation and sustainability currently are serious concerns. Coastlines around the world are facing increasing pressures and smarter coastal protection strategies are required that are more sustainable, adaptable and multi-functional (Gracia et al., 2018). Natural ways to approach issues like coastal erosion and salinity intrusion have proven to be more efficient over the long term (Walters et al., 2016). Natural approaches, or so-called Nature-based Solutions (NbS), can also create more co-benefits, which is highlighted in the IPCC land-based solution report (IPCC, 2019). Nature-based Solutions for climate and disaster resilience are already being implemented worldwide to a broad range of challenges and in various environments. NbS rely on the long-term sustainability and resilience of healthy coastal ecosystems, which have the capacity to recover after disturbance of external stressors. Moreover, NbS provide multiple benefits for biodiversity and human well-being (through ecosystem services). For successful implementation, bottom-up approach through local community participation is an important imperative (Cohen-Shacham et al., 2019; Seddon et al., 2021).

For example, a method to reduce hazard vulnerability is Ecosystem-based Disaster Risk Reduction (Eco-DRR); a sustainable management solution established through conservation and restoration of ecosystems, to provide facilities that reduce disaster risk and mitigate hazards, for example through restoration of coastal dune systems and active plantation near the shore. This will increase the resilience capacity of both ecosystems and social systems. The UN Framework Convention on Climate Change adopted an adaptation framework, which involved recommendations on how ecosystems can be adapted as a tool in coastal management strategies to deal with coastal erosion and encourages community participation in coastal erosion management processes (Gesing, 2019b). This section brings forward alternatives to grey engineering methods by introducing soft engineering methods; shoreline management practices using sustainable and ecological principles to restore shorelines and reduce coastal erosion.

Many people recognize that humans coexist in a multifaceted, interdependent relationship with the natural world. Social-Ecological-System (SES) thinking underlines this assumption by acknowledging that social and ecological systems are linked through complex feedback mechanisms, and that both systems can be adaptive and resilient when they are well attuned to each other (Ostrom, 2009). This principle emphasizes the importance of including natural resource protection in all decision-making processes and it needs to be realized that social acceptability is vital for the success of NbS. To support these approaches, integrated coastal zone management (ICZM) is a multidisciplinary framework that is being developed and adapted to local settings to promote sustainability in coastal zones. ICZM strives to find a balance between environmental, social, economic, cultural and recreational objectives for long-term periods (Cuong & Cu, 2014). For ICZM to be successful, it is important that all involved stakeholders are incorporated in the coastal development and co-operate in the process of decision making, implementation, management and monitoring (Anderson & Renaud, 2021). When successfully implemented, NbS will increase the resilience capacity of both ecosystems and social systems. Three methods will be described in more detail, including how they can be adopted in coastal areas of Taiwan and Vietnam and how these approaches can contribute to multiple benefits for both human and nature, following the principles of ICZM and NbS.

Setback Zoning

This is the principle of restricting or limiting coastal development in designated protection zones adjacent to the shoreline. These zones can be used as buffer zones for storm tides, coastal erosion and adaptation to sea level rise (Sanò et al., 2011). Setback policy might also apply in already inhabited areas. To manage continuous coastal erosion, sea level rise and the increasing risk of extreme weather events and floods in the future, this policy may be applied to protect coastal communities. This rigorous measure might be needed in the future with the predicted sea level rise in some areas (Furukawa, 2007). And to prevent damages or disasters, wider coastal setbacks should be considered in future coastal development plans. While it is easy to restrict development plans in already existing buffer zones, managed retreat is another way of setback zoning. Rather than construct more costly coastal defences, farm or residential land will be allowed to erode and in a designated buffer zone between the shoreline and coastal communities. In this case people should be compensated for their losses. This measure is adopted when compensational costs for the affected people are significantly lower than the costs of building coastal defences. Setback zoning can be beneficial for ecosystems, since developments will be restricted and new habitats are provided. Managed retreat is a controversial method though, because farmers or people actively have to leave their land and/or homes (Sanò et al., 2011; Lin et al., 2017)

Coastal Sand Dune Restoration

Dunes are a great coastal defence to protect the coastline as they can buffer against storm and wave energy and sea-level rise. Dune restoration or enhancement therefore can opt as a natural solution to coastal erosion and severe storms (Gracia et al., 2018). Nature-based Solutions concerning coastal dune restoration can have different levels of intensity and human involvement; one approach is that the dunes are conserved and protected from harmful activities such as sand-mining or industrial developments and natural dynamics can take over the dune formation process if there is sufficient sediment supply (De Jong et al., 2014). Dunes can also be actively enhanced by beach nourishment for extra sediment supply or planting marram grass in the foredunes to trap sand and grow larger dunes (De Vriend & Van Koningsveld, 2012). Another strategy is dune regeneration, when new sand dunes are artificially created to act as a buffer between land and sea. These are all types of Eco-DRR, a type of Nature-based Solution that involves maintaining and protecting the natural ecosystem, whilst offering protection, and it provides benefits for both biodiversity and human well-being. Both Taiwan and Vietnam have areas with coastal sand dunes along the coastline, which are now threatened by human activities. To preserve these areas, while still supporting economic development, coastal dune landscapes can be for instance turned into a natural park or local geopark. This is seen at the Caota Dunes in Taiwan, a newly opened coastal dune geopark (Taoyuan Office of Coast Administration Construction, 2021). In this way the dunes will be protected from further developments and local communities can make profit from the tourists that the area will attract and help in maintaining the site.

Restoring Mangrove Forests

Mangroves are typically found in shallow muddy grounds of intertidal zones and estuarine margins and they can grow in harsh environmental conditions (Giri et al., 2011). The extensive mangrove roots allow them to grow in unstable, dynamic settings; they grow in soils without oxygen and are regularly flooded with salt seawater. Wave and wind energy is drastically decreased when it passes through the network of mangrove branches and roots (Gracia et al., 2018). Along the Central Vietnam coastline and in West Taiwan, mangrove forests have been destroyed to make place for new land developments or the development of aquaculture or agriculture and at these sites coastal erosion has, coastal erosion has increased significantly (Nehren et al., 2017). Therefore, by actively replanting mangroves near the shore, erosion and also salinity intrusion can be prevented to some extent, while important ecological functions such as increasing biodiversity and providing shelter for many species are also fulfilled (Powell et al., 2007). During afforestation and reforestation projects the planted species should be carefully considered, since invasive species can become a major problem when introduced in vulnerable habitats. Therefore, it is easier to restore existing mangrove systems than starting a new mangrove habitat in another ecosystem, such as for coastal wetlands that have been free of mangroves before. Several mangrove reforestation projects in Taiwan eventually led to the removal of mangroves, because the newly planted mangroves acted as an invasive species and changed the original habitat while posing a threat to vulnerable endemic species (Chen & Shih, 2019). However, mangroves have proven to be efficient in reducing storm surge levels and can be a good ecosystem-based solution to coastal erosion problems in certain areas.

When addressing issues like coastal erosion through ecosystem-based management approaches, the political will of the local authorities and national government play an important role; when development is prioritized over natural resource protection, short-term solutions become prevalent. The protection or enhancement of coastal ecosystems could be incorporated in development processes when the decision makers will realize the long-term benefits, also on social and economic level, of more sustainable solutions to coastal erosion. Likewise, for long term sustainability coastal communities should be included in the decision-making process though bottom-up approaches, since NbS can take several years to implement and become beneficial, local community support will increase the number of benefits and long-term effectiveness (Mekonen, 2017). Lastly, before any decisions are made on which solution is the best approach for a local setting, it is important to make an impact assessment and analyse the natural dynamics, current threats and connect the involved stakeholders to realize more humanitarian and all-inclusive engineering solutions.

DISCUSSION AND RECOMMENDATIONS

In many cases human activities and interference in natural environments are pointed out as the most contributing factor to coastal erosion. In central Vietnam, the tourism industry is rapidly evolving, while the coastal zone attracts more and more people, who are vulnerable to the threats of natural hazards and ongoing coastal erosion. Due to the huge population growth, the land shortage for future urban and industrial development starts to limit the socio-economic development. But new land reclamation projects are initiated in both countries, reshaping the coastline and resulting in the loss of important marine ecosystems and overall biodiversity. And although hard engineering structures designed to deal with coastal erosion seem a quick and efficient solution, they also severely alter the natural environment and can even increase erosion at other places along the coastline. Both population growth and economic growth are expected to continue in Taiwan and Vietnam and as a result, the development of coastal areas will be further accelerated. The appropriate management of coastal zones is required to reduce the coastal disaster risks. The impacts of global climate change will put a growing pressure on these coastal environments, while creating more inequities for vulnerable people.

To protect coastal ecosystems and efficiently utilize the potential of coastal areas, more sustainable and integrated solutions are needed. The ICZM framework underlines the importance of integration socio-economic development and environmental protection, and has been adapted in some policies or guidelines in Vietnam and Taiwan, but is not yet fully incorporated in new development plans. Improper planning, poor implementation and enforcement of policies and lack of knowledge results in ecosystem degradation. It is crucial for ICZM to build knowledge and increase awareness at all levels and the recognition that local needs also need to be incorporated in local policies is growing. Without sufficient local support it will be hard to guarantee long-term sustainability of these kinds of solutions in the coastal zone. Once public understanding is enhanced of the values of the coastal zone, people could adapt more responsible ways to sustainable use natural resources. Education events and workshops can help raise awareness about environmental degradation and sustainable solutions to coastal threats. Projects and movements to increase knowledge will hopefully lead to well-informed decisions to work towards a more sustainable future for these heavily degraded coastal zones.

However, the fragmented approaches and restricted exchange of information between responsible authorities and different stakeholders is still not sufficient in the coastal management processes of both Taiwan and Vietnam. Especially in Vietnam, the amount of data on coastal processes and biodiversity in the coastal zone is very limited and the government is not transparent in sharing information (Mekonen, 2017; Thang et al., 2011). The use of scientific research data together with local knowledge can provide better solutions in line with the sustainable development goals. Since coastal processes and erosion problems can be very site specific, it is crucial to study these issues and collect data at a local level and then combine data from all over the coastline in a dynamic database. This can help to correctly implement new solutions and support integrated decisions on coastal development plans, while minimizing the impacts of human disturbances on the coasts. Therefore, it is necessary to invest in these types of research and strengthen the legal framework and prioritize plans for the protection of natural resources. International organizations that are involved in adaptation and mitigation efforts to climate change, can support each other with research data and with the development of appropriate action programs and best practices from soft engineering solutions should be shared worldwide in coastal management programs.

For future coastal development plans, governments can consider different strategies. In the case of Vietnam and Taiwan, large parts of the coastline are already cemented and to enhance coastal ecosystems, deconstruction of non-maintained or failing coastal engineering structures is suggested. This will make room to introduce soft engineering approaches and help increase the biodiversity, number of ecosystem services and provide societal benefits. If it is not possible to deconstruct coastal engineering structures, hybrid strategies are another possibility; by combining artificial engineering approaches and nature-based approaches, so called 'hybrid' or 'green-grey' options can be implemented. This will help restore degraded ecosystems or increase natural resources while enhancing the protection function, and this approach might be a useful alternative when NbS cannot be fully implemented. The restoration of ecosystems can be initiated along the coastline by planting trees in the right habitats, or restoring coastal sand dunes systems. Furthermore, policies should incorporate plans for permanent relocation in the future, to move communities away from danger zones and implement setback zoning where possible to create buffer zones. New innovative strategies are still being developed and tested, and more long-term experiments and evaluation data need to point out the long-term effects or large-scale dynamics of soft engineering strategies. Each country and each region will have different interests for the management of the coastal zone, and local factors play an important role in the management plans. However, increasing biodiversity and enhancement or restoration of ecosystems have demonstrated positive effects for more resilient and climate adaptable environments.

CONCLUSION

Vietnam and Taiwan face similar future threats; natural hazards have increased due to global climate change and the highly populated West coast of Taiwan and Central coastline of Vietnam are threatened by sea level rise and coastal erosion. Socio-economic developments along these coastlines have resulted in the loss of biodiversity and important coastal and marine ecosystems. Due to the growing tourism industry, countless hotels and resorts that have been established close to the shoreline and are now collapsing due to severity of coastal erosion. To counteract coastal erosion, grey engineering structures like groins, seawalls and breakwaters have been established along these coastlines. These 'hard structures' seem to be merely short-term solutions and they often quickly erode or are washed over by waves and sediments.

Many studies point out the severe side-effects of these hard structures, such as downdrift erosion, alteration of long-term shore dynamics and dangerous currents near the shoreline. To avoid negative impacts of these coastal defence structures, it is important to carefully assess and evaluate the shoreline dynamics and processes before construction and increase data collection. Moreover, it is recommended to include more natural ways of coastal defence strategies, using healthy ecosystems as part of the solution. For instance, restoration of coastal dune environments will help create a natural buffer to storm surge and sea-level rise, or mangroves decrease wave and wind energy and help decrease coastal erosion. These solutions will offer multiple benefits, both socio-economic in the form of ecosystem services and benefits for biodiversity.

Awareness and education about environmental issues and benefits of Nature-based Solutions is an important factor; it can help reduce or find alternative solutions for the number human activities that are underlying factors to coastal erosion. Furthermore, with local support these solutions will be more effective, also for longer periods of time and to reduce the adverse effects of human activities. Data on the long-term effects of soft engineering solutions is still being collected, but studies show promising results for the function of coastal protection and increased resilience of coastal environments. It is a challenging task to plan a more sustainable future in this age of the 'Anthropocene'. This will need global cooperation, taking into account that the effects of climate change are different in each country. Taiwan and Vietnam face many challenges related to this issue, but there are many future opportunities to work towards more humanitarian engineering options by incorporating soft engineering solutions in coastal management plans. With global cooperation and increasing awareness and data-collection, new ways will be found and implemented to adapt and mitigate to climate change through sustainable solutions.

REFERENCES

Alcántara-Ayala, I., & Goudie, A. S. (2011). *Geomorphological Hazards and Disaster Prevention*. Cambridge University Press.

Alves, B., Angnuureng, D. B., Morand, P., & Almar, R. (2020). A review on coastal erosion and flooding risks and best management practices in West Africa: What has been done and should be done. *Journal of Coastal Conservation*, *24*(3), 38. Advance online publication. doi:10.100711852-020-00755-7

Anderson, C. C., & Renaud, F. G. (2021). A review of public acceptance of nature-based solutions: The 'why', 'when', and 'how' of success for disaster risk reduction measures. *Ambio*, *50*(8), 1552–1573. Advance online publication. doi:10.100713280-021-01502-4 PMID:33606249

Bruun, O. (2012). Sending the right bill to the right people: Climate change, environmental degradation, and social vulnerabilities in central Vietnam. *Weather, Climate, and Society*, *4*(4), 250–262. doi:10.1175/WCAS-D-11-00040.1

Bui, V. T., & Ly, H. M. (2016). Detection of shoreline changes: A geospatial data analysis in Vung Tau City, Southern Vietnam. *Science and Technology Development Journal*, *19*(1), 131–139. doi:10.32508tdj.v19i1.567

Chang, H.-K., Chen, W.-W., Tsai, F.-C., Liou, J.-C., & Kuo, S.-G. (2018). Annual and Seasonal Shoreline Evolution of Tainan Gold Coast Using Multiband Satellite Images. *IACSIT International Journal of Engineering and Technology*, *10*(1), 52–56. doi:10.7763/IJET.2018.V10.1034

Chen, Y. C., & Shih, C. H. (2019). Sustainable management of coastal wetlands in Taiwan: A review for invasion, conservation, and removal of mangroves. *Sustainability (Switzerland)*, *11*(16), 4305. Advance online publication. doi:10.3390u11164305

Cohen-Shacham, E., Andrade, A., Dalton, J., Dudley, N., Jones, M., Kumar, C., & Walters, G. (2019). Core principles for successfully implementing and upscaling Nature-based Solutions. *Environmental Science and Policy, 98*(June), 20–29. doi:10.1016/j.envsci.2019.04.014

Cuong, N. Q., & Cu, N. V. (2014). Integrated coastal management in Vietnam: Current situation and orientation. *Journal of Marine Science and Technology*, *14*(1), 89–96. https://www.vjs.ac.vn/index.php/jmst

De Jong, B., Keijsers, J. G. S., Riksen, M. J. P. M., Krol, J., & Slim, P. A. (2014). Soft engineering vs. a dynamic approach in coastal dune management: A case study on the North Sea barrier Island of Ameland, the Netherlands. *Journal of Coastal Research*, *30*(4), 670–684. doi:10.2112/JCOASTRES-D-13-00125.1

De Vriend, H. J., & Van Koningsveld, M. (2012). *Building with Nature: Thinking, acting and interacting differently*. Ecoshape. doi:10.1080/02513625.2014.925714

Furukawa, M., & Hanh, P. T. T. (2007). Impact of sea level rise on coastal zone of Vietnam. *Proceedings of the 10th International River Symposium and Environmental Flows Conference*, *59*(84), 1–9.

Gesing, F. (2019a). The politics of artificial dunes: Sustainable coastal protection measures and contested socio-natural objects. *Die Erde*, *150*(3), 145–157. doi:10.12854/erde-2019-423

Gesing, F. (2019b). Towards a more-than-human political ecology of coastal protection: Coast Care practices in Aotearoa New Zealand. *Environment and Planning E: Nature and Space*. doi:10.1177/2514848619860751

Giri, C., Ochieng, E., Tieszen, L. L., Zhu, Z., Singh, A., Loveland, T., & Duke, N. (2011). Status and distribution of mangrove forests of the world using earth observation satellite data. *Global Ecology and Biogeography*, *20*(1), 154–159. doi:10.1111/j.1466-8238.2010.00584.x

Gowing, J. W., Tuong, T. P., Hoanh, C. T., & Khiem, N. T. (2006). Social and environmental impact of rapid change in the coastal zone of Vietnam: An assessment of sustainability issues. *Environment and Livelihoods in Tropical Coastal Zones: Managing Agriculture-Fishery-Aquaculture Conflicts*, 48–60. doi:10.1079/9781845931070.0048

Gracia, A., Rangel-Buitrago, N., Oakley, J. A., & Williams, A. T. (2018). Use of ecosystems in coastal erosion management. *Ocean and Coastal Management*, *156*, 277–289. doi:10.1016/j.ocecoaman.2017.07.009

Hoang, V. C. (2020). Long-Term Evolution of Morphology at Loc An Estuary, Vung Tau, Vietnam. *Marine Geodesy*, *43*(2), 163–188. doi:10.1080/01490419.2019.1606125

Hsu, T., Lin, T., & Tseng, I. (2007). Human Impact on Coastal Erosion in Taiwan. *Journal of Coastal Research*, *234*, 961–973. doi:10.2112/04-0353R.1

Hsu, T. W., Tseng, I. F., Lin, T. Y., Shin, C. Y., & Ou, S. H. (2008). Review of countermeasures against beach erosion on the Taiwanese coast. *Coastal Management*, *36*(3), 274–293. doi:10.1080/08920750801968678

Hung, N. T., Vinh, B. T., Nam, S. Y., & Lee, J. L. (2017). Cause Analysis of Erosion-Induced Resort Washout on Cua Dai Beach, Vietnam. *Journal of Coastal Research*, *79*, 214–218. doi:10.2112/SI79-044.1

IPCC. (2019). *Climate change and land - IPCC, Summary for policymakers*. doi:10.4337/9781784710644

Lin, J. C. (1996). Coastal modification due to human influence in south-western Taiwan. *Quaternary Science Reviews*, *15*(8–9), 895–900. doi:10.1016/S0277-3791(96)00060-1

Lin, T.-Y. (2016). Morphological response to the installation of detached breakwaters along the Cigu coast of Tainan, Taiwan. *Journal of Coastal Research*, *1*(75), 887–891. doi:10.2112/SI75-178.1

Lin, T.- Y., Onselen van, V. M., & Vo, L. P. (2021). Coastal erosion in Vietnam: Case studies and implication for integrated coastal zone management in the Vietnamese south-central coastline. *IOP Conference Series: Earth and Environmental Science*, *652*(1), 0–11. doi:10.1088/1755-1315/652/1/012009

Lin, T.-Y., Vo, L. P., Thao, N. D., & Bao, N. G. (2017). Case Studies of Coastal Erosion in Central Vietnam and Their Application in Hazard Management. *Proceedings of the 39th Ocean Engineering Conference in Taiwan*, 431–436.

Long, B. T., Truong, C. D., & Loi, L. L. (2006). Building up a tool for integrated coastal management in Da Nang City. *International Symposium on Geoinformatics for Spatial Infrastructure Development in Earth and Allied Sciences*.

Mekonen, S. (2017). Roles of Traditional Ecological Knowledge for Biodiversity Conservation. *Journal of Natural Sciences Research, 7*.

Nehren, U., Ho, D. T. H., Trung, N. D., Raedig, C., & Alfonso, S. (2017). *Sand Dunes and Mangroves for Disaster Risk Reduction and Climate Change Adaptation in the Coastal Zone of Quang Nam Province*. Water Resources Development and Management. doi:10.1007/978-981-10-2624-9_13

Nguyen, H. (2019). *Vietnam's Tourism Sector: Opportunities for Investors in 2020*. Retrieved from: https://www.vietnam-briefing.com/

Nguyen, K., Liou, Y., & Terry, J. P. (2019). Science of the Total Environment Vulnerability of Vietnam to typhoons: A spatial assessment based on hazards, exposure and adaptive capacity. *The Science of the Total Environment*, *682*, 31–46. doi:10.1016/j.scitotenv.2019.04.069 PMID:31121354

Noshi, Y., Uda, T., Kobayashi, A., Miyahara, S., & Serizawa, M. (2015). Beach changes observed in Phan Rang City in southeast Vietnam. *Procedia Engineering*, *116*(1), 163–170. doi:10.1016/j.proeng.2015.08.278

Ostrom, E. (2009). A General Framework for Analyzing Sustainability of Social-Ecological Systems. *Science*, *35*(5939), 419–422. doi:10.1126cience.1172133 PMID:19628857

Pearson, J., Bruce, T., Allsop, W., Kortenhaus, A., & van der Meer, J. (2004). Effectiveness of recurve walls in reducing wave overtopping on seawalls and breakwaters. *Coastal Engineering*, *148*, 4404–4416.

Powell, N., Osbeck, M., Bach Tan, S., & Vu Canh, T. (2007). *Mangrove restoration and rehabilitation for climate change adaptation in Vietnam.* World Resources Report. Retrieved from http://www.worldresourcesreport.org

Saengsupavanich, C. (2020). Deconstructing a jetty to rectify the downdrift erosion. *Journal of Sustainability Science and Management*, *15*(2), 79–88.

Sanò, M., Jiménez, J. A., Medina, R., Stanica, A., Sanchez-Arcilla, A., & Trumbic, I. (2011). The role of coastal setbacks in the context of coastal erosion and climate change. *Ocean and Coastal Management*, *54*(12), 943–950. doi:10.1016/j.ocecoaman.2011.06.008

Seddon, N., Smith, A., Smith, P., Key, I., Chausson, A., Girardin, C., & Turner, B. (2021). Getting the message right on nature-based solutions to climate change. *Global Change Biology*, *27*(8), 1518–1546. doi:10.1111/gcb.15513 PMID:33522071

Shih, Y. C. (2016). Coastal Management and Implementation in Taiwan. *Journal of Coastal Zone Management*, *19*(4). Advance online publication. doi:10.4172/2473-3350.1000437

Takagi, H., Nguyen, D. T., Miquel, E., Tran, T. T., Hanne, L. K., Takahito, M., & Mitsuko, L. (2012). *Assessment of Disaster Vulnerability of Coastal Areas in Vietnam and Basic Study for Disaster Mitigation.* JST FY2011 SATREPS. doi:10.13140/RG.2.2.26388.30088

Taoyuan Office of Coast Administration Construction. (2021). *Caota sand dunes geopark*. https://www.tyoca.gov.tw/wp-content/uploads/2021/02/%E8%8D%89%E6%BC%AF%E6%B2%99%E4%B8%98%E6%89%8B%E5%86%8A.pdf

Thang, H. C., Misdorp, R., Laboyrie, H., Pos, H., van Zetten, R., & Huan, N. N. (2011). *Vietnam: a decade of Coastal Cooperation*. Climate of Coastal Cooperation.

Uda, T., & Yamamoto, K. (1994). Beach erosion around a sand spit -An Example of Mihono-Matsubara Sand Spit-. *Coastal Engineering*.

van Slobbe, E., de Vriend, H. J., Aarninkhof, S., Lulofs, K., de Vries, M., & Dircke, P. (2013). Building with Nature: In search of resilient storm surge protection strategies. *Natural Hazards*, *66*(3), 1461–1480. doi:10.100711069-013-0612-3

Vo, P. L., Lin, T.-Y., & Nguyen, B. G. (2018). Threat of coastal hazards to socio-economic development in the central coasts of Vietnam. *International Symposium on Lowland Technology*.

Walters, G., Janzen, C., & Maginnis, S. (2016). *Nature-based solutions to address global societal challenges*. IUCN Report. doi:10.2305/IUCN.CH.2016.13.en

Waryszak, P., Gavoille, A., Whitt, A. A., Kelvin, J., & Macreadie, P. I. (2021). Combining gray and green infrastructure to improve coastal resilience: Lessons learnt from hybrid flood defenses. *Coastal Engineering Journal*, 1–16. doi:10.1080/21664250.2021.1920278

KEY TERMS AND DEFINITIONS

Downdrift Erosion: In this chapter refers to erosion related to artificial engineering structures. A coastal engineering structure (such as jetties or groins) can block sediment that is spread by natural longshore currents along the shoreline. This results in sediment accumulation at the updrift side of the structure, while on the downdrift side erosion will occur.

Eco-DRR (Ecosystem-Based Disaster Risk Reduction): This method involves the management of ecosystems and ecosystem services to reduce the disaster risk (such as flooding, storm, or wave damage) for human communities.

Hybrid Engineering Approaches: Integration of grey (hard, artificial engineering structures) with green (nature-based solutions) to form a hybrid solution for coastal protection.

ICZM (Integrated Coastal Zone Management): This concept provides a framework for long-term sustainable coastal management. It involves stakeholder participation at all levels, throughout the process and seeks to find a balance between environmental, socio-economic, cultural and recreational objectives.

NbS (Nature-Based Solutions): Solutions to coastal hazards and environmental challenges though sustainable actions that enhance and work with nature and should have benefits for both society and biodiversity.

SES (Social-Ecological Systems): A concept that recognized that humans and the natural world are interlinked and affect each other through a complex feedback mechanism. Identifying these links and their consequences when working towards solutions is a key factor, based on a more holistic view.

Setback Zoning: Allowing natural dynamics to take place in a strip of land along the coastal zone (where development activities are prohibited or restricted), to function as a buffer zone for coastal erosion, storm tides and sea level rise.

Chapter 6
Community-Based Waste Management Model in Optimizing Waste Reduction:
Waste Bank Practices in Indonesia

Christia Meidiana
Brawijaya University, Indonesia

Tonni Agustiono Kurniawan
Xiamen University, China

Adipandang Yudono
Brawijaya University, Indonesia

Surjono Surjono
Brawijaya University, Indonesia

ABSTRACT

The concept of collect-transport-and-dump of waste in Indonesia has changed into reduce, reuse, and recycle (3R) since it causes environmental consequences that the Government of Indonesia (GoI) has banned including the open dumping practices since 2013. GoI targets 30% waste reduction by 2030, and some policies are introduced to achieve the target including promoting waste bank. Waste bank is an implementation of 3R involving community. The number of waste banks are increasing since waste banks offer many benefits. Currently, there are 7600 waste banks nationwide. Through waste banks, the community learns to sort the waste. It also raises public awareness about waste issues and the importance of 3R. It becomes an innovation program at the grassroots level to increase income and support the local governments to improve the performance of waste management. In this chapter, some cases presenting waste bank implementation and their related aspects, such as waste bank performance, public acceptance, and participation, as well as determinants of public participation, are discussed.

DOI: 10.4018/978-1-7998-9190-1.ch006

INTRODUCTION

Landfill is one of the cheapest and commonly used as final destination to dispose the waste In developing countries (Idrus, et.al., 2008). Therefore, it is the most preferred waste treatment in Indonesia that almost 69% waste ends up in the landfill, whereas 7% is for treatment and recycling purposes. About 24% of waste is unmanaged, causing illegal dumping, marine/river littering, and open burning (Ministry of Environment and Forestry, 2020). Many landfills in Indonesia are still open dumpsites. Open dump site causes environmental consequences such as pollution of surface water, surrounding soil, and groundwater due to the leachate (Gworek et al., 2016), high methane (CH4) emission contributing 1-2% of the total global greenhouse gas (GHG) emissions (Bogner et al., 2011; DEFRA, 2009), and several health risks due to hazardous pollutants leaks from landfill and penetration of leachates into the soil (Palmiotto et al., 2014, DEFRA, 2009).

In developing countries, SWM is one of the most crucial issues faced by the local authorities and characterized by inefficient organization, financial limitation and complexity (Burntley, 2007). Furthermore, the waste management processes are more straightforward, with a notable contribution from the informal sector (Ezeah et al., 2013).

In addition, open dumpsite is considered and acceptable method and part of the waste disposal process due to lack of infrastructure, financial resources, and management planning (Srivastava et al., 2015). Because of its adverse environmental effects, the Government of Indonesia (GoI) has banned open dumping practices in landfill through the enactment of Waste Law No. 8/2008. It mandates the local governments to close open dumping sites and operate the only environmentally friendly landfill, either controlled or sanitary. Currently, there are totally 388 landfills in Indonesia and 5.9% of them are equipped with biogas collection system and the landfill biogas is distributed to nearest villages for cooking. However, disposing of waste in landfills creates another problem that requires significant space demand.

Growing population leads to the increasing trend of municipal waste quantity. Thus finding the available place for expanding the landfill poses an additional challenge to consider, such as urban planning (Mallick, 2021), the possibility of groundwater pollution (Abdullah et al., 2018), soil pollution (Li et al., 2007), and agriculture (Thapa & Murayama, 2008). In 2011, GoI changed the old paradigm of waste management from the end-pipe-solution, which comprises waste collecting-transporting-disposing of, to the new one, waste minimization by source-reduction. There are various measures, approaches, and strategies to promote waste minimization, such as charging fees for waste disposal on each disposal facility to reduce the total waste generation potentially, "three Rs" principle of waste (reduction, reuse, recycling), as well as the zero-waste approach to motivates sustainable consumption and production (Yu et al., 2021). GoI has been promoting the new concept by introducing waste separation and reduction in the community through the enactment of Government Regulation No. 81 of 2012, defining the importance of domestic waste reduction through reduction, reuse, and recycling (3R). The regulation is essential since the GoI targets to achieve 20% waste reduction by 2019. This target was updated in 2017 to be 30% waste reduction by 2025.

Promoting the 3R can be effective within the community scale since the community has essential roles in waste reduction by controlling waste in their environment, maintaining and improving the implementation of 3R (Abdullahi et al., 2016). They can implement sanitary behavior by maintaining the cleanliness of households and surroundings, then storing the waste in a designated container/bin (Malik et al., 2015), or taking resource recovery actions by managing the solid waste services (Sekito et al., 2013). Still, in many cases, the level of community participation in waste separation and recycling activity is low for some factors. Timlett and Williams (2001) stated that the efficiency of a waste recycling program involving the community depends mainly on local conditions, cultural, situational or demographic factors. Community participation plays an important role in both formal and informal waste sector.

Informal waste sector in waste segregation and recycling has founded and developed itself in as a response to a demand of recyclable materials from the industrial sector because of increasing prices of imported recyclable materials, i.e. plastic or paper (Matter, 2013). Informal groups such as scavengers are commonly involved in 3R activities in many developing countries helping to separate recyclable materials in the transfer station and landfill (Asim et al., 2012; Wilson, et.al. 2009). The recyclable materials are sold to the middleman or recycling centers (Miranda et al., 2020). Therefore, the 3R helps involve the community scale as the primary producer of solid waste in waste reduction practices. The NIMBY (Not In My Back Yard) syndrome (Shen & Yu, 1997) is a phenomenon of the strong opposition of a community to some public interventions due to undesirable things such as waste. That could be potentially changed into YIMBY (Yes In My Back Yard) by recognizing the potential value of the waste and being responsible for the waste produced. YIMBY can stimulate community participation in waste reduction. Other informal waste sector is waste bank established typically as an organization. The good organization is one of the factors for effective waste reduction at source besides waste separation (Figueroa, 1998).

Waste Bank (WB) is a community participation approach for solid waste recycling through bank projects (Singhirunnusorn et al., 2017). WB could potentially stimulate solid waste management regarding its source separation, recovery attempt of recyclable materials, and storing potential materials for collection purposes (Visvanathan & Glawe, 2006). Successful MSW requires waste at source where community participation is necessary (Dehghani, et.al. 2014). According to Safdari, et.al, (2013) some economic and environmental benefits are offered by waste separation at source such as reduced recycling costs, increased efficient system for circular process of a product and decreased material input thus decreased raw material extraction from the nature giving environmental pressure.

WB could also act as an intermediary of selling sorted recyclable materials for more significant net revenues to benefit the members and officers alike, which is possible due to the collaboration between waste generators and other entities for hosting the activities such as schools or community centers (Mongkolnchaiarunya, 2005; Suttibak & Nitivattananon, 2008).

WB approach in developing countries like Indonesia has encouraged people to earn additional income by joining active membership. This approach has changed the perspective of waste as a resource or income generator rather than something useless, supported by the waste separation policy implemented to the local community (Meidiana et al., 2017). Ministry of Environment and Forestry (MoEF) encourages applying the YIMBY strategy by implementing the waste bank management model by issuing a ministerial regulation on 3R through a waste bank, namely the Regulation of the Minister of Environment of the Republic of Indonesia No.13/2012. The new regulation, Ministry of Environment and Forestry Regulation No. 14/2021 regarding on Waste Management in Waste Bank, was enacted in 2021 to replace the old one.

The WB progression in Indonesia showed an increasing trend from 1172 units and 151.419 membership in 2014 to 9106 units and 419.204 membership in 2019 (MoEF, 2021). The number of WB is expected to grow in response to achieving national waste reduction by 2025. On the other hand, a preliminary study revealed the inactivity of WB in some cities. One example is that the WB average activity rate in East Java Province accounted for 48,2%, showing that only about half of the registered WBs are actively operating. Therefore, it is essential to improve WB performance through community participation as a prerequisite for the WB approach. Singhirunnusorn et al. (2017) revealed a significant positive correlation between recycling knowledge and behavior to their involvement in the WB. However, the information

Figure 1. Waste Bank's role within waste management

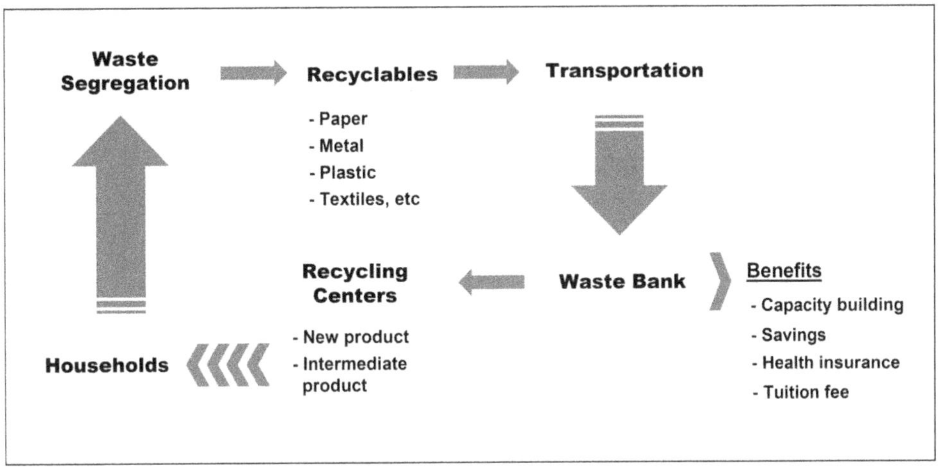

about WB's recycling capacity, mass, collection systems, and cash flow is limited, thus assessing the WB's performance poses another challenge. Therefore, this paper aims to present some cases about the various aspects of the waste bank system, including measurement of WB's performance, participation level, public acceptance, and its significant factors.

BACKGROUND

WB, by principle, is an institution like a bank founded by a community as core operators, which considers the social aspect that influences WB's performance. WB is regarded as a solution to the waste problem by earning money from the recyclable waste to deposit instead of money in a banking system (Pariatamby & Tanaka, 2013). Unlike commercial banks, which operate in the public domain, the WB working system is based on household-level. It compensates for the sorting and wastes deposit process in the form of money.

The private sector, for instance, is the recyclable industry, sets the prices and acceptable kinds of recyclable waste. Thus WB could keep enough money to cover the operating costs and deposit fees for the customers. In general, WB receives recyclable waste such as plastics (bottles, glasses, bags), papers (newspapers, magazines, books, office paper), aluminium cans, electrical wires, and other potential recyclable waste. The recyclable waste deposited from the member will be priced based on its type and weight, then the value is recorded in the account book. The member has the right to claim their deposit in money value or exchange it with available programs offered by WB, such as electricity or water services coupons, staple foods coupons, health insurance, school tuition, and other potential programs. The WB can produce value-added products instead of selling them directly to the recyclable industry (Purwaningsih, 2015). For instance, partnerships could be formed between WB and women initiatives for making value-added products such as handicrafts. Figure 1 describes the role of WB.

WB could also help to finance the education institutions in sustainable manners. Indrianti (2016) revealed that a non-formal voluntary religious education institute could sustain its financial needs by establishing a WB through community empowerment processes (training, counseling, and system establishment). The absence of regular financial supports from the government or charity programs motivates the committee members to achieve economic sustainability from WB. It is also beneficial by giving actual practice to the students on how to promote sustainable waste management and the traditions of the religious practice they learned. Therefore, the waste bank management model increases public awareness of the waste problem and the economic value. It can motivate people to involve more in waste separation, reduction, reuse, and recycling.

Approaches for Waste Bank Development and Evaluation

Some cases related to waste bank implementation in some cities in Indonesia were chosen for this study. These cases could provide compared understandings into how the developing countries (Indonesia-) may apply powerful technologies and innovations to solve solid waste management using the waste bank management model.

Measurement of Waste Bank Performance

In this case, Meidiana et al. (2018) measured Waste Bank Performance with two parameters, i.e. the efficiency and effectiveness. The research was conducted in five waste banks located in Batu District, Batu City. A survey concerning waste bank performance was conducted for about five months. The questionnaires were distributed during the study for effectiveness assessment, while the in-depth interview was conducted on waste bank staff for attaining waste bank characteristics. Overall Equipment Effectiveness (OEE) and Elasticity index were used to measure the waste bank's effectiveness and efficiency. These two parameters determined the waste bank performance. Table 1 presents the variables used for both measurements.

OEE is a product of availability (A), quality (Q) and performance (P). Therefore, OEE is calculated using Eq. 11.

$$OEE = A * Q * P \quad (11)$$

OEE value is classified into five groups with each remark from outstanding to low and whether the group indicates high, moderate, or, inadequate effectiveness. **Table 2** shows the classification for OEE value and its interpretation indicating higher OEE is higher effectiveness of WB.

Meanwhile, the elasticity index (E) was calculated to measure WB's efficiency monthly. Each value indicates the efficiency level as described in **Table 3**.

A matrix is generated by combining effectiveness and efficiency value presenting WB's performance. The results are categorized into nine groups shown in **Table 4.**

WB's efficiency is reflected by its monthly elasticity calculated using income and number of active members. Using those methods, the efficiency and the effectiveness of the study case in Batu City was calculated. The following sections discuss the result of the study.

Table 1. Variables for WB's effectiveness and efficiency measurement

No	Parameter	Variables	Equations		Remarks
1.	Efficiency	a. Income b. Number of active members	$APN = \frac{Q}{N}$	(1)	Q: Income (Rp/event) N: # active member per event ΔQ_n: income difference between event
			$MPN = \frac{\Delta Q_n}{\Delta N_n}$	(2)	ΔN_n: the difference between # active member per event
			$\Delta Q_n = \frac{Q_{n+1}}{Q_n}$	(3)	APN: average income
			$\Delta N_n = \frac{N_{n+1}}{N_n}$	(4)	MPN: marginal income of member
			$E = \frac{MPN}{APN}$	(5)	E= Elasticity index
2	Effectiveness	Availability	A = Aa/Ra x100%	(6)	Aa:WB's current operasional time (event/year)
					Ra: planned operational time (event/year)
		Quality	Q = Aq/Tq x 100%	(7)	Aq: # of program implemented
					Tq: total # planned WB's program
		Performance	$CT_a = \frac{O}{T_{op}}$	(8)	T_{op}:operasional time (minutes)
			$CT_t = \frac{IO}{T_{st}}$	(9)	O: Output (# active member
			$P = \frac{CT_a}{CT_t}$	(10)	CT_a: Actual cycle time (minutes)
					T_{st}: Operasional time standard (minutes)
					IO: Ideal Output
					CT_t:theoretical cycle time (minutes)

Source: (Meidiana, 2017) (modified according to the local context in Batu City)

Table 2. Classification for OEE value

OEE Value	Effectiveness
100%	High
85% ≤ OEE <100%	
60% ≤ OEE <85%	Moderate
40% ≤ OEE <60%	Low
OEE< 40%	

Source: (Meidiana, 2017)

Table 3. Elasticity index of waste bank

Elasticity index	Remarks	Efficiency
> 1	High efficiency	High
= 1	Average productivity (income) of members	Moderate
< 1	Need more members' involvement	Low

Table 4. Matrix for measurement of waste bank's performance

Efficiency Effectiveness	High Efficiency	Moderate Efficiency	Low Efficiency
High Effectiveness	High performance	High performance	Moderate performance
Moderate Effectiveness	High performance	Moderate performance	Low performance
Low Effectiveness	Moderate performance	Low performance	Low performance

Table 5. Value of waste bank's efficiency

No	Waste Bank	Elasticity Index	Output	Efficiency
1	Kartini Mandiri	0.57	E < 1	Low
2	Saras Asri	1.13	E > 1	High
3	Cahaya	0.96	E < 1	Low
4	RW 03 Krajan	2.74	E > 1	High
5	Kampung Damai	1.20	E > 1	High

Table 6. Values of each parameter of OEE and effectiveness

Waste Bank	A	P	Q	OEE	Effectiveness
Kartini Mandiri	92%	78%	33%	24%	Low
Saras Asri	88%	68%	50%	29%	Low
Cahaya	67%	59%	33%	39%	Low
RW 03 Krajan	100%	55%	100%	55%	Low
Kampung Damai	33%	65%	50%	24%	Low

Waste Bank Efficiency

Firstly, referring to the equations presented in **Table 1**, the efficiency of 5 WBs in Batu City was conducted. **Table 5** shows the result.

The efficiency of two WBs is low, while three WB's are high. Low efficiency (E<1) indicates that the activity of members within the programs offered by WB is low. Their productivity should be more enabled by increasing member income and maintaining the number of active members. For instance, the elasticity index of Kartini Mandiri is low, indicating low efficiency caused by a significant decrease of members' marginal income. Referring to Eq. 2, marginal revenue is determined by the difference of the number of active members and income between two opening events. Based on the data, one more member who sold the waste in the 15th event. At the same time,the income remained relatively constant from the 14th to 15th event leading to the member's marginal revenue. Referring to Eq.5, a member's marginal revenue while the average income is constant will decrease the elasticity index to -4.26 in the respective round. Therefore, encouraging the member to be more active in selling the waste is necessary to raise the average income leading to higher efficiency. Meanwhile, the high efficiency of the other three WBs is mainly caused by steadily increasing revenue and the number of members every round.

Waste Bank Effectiveness

WB's effectiveness is represented by the *Overall Equipment Effectiveness (OEE)* value calculated using three parameters, i.e., *availability (A), performance (P),* dan *quality (Q).* Eq. 6 to Eq. 10 in Table 1 are used to generate the OEE value. Each parameter value and the OEE value is shown in **Table 6.**

Table 7. Matrix of WB's performance in Batu District

Waste Bank	Efficiency		Effectiveness		Performance
	Elasticity Index	Efficiency	OEE	Effectiveness	
Kartini Mandiri	0.57	Low	24%	Low	Low
Saras Asri	1.13	High	29%	Low	Moderate
Cahaya	0.96	Low	39%	Low	Low
RW 03 Krajan	2.74	High	55%	Low	Moderate
Kampung Damai	1.20	High	24%	Low	Moderate
Mean	1.32	High	34%	Low	**Moderate**

Table 6 indicates that all WBs have low effectiveness because the value of OEE is lower than 60% representing the low and marginal group.

Referring **to Table 3,** the values in Table 4 dan Table 5 are correlated, resulting in each WB's performance in Batu City (**Table 7**). Generally, WB's performance can be categorized into two, which are low and moderate.

The results showed that three WBs (Sarah Asri, RW 3 Krajan, Kampung Damai) have high efficiency caused by steady income and, a stable number of members and the others (Kartini Mandiri and Cahaya) have high low efficiency. Still, all of them have low value because three variables of effectiveness, i.e., availability, performance, and quality, have low value. To sum up, WBs in Batu City have moderate performance, but improvement is required since WBs have low effectiveness and efficiency.

Measuring Public Acceptance towards Waste Bank

According to the discussion in the previous section, WB's performance is influenced by the number of active members. WB can attract many people if the public accepts it as an institution that can improve their wealth. Therefore, public acceptance plays an essential role in enhancing WB's performance. Meidiana et al. al. (2016) observed the public acceptance towards a WB (named Bintang Sejahtera) in a rural area in Lombok, West Indonesia.

Bintang Sejahtera WB accepts recyclable materials from households, including plastic bottles/glass, aluminium tin, plastic bags, paper, and cardboard. In 2016, WB's capacitywas 4.43 tons/day (25% of total waste generation on the island). This amount remains relatively constant until 2019. The most fraction of the waste is plastic which is about 3.145 tons/day. The WB sells the waste to some industries outside the island, and the price varies depending on the waste type.

Using the same procedures, the OEE value is calculated, and the result is 12.67% which is below 60%, indicating low effectiveness caused by the low number of implemented programs. There only one out of 6 programs implemented because of insufficient public participation. The reason is that waste separation is not typical for the villager leading to low membership. Only 37.5% of the households is member of WB. The main reason is unacceptable waste price. Based on this result, WTA is measured to determine the optimum price of recyclable waste sorted by the people and sold to the waste bank, asshown in Table 8.

WTA of HHs is measured to determine the expected compensation to separate waste and sell it to the WB. Furthermore, WTA may reflect the public by eliciting questions in questionnaires. Villager accepting the WB program is asked further for an acceptable price for the waste transporting to WB. Table 8 explains the acceptable price for each waste type for 94 respondents representing the whole HHs in the village.

Factors Influencing Public Participation in Waste Bank

The study case in Batu City aims to examine public participation in the waste bank and the determinant factors influencing people's decision to involve in waste reduction through WB. The research was conducted in three WBs located in Junrejo District, Batu City. A survey for public participation was conducted in 372 households using a random sampling technique. The questionnaires were distributed for related data required for logistic regression.

Table 8. The selling price of recyclable waste from different actors

Waste types	Acceptable price (US$/kg)	WB's Selling price (US$/kg)	Middleman's selling price (US$/kg)
Plastic bottle	0.118-0.234	0.157	0.157
Plastic cup	0.118-0.234	0.157	0.234
Small beer bottle (glass)	0.008-0.039	0.000	0.024
Big beer bottle (glass)	0.039-0.118	0.000	0.063
Ketchup bottle (glass)	0.024-0.078	0.000	0.046
Aluminium tin	0.391-1.015	0.703	0.781
Cardboard and paper	0.078-0.157	0.078	0.118
Plastic bag	0.008-0.039	0.015	0.000
Tetra Pack	0.008-0.046	0.024	0.000
*1 US$ equals Rp 13.606			

There are about 12,400 households (HHs) in the study area, but only 478 HHs are registered as a member of WB, indicating that public participation in WB is only 3.8%. This percentage is far below the target of the local government, which is 10% waste reduction through WBs (out of a total target of 30% reduction by 2025). However, this percentage does not necessarily imply that the public acceptance WB is low. Many factors contribute to low participation in an informal organization such as Waste Bank, and these factors can be determined through regression analysis. Regression analysis was applied to explain the relationship between public participation as a response variable and some predictor variables. There are 22 predictors variables used in this study, shown in Table 9.

Those variables are used for logistic regression conducted based on the procedures presented in Figure 2. Logistic regression is used primarily when the response variable is a dichotomy and the predictor variables are of any type. It utilizes maximum likelihood estimation after transforming the response into a logit variable (Garson, 2008). Hosmer and Lemeshow (2000) proposed the logistic regression model as described in Eq. 12.

$$g(x) = \beta_o + \beta_1 X_1 + \ldots + \beta_n X_n \quad (12)$$

where β_o, β_n, X_n, is the vector of the parameter to be estimated, the coefficient of predictor variables, and a vector of predictor variable observations, respectively.

Results of the logistic regression model of public participation are presented as follows:

$$g(x) = -12.70 + 1.03_{X1} - 1.60_{X3} + 2.07_{X13} + 1.75_{X18} + 2.65_{X19} + 1.10_{X20} + 2.54_{X21} + 6.48_{X22}$$

Referring to the above model, factors influencing public participation in waste reduction through WB ((x)) are the availability of waste bin (X_1), occupation (X_3), knowledge of sustainable development (X_{13}), local norms (X_{18}), public opinion about waste (X_{19}), motivation (X_{20}), sense of belonging (X_{21}),

Table 9. Variables for the regression analysis

No	Parameter	Variable	Sub-variable	References
1	Environmental Attitude Response variable (Y)	• Participation in WB		
2.	Situational characteristics	• Behavioural context	• Availability of waste bin (X_1)	
		• Individual characteristics	• Age (X_2) • Occupation (X_3) • Income (X_4) • Education (X_5) • Political mindset (X_6)	
		• Individual knowledge and experience of the behaviour	Knowledge about environment • Global warming (X_7) • Ozone (X_8) • Water pollution (X_9) Knowledge about the waste management • Domestic waste production (X_{10}) • Urban waste production (X_{11}) • Waste problems (X_{12}) Knowledge about policy • Sustainable development (X_{13}) • National environmental program (X_{14}) • Personal experience of the behaviour (X_{15})	• Barr, Stewart. 2007. • Dhokhikah et.al, 2015. • Ajzen, I. 1991. • Katz et.al. 1992..
3.	Psychological factors	• Awareness about waste issues (X_{16}) • Motivation to overcome waste problems (X_{17}) • Acceptance of related norms (X_{18}) • Opinion about waste problems (X_{19}) • Self-confidence (X_{20}) • Sense of belonging (X_{21})		
4.	Technical Factor	• Distance from house to WB (X_{22})		Sterner dan Baterlings. 1999.

and distance from HH to WB (X_{22}). All factors are proportional to public acceptance except occupation. It indicates that households with higher incomes participate less in the waste bank because of the relatively low recyclable materials.

SOLUTIONS AND RECOMMENDATIONS

The three case studies presented in the previous sections describe the analysis that can be implemented to measure the waste bank performance and determine the factors influencing community participation.

Figure 2. Procedures for logistic regression

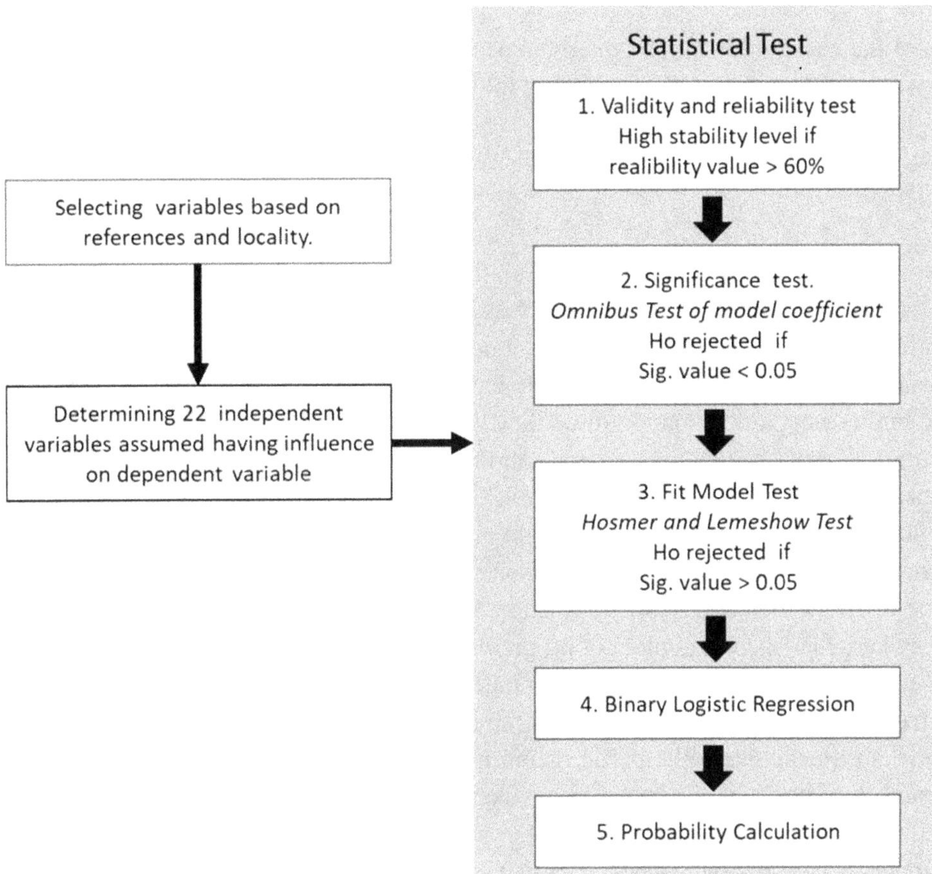

The result of the first case study showed that generally, WBs in Batu City have moderate performance. However, improvement is necessary because some WBs have low effectiveness and efficiency. A waste bank will have high performance if many members collect the waste and sell it to the WB. The analysis shows that WBs with low membership have relatively low service quality (long processing time, unattractive programs, and irregular opening time). These conditions should be avoided by providing better service than households are willing to register as a member. In the second case study, public acceptance towards WB is measured using willingness to separate the waste. Households are interested in separating and expecting specific prices for each waste type. There are mainly two actors interested in buying the waste from households, i.e. waste bank and intermediaries. The result shows that both actors offer various prices for selling the waste, giving rivalry consequences affecting the price stability because the household prefers to sell the waste to the one who provides a higher price. That is one of the reasons for low membership and small waste amount selling to the waste bank. Therefore, cooperation between WB and intermediaries may be initiated to overcome this problem. In the third case, the result of regression analysis shows factors affecting community participation in waste reduction through the waste bank. Significant variables should be the primary concern for development to improve the WB's performance.

FUTURE RESEARCH DIRECTIONS

The results of the case studies can be the reference for the following research, which will focus on developing a waste bank information system that the WB members and the local government can access.

CONCLUSION

Innovative solid waste management practices should be developed in developing countries. It should be robust, easy to implement, not expensive and sustainable. Implementation of high waste technology is not necessarily suitable for many reasons., As a developing country, Indonesia has developed a waste bank concept to reduce waste and support the national target of 30% waste reduction by 2025. The waste bank implements a new paradigm of the "waste as a resource or income generator" approach. It encourages household behavior to separate and reduce waste at source through their involvement in a waste bank that works at the community level. Community-based solid waste management involves close consultation with community organizations and community members. As an organization, a waste bank requires regular evaluation to improve its performance. As found in the study case, waste bank's performance is determined two main parameters, i.e., efficiency and effectiveness. Each parameter is influenced by variables such as number of programs, number of members, member's activity, and waste volume. The other case shows that public acceptance toward waste banks can be measured by their willingness to pay to separate the waste and adapt the behavior change to separate the waste. The last case presents factors influencing public participation in waste reduction through the waste bank. It shows that situational characteristicsss, psychological factors, and technological factors can influence people's perceptions, awareness, and susceptibility to the waste bank. The continuous informing process from the government and sustainable support for raising awareness on environmental issues and appropriate solid waste management are determinants to the success of the waste bank project.

ACKNOWLEDGMENT

Overall, the authors want to remark the clear contribution of the Institute of Research and Community Services Brawijaya University (LPPM) Brawijaya University, which has supported this research.

REFERENCES

Abdullah, T. O., Ali, S. S., Al-Ansari, N. A., & Knutsson, S. (2018). Possibility of Groundwater Pollution in Halabja Saidsadiq Hydrogeological Basin, Iraq Using Modified DRASTIC Model Based on AHP and Tritium Isotopes. *Geosciences*, *8*(7). Advance online publication. doi:10.3390/geosciences8070236

Abdullahi, Y., Badamasi, A., Galadimawa, N., & Abubakar, G. (2016). *Community based solid waste management strategy: a case study of Kaduna metropolis*. doi:10.2495/SDP160641

Asim, M., Batool, S. A., & Chaudhry, M. N. (2012). Scavengers and their role in the recycling of waste in Southwestern Lahore. *Resources, Conservation and Recycling*, *58*, 152–162. doi:10.1016/J.RESCONREC.2011.10.013

Bogner, J. E., Spokas, K. A., & Chanton, J. P. (2011). Seasonal Greenhouse Gas Emissions (Methane, Carbon Dioxide, Nitrous Oxide) from Engineered Landfills: Daily, Intermediate, and Final California Cover Soils. *Journal of Environmental Quality*, *40*(3), 1010–1020. doi:10.2134/jeq2010.0407 PMID:21546687

Burntley, S. J. (2007). A review of municipal solid waste composition in the United Kingdom. *Journal of Waste Management*, *27*(10), 1274–1285. PMID:17011771

Dehghani, M., Hashemi, H., Abedi, T., Shamsedini, N., Khodabakhshi, A., & Ghasemi, R. (2014). The survey of knowledge, attitude, and Behavior of students in Shiraz University of Medical Sciences about the recycling of solid waste in 2012. *Health System Research*, *10*, 821–829.

Department for Environment, Food and Rural Affairs. (2009). *Good practice and regulatory guidance on composting and odour control for local authorities*. Author.

Ezeah, C., Fazakerley, J. A., & Roberts, C. L. (2013). Emerging trends in informal sector recycling in developing and transition countries. *Waste Management (New York, N.Y.)*, *33*(11), 2509–2519. doi:10.1016/J.WASMAN.2013.06.020 PMID:23886489

Gworek, B., Dmuchowski, W., Koda, E., Marecka, M., Baczewska, A. H., Bragoszewska, P., Sieczka, A., & Osiński, P. (2016). Impact of the municipal solid waste lubna landfill on environmental pollution by heavy metals. *Water (Switzerland)*, *8*(10). Advance online publication. doi:10.3390/w8100470

Idrus, M. M., Mohd Baharudin, R., Shahabudin, M., & Rashidah, A. (2008). *An overview of landfill management and technologies: A Malaysian case study at Ampar Tenang*. 1st National Seminar on Environment, Development and Sustainability, Selangor, Malaysia.

Li, Z.-W., Zeng, G.-M., Zhang, H., Yang, B., & Jiao, S. (2007). The integrated eco-environment assessment of the red soil hilly region based on GIS—A case study in Changsha City, China. *Ecological Modelling*, *202*(3), 540–546.

Malik, N. K. A., Abdullah, S. H., & Manaf, L. A. (2015). Community Participation on Solid Waste Segregation Through Recycling Programmes in Putrajaya. *Procedia Environmental Sciences*, *30*, 10–14. doi:10.1016/J.PROENV.2015.10.002

Mallick, J. (2021). Municipal solid waste landfill site selection based on fuzzy-ahp and geoinformation techniques in ASIR region Saudi Arabia. *Sustainability (Switzerland)*, *13*(3), 1–33. doi:10.3390u13031538

Matter, A., Dietschi, M., & Zurbrügg, C. (2013). Improving the informal recycling sector through segregation of waste in the household. The case of Dhaka Bangladesh. *Habitat International*, *38*, 150–156.

Meidiana, C., Yakin, H., & Wijayanti, W. (2017). Household's Willingness to Accept Waste Separation for Improvement of Rural Waste Bank's Effectivity. doi:10.5772/intechopen.69428

Ministry of Environment and Forestry. (2020). *National Plastic Waste Reduction Strategic Actions for Indonesia, Republic of Indonesia*. https://wedocs.unep.org/bitstream/handle/20.500.11822/32898/NPWRSI.pdf?sequence=1&isAllowed=y

Miranda, I. T. P., Fidelis, R., Fidelis, D. A. de S., Pilatti, L. A., & Picinin, C. T. (2020). The integration of recycling cooperatives in the formal management of municipal solid waste as a strategy for the circular economy—The case of Londrina, Brazil. *Sustainability (Switzerland), 12*(24), 1–22. doi:10.3390u122410513

MoEF. (2021). *Performance Report of 2020.* Ministry of Environment and Forestry.

Mongkolnchaiarunya, J. (2005). Promoting a community-based solid-waste management initiative in local government: Yala municipality, Thailand. *Habitat International, 29*, 27–40. doi:10.1016/S0197-3975(03)00060-2

Palmiotto, M., Fattore, E., Paiano, V., Celeste, G., Colombo, A., & Davoli, E. (2014). Influence of a municipal solid waste landfill in the surrounding environment: Toxicological risk and odor nuisance effects. *Environment International, 68*, 16–24. doi:10.1016/J.ENVINT.2014.03.004 PMID:24685488

Pariatamby, A., & Tanaka, M. (2013). *Municipal Solid Waste Management in Asia and the Pacific Islands: Challenges and Strategic Solutions.* Springer Singapore. https://books.google.co.id/books?id=ZfsUswEACAAJ

Purwaningsih, A. (2015). Analysis of customer mindset change and accounting practice of garbage bank as medium of edupreneurship. Int. J. Monetary Economics and Finance, 8(4).

Safdari, M., Mirzaei Alavijeh, M., Ehrampoush, M. H., Qhaneyan, M. T., & Morowatisharifabad, M. A. (2013). Knowledge, attitude and performance students of shahid sadoughi university of medical ciences-yazd about recycling solid material: A short report. *Majallah-i Ilmi-i Danishgah-i Ulum-i Pizishki-i Rafsanjan, 12*, 157–164.

Sekito, T., Prayogo, T. B., Dote, Y., Yoshitake, T., & Bagus, I. (2013). Influence of a community-based waste management system on people's behavior and waste reduction. *Resources, Conservation and Recycling, 72*, 84–90. doi:10.1016/J.RESCONREC.2013.01.001

Shen, H.-W., & Yu, Y.-H. (1997). Social and Economic Factors in the Spread of the NIMBY Syndrome against Waste Disposal Sites in Taiwan. *Journal of Environmental Planning and Management, 40*(2), 273–282. doi:10.1080/09640569712227

Singhirunnusorn, W., Donlakorn, K., & Kaewhanin, W. (2017). Household Recycling Behaviours and Attitudes toward Waste Bank Project: Mahasarakham Municipality. *Journal of ASIAN Behavioural Studies, 2*(5), 17–26. doi:10.21834/jabs.v2i5.215

Srivastava, V., Ismail, S. A., Singh, P., & Singh, R. P. (2015). Urban solid waste management in the developing world with emphasis on India: Challenges and opportunities. *Reviews in Environmental Science and Biotechnology, 14*(2), 317–337. doi:10.100711157-014-9352-4

Suttibak, S., & Nitivattananon, V. (2008). Assessment of factors influencing the performance of solid waste recycling programs. *Resources, Conservation and Recycling, 53*, 45–56. doi:10.1016/j.resconrec.2008.09.004

Thapa, R. B., & Murayama, Y. (2008). Land evaluation for peri-urban agriculture using analytical hierarchical process and geographic information system techniques: A case study of Hanoi. *Land Use Policy*, *25*(2), 225–239.

Timlett, R. E., & Williams, I. D. (2001). Public participation and recycling performance in England: A comparison of tools for behaviour change. *Resources, Conservation and Recycling*, *31*, 149–170.

Visvanathan, C., & Glawe, U. (2006). *Domestic Solid Waste Management in South Asian Countries-A Comparative Analysis*. Academic Press.

Wilson, D. C., Araba, A. O., Chinwah, K., & Cheeseman, C. R. (2009). Building recycling rates through the informal sector. *Waste Management (New York, N.Y.)*, *29*, 629–635. PMID:18701272

Yu, A. T. W., Wong, I., Wu, Z., & Poon, C. S. (2021). Strategies for effective waste reduction and management of building construction projects in highly urbanized cities— A case study of hong kong. *Buildings*, *11*(5). Advance online publication. doi:10.3390/buildings11050214

Chapter 7
Lesson Learned From Public Participation in the Urban Infrastructure Projects for Engineering:
Case Studies in Ho Chi Minh City, Vietnam

Pham Thi Anh
Ho Chi Minh City University of Transport, Vietnam

Tien Thuy Nguyen
Ho Chi Minh City University of Transport, Vietnam

Tuan Anh Nguyen
Ho Chi Minh City University of Transport, Vietnam

Dong Doan Van
Ho Chi Minh City University of Transport, Vietnam

ABSTRACT

Raising public awareness and participation of the local communities can have an important input for public involvement in the project design and implementation process, and it has become a major objective of sound development programs. This chapter focuses on (1) public participation in the design and adjust boundary of the underground station of the Metro line 2 in Ho Chi Minh City and (2) public awareness and participation in canal environmental protection. Research is mainly based on the results of community meetings on project information, communication to raise public awareness on environmental protection, and survey questionnaires. The study will add to several recent studies on the roles of public participation in infrastructure projects for sustainable development. There are three important lessons that stand out from the experiences using public consultants in infrastructure projects.

DOI: 10.4018/978-1-7998-9190-1.ch007

INTRODUCTION

Since Sherry Arnstein, writing in 1969 about citizen involvement in planning processes in the United States, described a "ladder of citizen participation" (Arnstein, 1969), it is known that significant degrees exist in stakeholder involvement and participation and that the extent of their influence during decision-making processes is a crucial factor in determining their future stakeholders. Nowadays, stakeholder support is recognized as essential for the successful implementation of many policies and programs(Jonsson et al., 2007) (Hendriksen et al., 2012).

The roles of stakeholders and public participation in the implementation of policies and programs had been described in many aspects and spread from country to countries such as in the multi-level environmental governance of aquaculture (Anh et al., 2011) or food risks (de Krom et al., 2013), sanitation, and urban water sector (L. Zhong et al., 2008) (Van Buuren & Hendriksen, 2010), or transportation and other infrastructure projects (Pham Thi Anh & Nguyen Thi Bao Ngoc, 2019) (Li, Terry H.Y. Y et al., 2012).

The urban infrastructure projects often aim to bring benefits for the development of economic, social, and environmental to the invested area, but in other aspects, they also face the problems of investment, resettlement, and several impacts during the construction and operation of these projects. Raising public awareness and public consultation effectively can have an important input for public participation in the project design and implementation process.

Public participation in engineering has been more limited than in other fields. Meetings, where the community has the opportunity to comment on public works projects, are commonplace and the literature in this area focuses on the best way to gain support for projects (Rob Lawlor, 2016).

HCMC (Ho Chi Minh City) currently has 19 urban districts and 5 rural districts with an area of 2,095.01 km2, a population of 7,995 million inhabitants, and an average population density of 3,401 people/km2. HCMC is the economic center of Vietnam and accounts for a large proportion of Vietnam's economy. Thanks to its favorable natural conditions HCMC became an important traffic hub for Vietnam and Southeast Asia, including road, railway, waterway, and airway. In recent years, Ho Chi Minh City had invested several projects on transportation and flooding improvement.

This paper focus on two groups of the project as case studies in Ho Chi Minh City; (1) Projects on public transportation in Ho Chi Minh City, the authors analyzed public participation in the design and adjusted the boundary of the underground station of the Metro line 2 - in Ho Chi Minh City, and (2) projects on the canal and sewerage improvement in the effort to reduce flooding in Ho Chi Minh City, the authors focused on public awareness and participation in canal environmental protection.

PUBLIC PARTICIPATION IN THE DESIGN AND ADJUSTING OF THE UNDERGROUND STATION OF METRO LINE 2 – IN HO CHI MINH CITY

Ho Chi Minh City People's Committee (PC) has drawn out a plan of a Metro Rail System with six main lines to link the city center of Ho Chi Minh City, Vietnam to outlying areas of the City. The six metro lines will be the backbone of the new HCMC public transport system and Line No.2 is a priority in the system development. It is predicted to transport over 600,000 people/day, and to provide a more environment-friendly and safer means of public mass transport in the city (MVA Systra Group, 2013).

The Metro line 2 (Ben Thanh - Tham Luong) is a dual-track line with a length of approximately 11.3 km, of which approximately 9 km consists of a Twin Bored Tunnel and the rest of an Elevated Section.

It is planned to have 10 Stations, 1 km of Depot link, a new Workshop, and twelve new 3-car trains to operate the line.

Encouraging the active participation of local communities in development projects, through capacity-building and environmental and social education, has become a major objective of sound development programs (Borgstede et al., 2013) (Li et al., 2012) (Tran, 2006). In the framework of the Metro Railway Transit Line 2 (MRT 2) Project, Organizer had included the local communities in the program to ascertain the views and perceptions of the Public by collecting additional comments and concerns regarding the new concept Design Adjustment and Station Boundary, and the options for the location of the Technical Buildings. A face-to-face, a structured survey was used to evaluate attitudes and perceptions. The contribution in the consultation has made a positive step towards providing the local community with essential knowledge regarding development, environmental and social issues, thereby enabling them to become actively involved in the comments to the project (Pham Thi Anh, 2017a).

During the Feasibility Study, many public consultations had been organized. They are open forums; the participants were given the opportunity to raise their questions and concerns either orally or in writing. The open forum enabled direct communication and exchange of information between MAUR, concerned or relevant HCMC District PC, affected households, and Relevant Agencies. The issues and concerns, suggestions, and recommendations were recorded during the public consultation.

Participants are the people in the area around the project, affected by relocation, or during the construction and operation of the project, representatives of the Women's Union, Veterans' Association, Youth Union, Vietnamese Fatherland Front, representatives of non-governmental organizations, representatives of donors Asian Development Bank - ADB. Chairing the consultation are representatives of local authorities and representatives of the Metro Line 2 Project Management Board, along with a group of experts to provide consultation and technical assistance.

The first public consultations were organized in 2013. The result of this period was drawn from 6 public consultations (Marcelo Caleda & Pham Thi Anh, 2014). The study recorded 113 direct comments. These comments were synthesized, analyzed, and classified into 4 main topics: (1) Compensation and Resettlement, (2) Adjusted boundary of the Station, Ventilation Shaft and Technical Buildings, (3) Comments and concerns addressed the need for Project Information and Clarification, and (4) Suggestions and Recommendations from the participants.

During the first public consultation, most comments (37/113 comments, rated 32.74%) were requesting to minimize the clearance or resettlement of resident's houses and land, make use most to public. There are also many opinions that the distance between the stations is quite short, it could be made longer distance to reduce the number of stations and reduce the land to be cleared and resettlement (see table 1).

The Management Authority for Urban Railways (MAUR) and designed consultant had considered these comments and the boundary and design of the Station, Ventilation Shaft, and Technical Buildings had been revised and adjusted accordingly.

The second public consultations were organized in the period from September 2014 to March 2015 (Joselito Losaria & Pham Thi Anh, 2015). The purpose of the public consultation is two folds. First, the public consultation intends to enable MAUR to directly provide to the public information regarding the Underground Station Concept Design and the boundary adjustments of the Stations, and the associated Technical Buildings under the Metro Line 2 Project. Secondly, the consultation wants to ascertain the views and perceptions of the public by collecting comments and concerns regarding the revised concept Design Adjustment and Station Boundary, and the options for the location of the Technical Buildings.

Table 1. The results from 2 public consultations

	The first public consultation period (2013)	**The second public consultation period (2014-2015)**
Participants	258	494
Number of questions, issues, and concern	113	195
Compensation and Resettlement	There were 25/113 (22.12%) opinions related to compensation and *Resettlement*, most of which are concerned about compensation price, time of *Resettlement*, procedures for house rebuilding and resettlement; and the need for people's representation on the compensation committee.	There were 39 comments (20%) relating to the Compensation, and Resettlement. Most of these comments asserted that "this revised design had been improved, the affected houses were less than in the design in 2013".They were also concerned about the compensation prices, resettlement and the procedure for rebuilding the houses; residents also proposed to have a representative of the residents among the compensation committee's members.
Adjusted boundary of the Station, Ventilation Shaft and Technical Buildings	There were 37/113 (32.74%) opinions regarding station boundaries, ventilation towers, and technical buildings. The comments focused on the request to minimize the clearance of people's houses, make use of public land to reduce damage to resident' houses and land. There are also many opinions that the distance between the stations is quite short, it may be longer distant to reduce the number of stations and reduce the land to be cleared.	There were 56 comments (28.7%) relating to the adjusted boundary of the Station, Ventilation Shaft, and Technical Buildings. Most of the comments said: "This design was better than the previous design from 2013. Residents were very pleased that authorities had done their best to reduce the land taken from them for the project locating the vent shaft and chiller in one place was highly appreciated
Comments and concerns addressed the need for Project Information and Clarification	There were 23/113 (20.35%) comments regarding the need for more detailed information on station boundaries, project implementation time, and procedures for housing reconstruction.	There were 56 comments (28.7%) relating to the necessity to have further detailed information on the station border, land to be cut and the procedure to rebuild the houses.
Suggestions and Recommendations from the participants	There are 28/113 (24.78%) of the community's opinion that are concerned about issues of construction quality, management, operation, maintenance, maintenance, and safety. Many opinions suggested that by the way of a metro project, it is necessary to study and redesign the infrastructure, connect and synchronize the drainage system, footpaths, create a green corridor, etc. to organize the construction well to less impact on the environment and people's lives.	The meeting also received 44 comments (22.6%) relating to the technical aspects, quality, safety, operation, and mainternance in construction. Many of them have proposed to redesign the infrastructure, such as the connection between the sewerage system, roads, and green spaces around the project area.

Source: (Pham Thi Anh, 2017a).

The following table will show how the community was involved and what responses they played, and what role they played in realigning the boundaries and design of the stations.

The results of these public consultations showed that residents were really concerned and interested in the project. Residents support and concur with the adjusted design which is the result of the public consultation in 2013. There is no objection to the implementation of the project. Most of them are comments that the project will have less impact on the residential houses and few families have to be resettled. In the case of resettlement, the residents are most concerned about the compensation price and the living condition after resettlement. The residents also need to have transparency and clear information about the project. Many comments contributed to technical and safety issues in the design, construction, and implementation of the project. It shows the effectiveness of public awareness and participation of the public in the project.

Effective public consultation can add substantial value to the design and implementation of the Project. The information gained through public consultation on the stakeholders' concerns and interests can help identify key problems. This can be used to evaluate the issues and concerns regarding the underground Stations, Ventilation Shaft and the location of the Technical Buildings and the Boundary. The public consultation meetings conducted were satisfactorily accomplished in providing Project updates to the public.

In encouraging and developing public participation in governing social and environmental issues, such as in infrastructure projects, resource conservation and environmental protection, public hearings are relatively 'mild' forms of participation. Public hearings do not necessarily threaten the fundamental position of the Vietnamese Government. But public hearings do indicate that enhanced opportunities are provided to citizens to play a more formative role in environmental governance.

Public participation via hearings is also widely applied in Western Countries but has a different outlook and performance in Vietnam due to its different economic, social, and political contexts. But the objectives of public hearings in both contexts are not too different: to provide the public insight into, access to, and influence on governmental policy-making processes.

Participative policy-making is still at an early stage in Vietnam, and public hearings in the environmental and social aspects are arguably the first institutionalized form of participative policy making. In principle, public hearings can be an important contribution to more public participation in policy and decision-making (L. J. Zhong & Mol, 2008). But whether these hearings in Vietnam's infrastructure projects really live after these participatory promises depends very much on how the system of public hearings is designed, implemented, and executed in practice.

PUBLIC AWARENESS AND PARTICIPATION IN CANAL ENVIREONMENTAL PROTECTION

Results from Public Awareness Program

Due to rapid population growth, inadequate and poorly functional infrastructure, low public awareness, HCMC is challenged with environmental pollution, the causes of which include direct discharge of untreated wastewater to canals, creeks, and rivers.

Several organizations, in recent years, have supported the HCMC government to construct and operate facilities relating to drainage and canal protection, flood control, and pollutants removal.

To maintain the obtained results from these projects to the environment and flooding prevention of the city, it needs to have the participation and commitment from the whole society.

The public awareness program was developed by the city government under the HCMC Steering Center of Urban Flooding Control. The program had been organized in various forms such as public meetings, brochures, leaflets, posters, propaganda gifts, face-to-face discussions, study visits, and survey questionnaires. There were 32 public awareness meetings for 14 districts with about 2,726 participants from the year 2012 to 2015.

Especially, in the public meeting, there is an open forum; the participants were given the opportunity to raise their questions, concerns, and comments either orally or in writing. The raised issues and concerns, suggestions, and recommendations were recorded during the public meetings.

In the results, the research team collected in total 596 proposes from the participants via a direct speak which were noted into the meeting minutes and written on paper. In there, 101 comments (17%) relating to the improving regulation; 125 comments (21%) relating to improving the enforcement; 191 comments (32%) are relating to technical solutions and remains 179 comments (30%) relating to Improving awareness. With a diverse social, political and educational background, the interests or concerns of every stakeholder involved in a public participation program could vary (S & A, 2005).

The detailed mentions of these comments are presented in (Pham Thi Anh, 2017b). There are many important positive outcomes from community involvement and citizen participation in these public meetings. The question for the government and policy advisors is how we can see a more systematic change in government decision-making from passively responding to public claims on an ad hoc basis to actively engaging the public and thus enhancing the legitimacy and mass support for public policies.

Results from the Public Survey

Every year, the city continuously spends tens of billions of dong (Vietnam currency) on garbage and water hyacinth removal projects along canal basins. In 2018, the city used about 6.3 billion VND to collect garbage along Nhieu Loc - Thi Nghe canal; 1.1 billion VND for Tan Hoa - Lo Gom canal and 14.4 billion VND for Tau Hu - Ben Nghe - Đoi Te canal with a volume of about 31 - 46 tons/day, peaking at about 68-85 tons/day (DONRE, 2018). As indiscriminate disposal of garbage is one of the reasons contributing to the flooding in the city, the mitigation of floods has an important behavioral aspect.

In 2018, the study team with support from the HCMC Steering Center of Urban Flooding Control had made an actual survey of the characteristics of the roads, the situation of littering in public places, the operation status of the drainage system. The survey focused on awareness and behavior of public garbage disposal of households living in flooded areas. The survey has been carried out as an open interactive exercise to create conditions conducive for people to contribute their ideas, without any restrictions. The survey included especially the issues related to management, technology, and public information campaigns to solve the city's flooding problems. The surveys involved 820 respondents and were conducted from June 23 to August 25, 2018. The detailed results of this public survey are presented in (Anh & Ngoc, 2021). The analysis had focused from the public perception to public awareness and action.

Regarding the situation in the surveyed area, 77% of the respondents in the area have often faced flooding. They said that the time of frequent flooding often occurs when it rains (99.5%) and during high tide (13.9%).

With regard to the influence of flooding on the health of the family, 50% of surveyed think that it is not affected, 28% of people think that flooding affects the health of the family but is not serious, 21% of people rate the impact as serious and only 1% of the remaining assesses the level of influence is very serious.

In the past year, 25% of families had health problems, of which 56% of people suffer from skin diseases, 29% of people with respiratory problems, 10% of people with dengue fever, 4% number of people suffering from digestive issues and 1% suffer from other diseases. According to the surveyed participants, the increasing flooding has seriously affected the lives and activities of families, especially difficulty in travelling.

Regarding domestic solid waste collection, along reconstructed roads, 48% of the surveyed choose to take the waste to the street irrespective of the garbage collection time, 50% of the surveyed store in the house until the collection time and then hand it over to the garbage collectors. Particularly along the roads that still have to, be reconstructed to solve flooding, 60% of the surveyed choose to put the waste at any time in the street, 35% of the surveyed will store in the house and hand over to the garbage collectors at the collection time for disposal and 5% of the people choose to take the garbage to the street front and put it in the grass/tree/storm sewer inlets.

Regarding questions whether people were familiar with the local cleaning up program of the city only 34% of survey respondents answered that they knew of this program. This local cleaning up means often: scraping walls cleaning neighborhoods, sorting garbage, collecting bottles, spraying flies and mosquitoes, distributing leaflets for public information, and participating in the green summer campaign. The study found that the participation of the central districts participated to the cleaning up program (42%) is higher than in the sub-urban districts (33%).

Regarding the attitude of citizens when they see people throwing garbage in a storm sewer inlet, 93% of the respondents said that they would feel uncomfortable and only 7% of the rest (mostly men) found it normal. In order to prevent littering, 57% of respondents would choose to remind these littering people about the ills of littering, 3% would choose to take photos and bring them to the notice of administration and media / or report to local security and 40% would not do anything.

When asked about the environmental impacts of waste disposal near the storm sewer inlets of respondents living along roads that still have to be reconstructed said that it would obstruct the drainage/ sewerage system, expressed that urban beauty is lost and 24% people think that it will cause environmental pollution.

For the question regarding access to sanctioning regulations for illegal garbage disposal, issued by the State, 57% of respondents said to have access to this information. 43% of respondents did not know/ could not access. Among those who have access to information 6% of respondents said the information had reached them through banners, street signs, 18% of people got access via local channels (message boards, leaflets, meetings, etc.), 36% via television, radio and 40% are self learn/see online/listen to others. 61% of the surveyed respondents in the inner districts of the city responded that they had access to sanctions regulations for illegal dumping, which was higher than respondents in other districts.

There is a difference in people's understanding of the causes of flooding, the impact of flooding and the sense of environmental protection as well as flood risk management among the residents in the area that has been flooded.

Propaganda solutions in various forms to the people are effective to raise people's awareness and understanding in environmental protection in general and drainage system in particular to contribute to flood risk management.

In addition to technological solutions, community awareness, solutions for management and sanctioning are also necessary. This is recommended to enhance quality of all the residents.

LESSON LEARNED FROM PUBLIC PARTICIPATION IN THE INFRASTRUCTURE PROJECTS

In this section, there are three important lessons that stand out from our experiences using public consultants in infrastructure projects:

1. The preparation and process activities employed for the public consultation and information were validated based on the principles and objectives of conducting the public consultation. They include:

(i) Information Dissemination - provides timely disclosure of relevant and adequate information that is understandable and readily accessible to affected people. Project information includes the technical, environmental, social, economic, and financial aspects of the project. Information regarding the benefits and negative effects of the project are made known to the public; (ii) Information solicitation - is engaging people in dialogue. Asking and listening to the local residents, and interested groups about their views and input into the project design; (iii) Incorporation of all relevant views of the affected people and stakeholders' issues and concerns in developing mitigation measures; and (iv) Coordination – A successful conduct of public consultation meeting depends on how well the individual Project team members understand their roles and responsibilities and cooperate with each other to achieve an effective and meaningful public consultation.

2. Community participation has a role to play in public projects. The information gained through public comments can help identify key issues and solutions for the authorities at the district and city levels. In the frameworks of these public awareness programs, it was determined that some of the most important outcomes from the public participation were educating the public, incorporating public values, assumptions, and preferences into the decision-making process, increasing the substantive quality of decisions fostering trust in institutions, reducing conflict and achieving cost-effectiveness as mentioned by Refs (T.C. Beierle, 1998)(Tran, 2006)(Anh et al., 2011). Without understanding and prioritizing the grievances of affected groups, the values to the entire society, the thoughts of the government, and the worries of the pressure groups, it is impossible to reach a consensus (S Thomas Ng et al., 2012). In view of this, more education for the community about the value of public participation, as well as the need to respect each other, is desirable.
3. Engineering students should recognize an initial insight into the profession of engineering, and idea of what their future career might look like. Engineering needs to be understood in the context of its role in society, and their role as an engineer has to be understood in the context of their work within a company and ultimately within society. As an engineer, they may be involved in negotiations; they may become a manager, supervising the work of a team of engineers; they may have special responsibilities to ensure that work is safe, or to ensure it is not damaging the environment.

CONCLUSION AND RECOMMENDATIONS

This paper will add to several recent studies on the roles of public participation in infrastructure projects for sustainable development.

The study focus on the role of public participation in these projects. It was determined that the most important outcomes from public participation were educating the public, incorporating public values, assumptions, and preferences into the decision-making process, increasing the substantive quality of decisions fostering trust in institutions, reducing conflict, and achieving cost-effectiveness. Without understanding and prioritizing the grievances of affected groups, the values to the entire society, the worries of the pressure society and environment, it is impossible to reach a consensus. In view of this, more education for the engineer and community about the value of public participation, as well as the need to respect each other, is desirable.

The consulting engineers responsible for these public infrastructure projects are accustomed to keeping the community informed about a project, seeking their input through public forums, and answering their questions, but using the citizens as a part of the team to collect the needed data is rare. The literature search results suggest that this is similar among engineering researchers and consulting engineers.

As Sherry Arnstein described a "ladder of citizen participation" (Arnstein, 1969), that showed participation ranging from information provision through consultation, placation, partnership, delegated power to citizen control; each with different implications for citizen participation and democracy.

ACKNOWLEDGMENT

This research was financially supported by several organizations due to the implementation of these projects, they include the steering center of the urban flood control program of Ho Chi Minh City, MVA Vietnam, MAUR. The skilled human resource was supported by the People Committee from Districts, Ho Chi Minh City University of Transport and Van Lang University. We would like to thank these organizations and people for their support.

REFERENCES

Anh. (2017a). How public Participation Contributed to the Infrastructure Projects in Vietnam: The case of MRT 2 project in Ho Chi Minh City. *Imperial Journal of Interdisciplinary Research, 3*(4).

Anh & Ngoc. (2019). Public Participatory Role in Urban Flood Risk Management of Ho Chi Minh City - Vietnam : From Awareness to Action. *South Asian Journal of Social Studies and Economics, 4*(4), 1–10. doi:10.9734/sajsse/2019/v4i430133

Anh, P. T. (2017b). Public Awareness and participation in Canal Environmental Protection: Case Studies in Ho Chi Minh City, Vietnam. *Journal of Shipping and Ocean Engineering*, *7*(3), 121–126. doi:10.17265/2159-5879/2017.03.005

Anh, P. T., Bush, S. R., Mol, A. P. J., & Kroeze, C. (2011). The multi-level environmental governance of Vietnamese aquaculture: Global certification, national standards, local cooperatives. *Journal of Environmental Policy and Planning*, *13*(4), 373–397. doi:10.1080/1523908X.2011.633701

Anh, P. T., & Ngoc, N. T. B. (2021). Promoting Public Participation in Environmental Protection and Flood Risk Management in Ho Chi Minh City - Vietnam. In *Insights into Economics and Management* (Vol. 11). Issue SE-Chapters. doi:10.9734/bpi/ieam/v11/2508E

Arnstein, S. (1969). A Ladder of citizen participation. *Journal of the American Institute of Planners*, *35*(4), 216–224. doi:10.1080/01944366908977225

Beierle, T. C. (1998). *Public participation in Environmental DEcisions: An Evaluation Framework Using Social Goals*. Resources for the Future.

Caleda & Anh. (2014). *Report on public consultation for Ho Chi Minh city urban mass rapid transit line 2*. Academic Press.

de Krom, M. P. M. M., Oosterveer, P., & Mol, A. P. J. (2013). Public Interests and Values in Multi-Level Food Risk Governance: European Responses to Avian Influenza. *Journal of Environmental Policy and Planning*, *15*(2), 161–177. doi:10.1080/1523908X.2012.746867

DONRE. (2018). *Annual report on solid waste management program in Ho Chi Minh City*. Author.

Hendriksen, A., Tukahirwa, J., Oosterveer, P. J. M., & Mol, A. P. J. (2012). Participatory Decision Making for Sanitation Improvements in Unplanned Urban Settlements in East Africa. *Journal of Environment & Development*, *21*(1), 98–119. doi:10.1177/1070496511426778

Jonsson, A., Andersson, L., Alkan-Olsson, J., & Arnheimer, B. (2007). How participatory can participatory planning be? Degrees of influence of stakeholder and expert perspectives in six dimensions of participatory modeling. *Water Science and Technology*, *56*(1), 207–214. doi:10.2166/wst.2007.453 PMID:17711017

Lawlor, R. (2016). Engineering in society - Beyond the technical...what every engineering student should know. Royal Academy of Engineering.

Li, T. H. Y. Y., Thomas Ng, S., & Skitmore, M. (2012). Public participation in infrastructure and construction projects in China : From an EIA-based to a whole-cycle process. *Habitat International*, *36*(1), 47–56. doi:10.1016/j.habitatint.2011.05.006

Li, W., Liu, J., & Li, D. (2012). Getting their voices heard : Three cases of public participation in environmental protection in China. *Journal of Environmental Management*, *98*, 65–72. doi:10.1016/j.jenvman.2011.12.019 PMID:22245866

Losaria & Anh. (2015). *Reports on public consultation of Ho Chi Minh City urban mass rapid transit line 2*. Academic Press.

MVA Systra Group. (2013). *Sustainable Urban Transport for Ho Chi Minh City MRT line 2 Project Integrated MRT2 Station Designs- Feasibility Study Report*. Academic Press.

S, O., & A, L. (2005). Evaluation of stakeholder influence in the implementation of construction projects. *International Journal of Project Management, 23*(4), 321–328.

Thomas Ng, S., Li, T. H., & Wong, J. M. W. (2012). Rethinking public participation in infrastructure projects. *Proceedings of the Institution of Civil Engineers: Municipal Engineer,* 101 - 113.

Tran, K. C. (2006). Public Perception of Development Issues: Public Awareness can Contribute to Sustainable Development of a Small Island. *Ocean and Coastal Management*, *49*(3), 67–83. doi:10.1016/j.ocecoaman.2006.02.005

Van Buuren, J., & Hendriksen, A. (2010). A learning and decision methodology for drainage and sanitation improvement in developing cities. Springer.

Von Borgstede, C., Andersson, M., & Johnsson, F. (2013). Public attitudes to climate change and carbon mitigation— Implications for energy-associated behaviours. *Energy Policy*, *57*, 182–193. doi:10.1016/j.enpol.2013.01.051

Zhong, L., Mol, A. P. J., & Fu, T. (2008). Public-private partnerships in China's urban water sector. *Environmental Management*, *41*(6), 863–877. doi:10.100700267-008-9070-1 PMID:18256780

Zhong, L. J., & Mol, A. P. J. (2008). Participatory environmental governance in China: Public hearings on urban water tariff setting. *Journal of Environmental Management*, *88*(4), 899–913. doi:10.1016/j.jenvman.2007.04.018 PMID:17573184

Chapter 8
Rohingya Refugee Crisis in Cox's Bazar:
A Great Humanitarian Challenge

Mohammad Shoeb
University of Dhaka, Bangladesh

Mithila Farzana Mumu
University of Dhaka, Bangladesh

ABSTRACT

About one million Rohingya Muslim minority people fled from their home on 25 August 2017 due to the recent violence in the Rakhine State of Myanmar and entered Bangladesh, a neighboring country. Many of them were the vulnerable people with 58% refugees under 18 children, 67% female, and 9% infants below one year. This was a great humanitarian challenge for the Government of Bangladesh and the local community. The mass influx of refugees caused a major humanitarian emergency, and the majority were reliant on humanitarian assistance including shelter, food, healthcare, clean water, sanitation, education, nutrition, safety, and security. They were given accommodation, food, emergency aids, etc. upon arrival, and finally, 890,276 Rohingya refugees settled in 34 refugee camps in Cox's Bazar and Teknaf area. Many houses, schools, shops, hospitals, and other service centers were built day by day in the camp area. Children are getting vaccines for different diseases and are going to learning centers for education. Still, there are several challenges for better management.

INTRODUCTION

Bangladesh, situated between 20'34 North to 26'38 North latitude and from 88'01 East to 92'41 East longitude, is a densely populated country with a population of 170 million in 147,570 square kilometers. It is mainly an agricultural country. Although the economy primarily depends on agriculture, small industries, for example garments, shrimp industry, leather are developing and contributing to the Gross Domestic Product (GDP) of the country. Bangladesh is fighting with natural disasters such as cyclones,

DOI: 10.4018/978-1-7998-9190-1.ch008

tornados, floods, monsoons and heavy rain, and is also vulnerable to climate change. The cultivable land is decreasing day by day due to these natural disasters and urbanization. In order to meet the excess requirements of food supply for a huge population, a large amount of pesticides and fertilizers are being used in crop cultivation. Because of the climatic condition of Bangladesh, there is no other alternative than the use of pesticides. Rice, pulses, lentils, grains and vegetables grow naturally and meet the supply of local needs. Fish and chicken are the main protein supply of the country. But a large number of chemicals including antibiotics are also used in fish culture, chicken and beef farming. Pesticides and fertilizers are usually discharged in the water body from the agricultural fields. Discharge of industrial effluents, tannery waste, fuels and surfactants in the river, lakes and canals also contaminate water bodies. Heavy metals, toxic chemicals, and their ions (phosphate, nitrate) reduce the water quality and, hence freshwater fishes and aquatic organisms are in danger.

Principal Administrative units in Bangladesh are known as divisions and there are eight divisions namely Barishal, Chattogram, Dhaka, Khulna, Rajshahi, Rangpur, Mymensingh and Sylhet and 64 districts (known as Zilas) are under these divisions. Cox's Bazar is one of the largest districts belonging to Chittagong and is situated between 21.43° North latitude and 92.00° East longitude with an area of about 2,492 square kilometers (Figure 1). There are eight Upazila in Cox's Bazar district and their names are Chakaria, Cox's Bazar Sadar, Kutubdia, Maheshkhali, Pekua, Ramu, Teknaf and Ukhiya (Banglapedia, 2021). Cox's Bazar is famous for its natural beauty with the World's largest sea beach, Himchari Water Fall, Inani Sea Beach etc. and attracted tourists from home and abroad. However, the recent Rohingya Refugee influx has posed a threat to the biodiversity of the Cox's Bazar district as it hosts the world's largest number of refugees in a crowded area.

ROHINGYA INFLUX

Rohingya people are an ethnic Muslim minority group who used to live in the Rakhine state formerly known as Arakan. The word "Rohingya" is derived from the *Rohang* which is one of the old names of the Rakhine state. The Rakhine State is a state in Myanmar with an area of 36,762 square kilometres (Rakhine State, 2021). Bangladesh is a neighboring country of the state along the Naf river which is around 64 km long. About 135 ethnic groups were officially recognized by the Myanmar Government except Rohingya Muslims who have been treated stateless and stripped of their citizenship in 1982 (Monwaza, 2020; Parmar et al., 2019). This led to failing their right to education and other basic rights which are usually a citizen can possess (Nare, 2021). Historically, the Rohingya Muslim minority have been deprived of basic rights for many years and fled to many neighboring countries for shelter and better lives.

The migration of the Rohingya ethnic minority to the neighboring countries started a long ago. Bangladesh has been giving shelter to Rohingya refugees from Rakhine State of Myanmar since 1978 by giving shelters to 250,000 Rohingya people in Cox's Bazar. Another large wave of Rohingya refugees arrived in Bangladesh in 1991–1992 when around 250,000 people crossed the border and entered Bangladesh (Mahmood et al., 2016; Parmar et al., 2019). They were recognized by the Government of Bangladesh (GoB) as refugees and about 300,000 to 500,000 Rohingya refugees were given shelter in camps in the Cox's Bazar district in Bangladesh (Khatun, 2017). The latest influx of Rohingya has restarted on 25 August 2017 due to recent violence in the Rakhine State of Myanmar. It was reported that around 379,000 people have taken shelter in Bangladesh within one week since the attacks (Babu, 2020). The Inter-Sector Coordination Group (ISCG) reported after one month on 27 September 2017 that

Figure 1. Cox's Bazar district

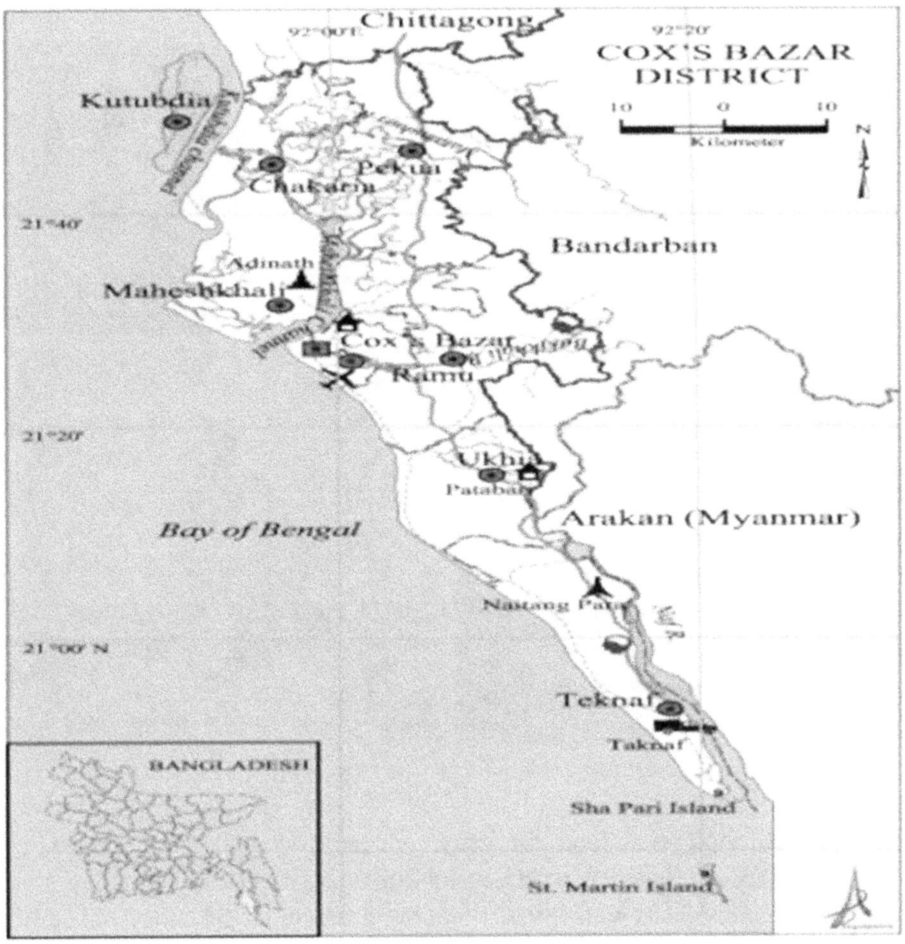

an estimated 501,800 people entered Bangladesh since the attacks (Bremner, 2020; UNICEF, 2017a). Of them, 67% are female, 58% people are under 18 children, 23% are children under five and 28% are children between 6-18 years, while 9% are infants below one year, and all are valuable people. The tortured and distressed people took different strategies to cross 271 km long Bangladesh and Myanmar border. Many of them walked 50 to 60 kilometers for up to six days, used boat or ferry to cross the river or sea and were looking for food, water, shelter and protection. However, the journey was not smooth as they had to travel without food and had to walk often carrying children or dependent on shoulder for many days (Figure 2, UNICEF, 2017d).

Under the leadership of Prime Minister Sheikh Hasina, the Government of Bangladesh allowed Rohingya refugees to enter Bangladesh and took shelter in Cox's Bazar and Teknaf area, and finally facilitated to settle down in 34 refugee camps. They were given accommodation, food, emergency aids etc. upon arrival with the help of host community and international donor agencies. Although the Government of Bangladesh helped in settling down hundreds of thousands of Rohingya refugees in Cox's Bazar, Teknaf and surrounding area, the situation is vulnerable due to land degradation, reduced avail-

Figure 2. Rohingya influx

ability of food, increased food price, food insecurity, reduced feed availability, reduced availability of water etc. The mass influx of refugees caused a major humanitarian emergency.

Until today, 890,276 individuals with 189,901 families are living in 34 camps at Ukhiya and Teknaf Upazila. Of them 854,128 individuals with 183,601 families are at Ukhiya and 36,148 individuals with 6,300 families are residing at Teknaf (Figure 3, UNHCR, 2021). Among them, children, adults, older persons and persons with disabilities are 51%, 45%, 4% and 1% accordingly (UNHCR, 2021). In general, 52% female and 48% are male out of 890,276 Rohingya refugees and the percentage within age range are <1, 1-4, 5-11, 12-17, 18-59 and 60+ are 1%, 6%, 11%, 7%, 25%, 2% and 1%, 6%, 12%, 7%, 20%, 2% for female and male, respectively (Figure 4, UNHCR, 2021).

Impact of Influx

The Rohingya influx has much impact on the environment, health, education, food security, WASH (Water, Sanitation and Hygiene), safety and security among refugee and host community. Government of Bangladesh had provided 2,000 acres of land upon their arrival. Around 7000 tons of firewood were collected by the refugees from the nearby forests which was highlighted in different newspapers and also discussed in several TV channels (Figure 5). About 60 bamboos per family were used for building their houses, about 1500 hectre of forest was destroyed while 14000 hectre forest were under threat. 6000 acre land were used for refugee camps which was extended later to 10000 acre. More than 1000 species of trees and wildlife are under threat and cultivable lands are also destroyed for this settlement

Figure 3. Families and individuals in camps

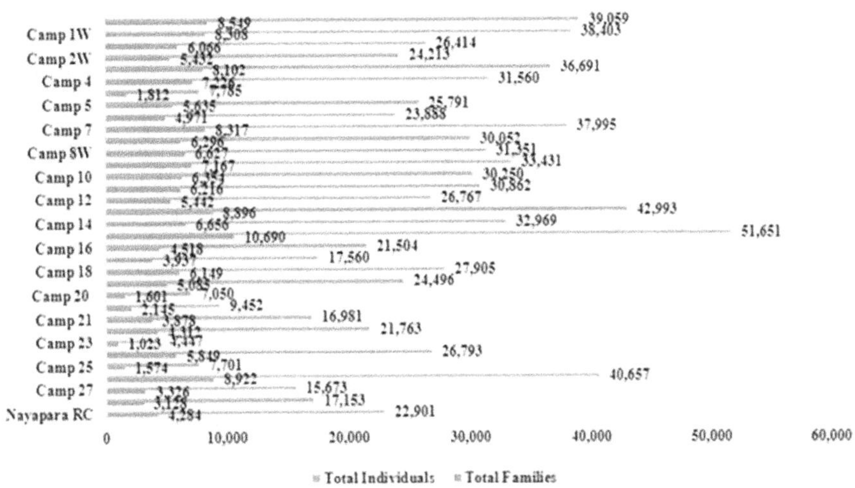

process (Figure 1) (Abrar, 2001; Baby, 2018). The Ministry of Environment and Forests, and United Nations Development Programme (UNDP) had jointly warned that the influx will affect natural forests, protected areas, vegetation, wildlife, marine and freshwater ecosystems (UNDP, 2017).

Shoeb reported the investigation of physic-chemical parameters in thirty water and thirty soil samples collected from Rohingya refugee camps. The range of pH and total suspended solids in water collected, were in the range of 6.34-7.54 and 34-255 mg/L, respectively. The range of As, Cr, Hg and Pb were 0.50-

Figure 4. Demographic profile of Rohingya refugee

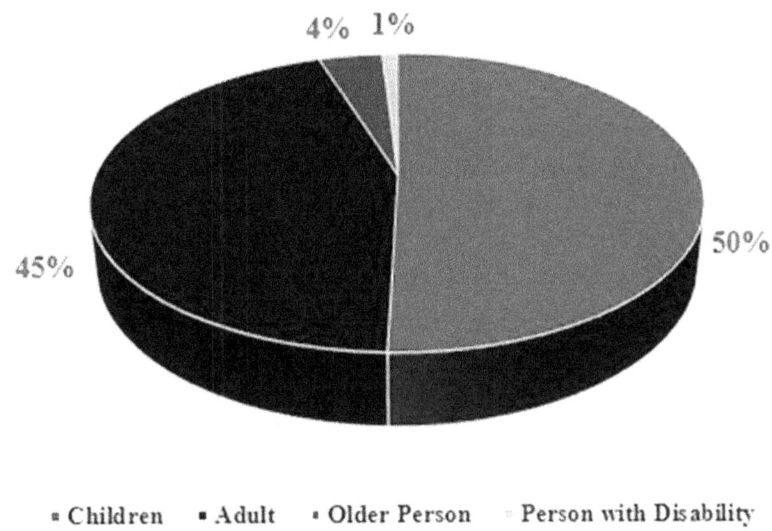

Figure 5. Uses of firewood and bamboo

2.62, 6.10-33.9, 0.12-0.80, and 6.98-22.3 mg/kg in soil samples, and below than 0.005, 0.005, 0.001, and 0.01 mg/L in water samples, respectively. These results indicated that water and soils samples were less contaminated from toxic metals (Shoeb, 2020).

HUMANITARIAN RESPONSE

Refugee Relief and Repatriation Commissioner (RRRC) is monitoring everything. The United Nations High Commissioner for Refugees (UNHCR), United Nations Development Programme (UNDP) Bangladesh, Action against Hunger (ACF), World Health Organization (WHO), United Nations Children's Emergency Fund (UNICEF), Food and Agriculture Organization of the United Nations (FAO), World Food Programme (WFP) are working with the government and other Nongovernmental organization (NGOs) for humanitarian response, and ISCG was established for monitoring and communication of all activities for humanitarian response to Rohingya refugee crisis. The major sectors including health, WASH, nutrition, food security, education, shelter, child protection, site management etc. were identified to share responsibilities among humanitarian partners (ISCG, 2019a). Bangladesh is a densely populated country with many obstacles and limitations. In this respect, Rohingya issue has added a new feather of problems in the existing situation. As the refugees continue to enter Bangladesh for shelter, Bangladesh is facing numerous challenges while providing basic human needs to those oppressed refugees. The local authority and humanitarian partners had faced several challenges for the management of health, nutrition, education, disaster, WASH facility and safety and security of the Rohingya Refugees in camps.

MANAGEMENT OF HEALTH

WHO reported on 01 November 2017 that the cumulative number of Rohingya refugees with 607,000 new arrivals and 300000 earlier migrated in Ukhiya, Teknaf, Cox's Bazar and Ramu was 907,000 (ISCG, 2017a). This was a great humanitarian challenge for the government and host community to offer health services to the large number of population where more than 50% were women and children. But Bangladesh could neither reject to give shelter and health facilities to those stateless people on the basis of humanitarian grounds. From the very beginning of the influx of Rohingya refugees into Bangladesh, public health concerned have been anxious about an innumerable health problems among the refugees. As imagined, mental condition of the refugees was deteriorating, food and water-borne diseases were spreading as a result of lacking hygiene and contagious diseases were evolving.

Contagious Diseases

Based on a report by Early Warning Alert and Response System (EWARS), the alarming health problems that are predominant among Rohingya refugees are diarrhea, acute respiratory infection and fever (EWARS, 2018). Several factors including poor living conditions due to overcrowded, malnutrition, lack of water and sanitation, poor cloth washing & shower facility and proper self-hygiene were responsible for these health risk. It was reported that acute respiratory infections represent the largest burden (32%) followed by unexplained fever (26%), acute watery diarrhea (20%) and skin diseases (9%) with 139 suspected measles and a total of 104 deaths and the majority of which (33%) are attributed to acute respiratory infections at the very beginning. The similar cases were reported with diphtheria, tetanus, acute jaundice, meningitis, malaria, Chronic Hepatitis B or C markers, measles or rubella (Bhatia et al., 2018; Chin et al., 2020). Acute respiratory infection, acute watery diarrhea, unexplained fever and skin diseases with 68769 (11.1%), 65092 (10.5%), 98523 (15.9%), 28693 (4.6%) cases, respectively and 1860 patients with suspected measles or rubella were the most frequently reported diseases/syndromes among Rohingya refugees population within 4 months of their arrival (WHO, 2021a).

Under the leadership of the Government of Bangladesh, the Ministry of Health and Family Welfare (MoHFW), office of RRRC, WHO, Civil Surgeon Office and other Health Sector partners tackled the situation by coordination and monitoring. The situation is improved now and children are getting immunization from the Expanded Programme on Immunization (EPI). Diphtheria surveillance started from the very beginning after the arrival of Rohingya Refugee. The total number of diphtheria cases reported 9296 in the 34 camps until today and it was 3016 in 2017, 5330 in 2018, 614 in 2019, 226 in 2020 and 110 in August 2021 (WHO, 2021b; WHO, 2021c). Diphtheria surveillance is still ongoing. Polio and Measles were another concerns of the congested Rohingya refugee camps. About 72,000 children were taken under an oral polio vaccination project (WHO, 2017). Rohingya women's are more vulnerable to the Rohingya refugee crisis. Early marriage is a common practice among the Rohingya population (Ainul et al., 2018). 50% of the Rohingya population that had arrived were women and it was reported that 64,000 women were pregnant when they arrived. Special care were taken for them with gynecological and obstetrical care services and there are some hospitals dedicated to look after pregnant women by midwives (Figure 6).

Figure 6. Health care in hospital in Rohingya refugee camp area

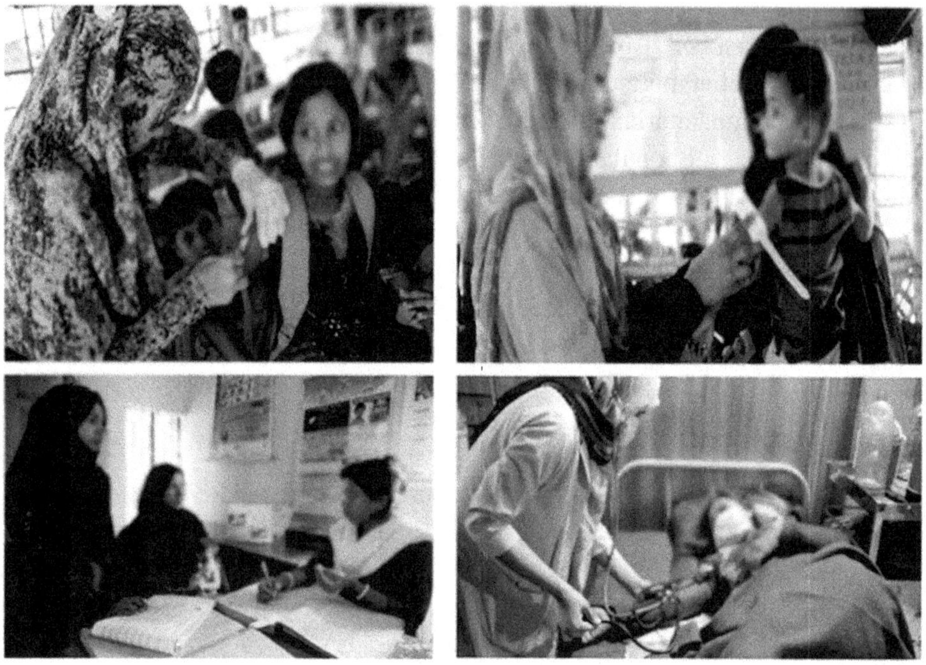

Respond to COVID-19

The MoHFW, office of RRRC, Civil Surgeon Office, WHO, other Health Sector partners and Community Health Workers (CHW) were working to tackle the COVID-19 pandemic situation. They took different programme including creating awareness about washing hands with soap and using hand sanitizer, using mask, maintaining social distance (WHO, 2021d). A total of 2677 COVID-19 cases with 49% male and 51% female were reported out of 56,109 tests conducted among Rohingya refugees in 34 camps. In the refugee camp targeting Rohingya refugees of 55 years and above, COVID-19 vaccination campaign started on August 10 and is currently ongoing (WHO, 2021b; WHO, 2021c).

MANAGEMENT OF NUTRITION

New arrivals of Rohingya refugees in Cox's Bazar lead to the health challenges including nutrition. Malnutrition is an alarming health problem for Rohingya refugee children. The refugee camps in Bangladesh, contain over 25% of Rohingya children that are undernourished as a result of lack of sufficient foods and nutrition. Infants, aged under 6 months, have malnutrition rates near 50% (Patinkin, 2018). A data of 2015–16 Myanmar Demographic Health Survey, Rakhine State represented the worst nutritional status of children in the entire Myanmar, about 38% of the children aged less than 5 years old were chronically malnourished, where 14% of the children were acutely malnourished (MoHS and ICF, 2017). After the recent massive influx in August 2017, a conducted survey at Kutupalong Refugee Camp provided evidence that about 24% of children in the camp were acutely undernourished which is above the threshold level (15%) given by WHO (Leidman et al., 2018; WHO, 2011). A total of 3,166 children among the new arrivals, aged under five, were screened and 143 were detected as severely acute malnourished (SAM). Until 08 October 2017, approximately 288,400 new arrived refugees required for nutrition assistance. Moreover, almost 0.15 million children under five and more than 50,000 pregnant and lactating women needed malnutrition prevention.

Rohingya people are largely dependent on emergency sustenance consisting of rice, lentils, oil, vegetables (WFP, 2017). House-to-house screening of children for acute malnutrition has enhanced. Recently, more latrines, water points and primary health centers along with other physical infrastructure, have been constructed.

Educational Management

Children, with a population of over 469,430 are the most vulnerable group in Rohingya refugee camps. Many of them had experienced extreme violence, lost either mother or father or both or siblings. Many of them were also suffering from malnutrition and having water borne diseases including measles, diphtheria, diarrhea etc. In this engraving situation it was a big challenge to offer them education after arrival. Many of them had lack of eagerness, physical disabilities, visual impairment, speech difficulties and involvement in families. However, under the leadership of the government of Bangladesh, UNICEF with other humanitarian partners and NGOs took the initiatives to provide education to the newly arrived children from the Myanmar after 25 August 2017, although preliminarily 71 out of 76 Learning Centers (LCs)/schools and 30 Madrasa/Moktab were used as shelters by the newly arrived refugees in Kutupalong makeshift settlement (UNICEF, 2017b). It was estimated on 14 October 2017 that a total of 453,000 children aged 4–18 years from Rohingya (including 270,000 newly arrived Rohingya children) and Bangladeshi host community urgently need access to education, and UNICEF supported 19,400 children along with 5,451 newly arrived children and they received pre-learning and non-formal basic education through 228 learning centers (Figure 7) (UNICEF, 2017c). UNICEF also trained teachers to give services to the children and helped to develop their skills. Teachers always visited camps and encouraged children to attend school regularly. In addition, teachers also counseled them about psychosocial support, food, primary health care, safety and legal support, shelter, WASH, women friendly space, child friendly space before and after school hours. There are still several challenges including inadequate learning centers, lack of necessary teaching materials, lack of qualified teachers etc.

Most of the learning centers in the camps are single-room bamboo structures that are vulnerable to loss during extreme weather conditions like monsoonal rains or cyclones. The limited space in the learning centers makes it difficult to accommodate more than 40 students. Furthermore, there is not that much space in the camps to build more learning centers. At present, there are around 8900 teachers working in the learning centers. But this number is inadequate regarding the massive need. In addition to the above inadequacy, the learning centers have lack of necessary learning materials. Even, most of the learning centers do not have sufficient WASH facilities, as a result normal learning process is hampered, especially for women. Many adolescents and young population aged between 15 and 24 do not receive any kind of primary education in these refugee camps (Figure 8). This group of population is highly vulnerable to early marriage, child labour, human trafficking, exploitation and child abuse.

Due to the outbreak of COVID-19 pandemic all educational institute were closed from 17 March 2017. Children in camps were not able to go school. Later, teachers visited camps to teach students and online classes were started. However, there were some challenges including monetary challenges, access to device, availability of learning materials, internet access, scarcity of electricity, lack of support from school, access to TV, access to radio for children in refugee camps in accessing distance learning modality (UNICEF, 2021).

Figure 7. Children in learning centers

Figure 8. Educational facility

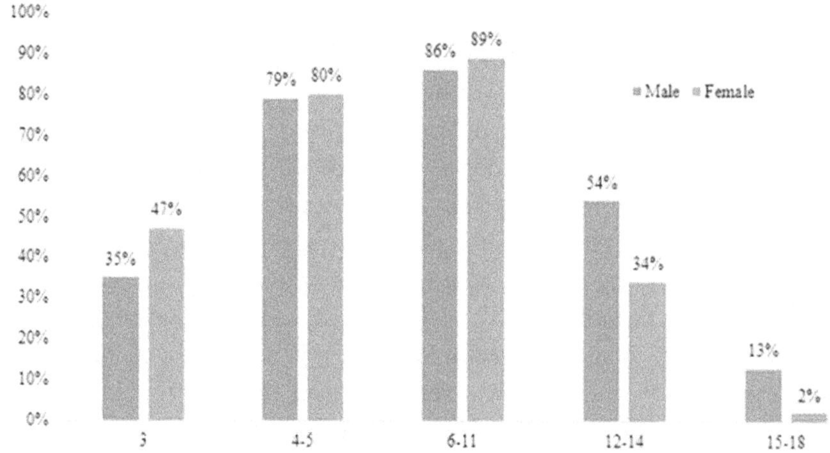

Figure 9. Disaster management

Disaster Management

Bangladesh is a land of natural beauty with six seasons, namely summer, rainy season, autumn, late autumn, winter and spring. But the country is prone to natural disaster due to its climatic condition. Bangladesh is most often cited as one of the most pervious countries to climate change for its geographic location, and flat and low-lying lands. It is in the tropical monsoon region. High temperature, heavy rainfall, sometimes excessive humidity are the climatic characteristics of Bangladesh.

From the very beginning to the massive influx in 2017, Rohingya refugees are residing in several camps in Cox's Bazar which with its long coastline on the Bay of Bengal is highly exposed to natural disasters and extreme weather including flash floods, cyclones, extreme rain, storm, landslides and extreme temperatures. A large number of exhausted and distressed Rohingya refugee took shelter in house made of wood/bamboo and turpail after 25 August 2017. However, heavy rain washed away their houses within one month of their arrival.

Monsoon season in Bangladesh, is parted by pre-monsoon hot season from March to May, the rainy summer monsoon season from June to mid-October, and a short mild autumn season from mid-October to mid-November (Ahmed, 2003a; 2003b). According to a report by ISCG, the monthly average rainfall in Cox's Bazar were recorded 802.38 mm and 915.45 mm, respectively in the months of June and July. On 25 July 2018, 463 mm of rain fell in a single day considered to be the worst rains in Cox's Bazar. Between 22 July and 10 August 2018, 979 mm of rainfall were recorded. The monsoon rain in 2018 terribly affected the Rohingya refugee camps and more than 55,000 refugees were affected by monsoon-related incidents such as landslides during mid-May to October 2018. As a result, over 6,000 were

Figure 10. WASH facilities

dislocated from the affected sides. One person died and another 55 were injured in the affected areas (ACAPS, 2018). During the monsoon in 2018, 1,577 shelters were fully ravaged across the Rohingya refugee camps. 110 learning centers were damaged by landslides and flooding damaged 70. The heavy rains from 24 to 28 July, damaged over 500 latrines, 300 of which had undergone repairing and 40 were dismantled, according to ISCG's report (ACAPS, 2018) (Figure 9).

According to a report by ISCG in response to the consequences of Monsoon in the Rohingya refugee camps between April and October 2019, about 83,517 people were affected during that time, including 17,240 people who were dislocated for a time being. 11 death were reported in the camps at that time. In total 12,800 households were damaged that led to the relocation of some of these households. About 82 shelters were completely ravaged. Moreover about 904 learning facilities were impinged. Sanitation facilities of a total of 98,750 people were affected in the camps including the total 2,199 latrines that were partly or entirely damaged.182 tube-wells of different Rohingya camps were damaged. These tube-wells required immediate repair and decontamination later (ISCG, 2019b).

Monsoon season of 2020 affected more than 118,000 Rohingya refugees that were about 41% more than in 2019 with 9 monsoon-related fatalities, according to a report by ISCG. About 8,700 refugees were primarily displaced. During the Monsoon of 2020, there were a large adverse impact of heavy rains and floods. In Ukhiya, 2,150 toilets, 111 tube-wells were impacted in several refugee camps. Damage for 24,106 households were reported. Most houses were partly and fully damaged; roofs and walls of several houses were blown away or damaged severely. From those, 121 shelters were completely unrepairable due to excessive winds and storms (ISCG, 2020).

From the afternoon of 20 May 2020 cyclone Amphan started to show its fury generating severe wind and heavy rainfall devastating several district. According to a report by UNICEF Bangladesh, on Cox's Bazar along with the Rohingya camps the impacts of the cyclone were not so significant. About 199 latrines were damaged along with 6 tube wells, causing difficulties for the over-crowded Rohingya camps.

MANAGEMENT OF WASH

Very limited and unstable sources of drinking water were available for the newly arriving Rohingya refugees in Cox's Bazar district (Figure 10, UNICEF, 2020). Due to sudden entry, most of them defecated in the open space which have a negative impact and potential threats for the environment and health. This was a major challenge for the host community as well. Mostly disabled people are susceptible to the spread of diseases. The risk of an outbreak increased with the increasing number of residing refugees in the areas where there was very limited access of pure water and wash facilities. The Department of Public Health Engineering (DPHE) managed two water treatment plants, installed 50 water tube wells, and delivered 9,000 litres safe water per day, planned for constructing latrines and distributing jerry cans for the refugees initially in Cox's Bazar with the support of UNICEF (UNICEF, 2017d).

DPHE, UNICEF partners and NGOs took initiatives to improve the WASH facilities day by day and more than 2,000 tube wells were installed in the overcrowded areas (UNICEF, 2017b). They upgraded safe water delivery to more than 60,000 liters per day, constructed 419 water points and 3741 latrines. Furthermore, DPHE has decided to improve the faecal sludge treatment facilities to provide more effective coverage. Solid waste management systems will be strengthen, by establishing the suitably-managed dumping sites, safe dumping and recycling. Partners of WASH will emphasize to meet the needs for hygiene supplies, including soaps, water containers, and other hygienic materials (ISCG, 2017b).

SAFETY AND SECURITY MANAGEMENT

Rohingya refugees have settled down completely and the safety and security look good as Bangladeshi security forces, including the army, BGB (Border Guard Bangladesh) and police are deployed in the camp locations to prevent crime and to provide safety to support the humanitarian response operation. However, there are some incidents and crimes such as robbery, intimidation, drug trafficking, exploitation and trafficking of children, abuse, violence, arbitrary killing and arrest, murder and abduction have been reported. The government is taking safety measures and has already deployed APB (Armed Police Battalion) 14 and Bangladesh Ansar to monitor such crimes.

CONCLUSION

890,276 Rohingya refugee individuals with 189,901 families are living in 34 camps at Ukhiya and Teknaf Upazila of Cox's Bazar. Among them, children, adults, older persons and persons with disabilities are 51%, 45%, 4% and 1%, respectively. They are getting all assistance and support from local people after arrival and their living standard have been improved. However, this is a great challenge for the Government and local community to provide health, nutrition, education, WASH facility and safety and security service to a huge population in a crowded area. Rohingya refugees are required to return their home in a voluntary, safe, dignified way.

ACKNOWLEDGMENT

Authors are grateful to the Ministry of Science and Technology, Government of the People's Republic of Bangladesh for financial supports from Special allocation for Science and Technology, and International Science Programme (ISP), Uppsala University, Sweden, UNICEF, UNHCR, UNDP, RRRC, WASH, Cox's Bazar for all co-operations.

REFERENCES

Abrar, C. R. (2001, January). On the Margin: Refugees, Migrants and Minorities, Refugee and Migratory Movement Research Unit. *Refugee Survey Quarterly*, *20*(1).

ACAPS. (2018). *Rohingya Influx Overview Key Changes During 2018 Monsoon Season*. Retrieved 12 August 2021 from https://reliefweb.int/report/bangladesh/rohingya-influx-overview-key-changes-during-2018-monsoon-season

Ahmed, R. (2003a). Climate of Bangladesh. In S. Islam (Ed.), *Banglapedia* (Vol. 3, pp. 45–48). Asiatic Society of Bangladesh.

Ahmed, R. (2003b). Climate of Banglaesh. In R. Ahmed (Ed.), *Abhaoa O Jalabayu Bijnan* [Meteorology and Climatology] (3rd ed.). University of Rajshahi.

Ainul, S., Ehsan, I., Haque, E.F., Amin, S., Rob, U., Melnikas, A.J., & Falcone, J. (2018). *Marriage and Sexual and Reproductive Health of Rohingya Adolescents and Youth in Bangladesh: A Qualitative Study*. Population Council. UNFPA.

Babu, K. E. K. (2020). Impacts and Challenges to Host Country Bangladesh Due to Sheltering the Rohingya Refugees. *Cogent Social Sciences*, *6*, 1770943. doi:10.1080/23311886.2020.1770943

Baby, B. (2018, January 1). Rohingya crisis: A postscript. *The Daily Star*. Retrieved 1 January 2018 from https://www.thedailystar.net/supplements/unpacking-2017/rohingya-crisis-postscript-1513045

Banglapedia. (2021, June 18). *National Encyclopedia of Bangladesh*. Retrieved 21 September 2021 from https://en.banglapedia.org/index.php?title=Bangladesh_Geography

Bhatia, A., Mahmud, A., Fuller, A., Shin, R., Rahman, A., Shatil, T., Sultana, M., Morshed, K.A.M., Leaning, J., & Balsari, S. (2018). The Rohingya in Cox's Bazar: When the Stateless Seek Refugee. *Health and Human Rights Journal, 20*(2).

Bremner, L. (2020). Sedimentary Logics and the Rohingya Refugee Camps in Bangladesh. *Political Geography*, *77*, 102109. doi:10.1016/j.polgeo.2019.102109

Chin, T., Buckee, C. O., & Mahmud, A. S. (2020). Quantifying the Success of Measles Vaccination Campaigns in the Rohingya Refugee Camps. *Epidemics, 30*, 100385. doi:10.1016/j.epidem.2020.100385 PMID:31951876

EWARS. (2018). *Epidemiological bulletin week 25*. Retrieved 21 June 2018 from http://www.searo.who.int/bangladesh/ewarsw252018.pdf

ISCG. (2017a, November 2). *ISCG Situation Update: Rohingya Refugee Crisis, Cox's Bazar*. Retrieved 21 September 2021 from https://reliefweb.int/report/bangladesh/iscg-situation-update-rohingya-refugee-crisis-cox-s-bazar-2-november-2017

ISCG. (2017b, October 15). *Bangladesh Humanitarian Situation Report No.6 (Rohingya influx)*. Retrieved 12 August 2021 from https://reliefweb.int/report/bangladesh/iscg-situation-report-cox-s-bazar-influx-15-october-2017

ISCG. (2019a). *Government of Bangladesh Humanitarian Stakeholders*. Retrieved 23 September 2021 from https://www.humanitarianresponse.info/en/operations/bangladesh/infographic/iscg-coordination-structure-2019

ISCG. (2019b). *Monsoon Response in Rohingya Refugee Camps Season Update | 21 April to 24 October 2019 Cox's Bazar, Bangladesh*. Retrieved 12 August 2021 from https://reliefweb.int/report/bangladesh/monsoon-response-rohingya-refugee-camps-season-update-21-april-24-october-2019

ISCG. (2020). *Monsoon Response in Rohingya Refugee Camps | 12 June to 15 October 2020 Cox's Bazar, Bangladesh*. Retrieved 12 August 2021 from https://reliefweb.int/report/bangladesh/monsoon-response-rohingya-refugee-camps-season-update-12-june-15-october-2020

Khatun, F. (2017). *Addressing Rohingya Crisis: Options for Bangladesh*. Center for Policy Dialogue.

Leidman, E., Humphreys, A., & Greene, C. B. (2018, April 10). Acute Malnutrition and Anemia Among Rohingya Children in Kutupalong Camp, Bangladesh. *Journal of the American Medical Association, 319*(14), 1505–1506. doi:10.1001/jama.2018.2405 PMID:29634821

Mahmood, S. S., Wroe, E., Fuller, A., & Leaning, J. (2016, December 1). The Rohingya People of Myanmar: Health, Human Rights, and Identity. *Lancet*. PMID:27916235

Ministry of Health and Sports—MoHS/Myanmar and ICF. Myanmar Demographic and Health Survey 2015–16. (2017). Myanmar: MoHS and ICF. Retrieved 20 August 2021 from https:// dhsprogram.com/publications/publication-fr324-dhs-final-reports.cfm

Monwaza, M. (2020). Politics of Nationalism of Myanmar an Analysis from the Perspective of Rohingya Crisis. *Journal of International Affairs, 22*(1 & 2), 1–24.

Nare, H. (2021). Political Restructuring, Bureaucratization and Minority Groups in Myanmar: The Case of Rohingya from 2011 to 2018. *Asian Journal of Social Science, 49*, 16–20.

Parmar, P. K., Leigh, J., Venters, H., & Nelson, T. (2019). Violence and Mortality in the Northern Rakhine State of Myanmar, 2017: Results of a Quantitative Survey of Surviving Community Leaders in Bangladesh. *The Lancet. Planetary Health, 3*(3), e144–e153. doi:10.1016/S2542-5196(19)30037-3 PMID:30904113

Patinkin, J. (2018, May 4). Rohingya Refugees Face Malnutrition. *VOA*. Retrieved 20 August 2021 from https://www.voanews.com/a/rohingya-refugees-facing-malnutrition/4377429.html

Rakhine State. (2021). Retrieved 12 July 2021 from https://en.wikipedia.org/wiki/Rakhine_State

Shoeb, M. (2020). *Project Report on Resilience in Cox's Bazar and Environmental Impact of Rohingya Influx*. Ministry of Science and Technology Government of Bangladesh.

UNDP. (2017). *Social Impact Assessment of the Rohingya Refugee Crisis into Bangladesh Key Findings and Recommendations*. Retrieved 12 July 2021 from https://www.humanitarianresponse.info/sites/www.humanitarianresponse.info/files/assessments/171207_social_impact_assessment_and_rapid_host_community_impact_assessment_summary.pdf

UNHCR. (2021, July 31). *Rohingya Refugee response/Bangladesh*. Joint Government of Bangladesh. Retrieved 23 September 2021 from https://reliefweb.int/report/bangladesh/rohingya-refugee-response-bangladesh-joint-government-bangladesh-unhcr-15

UNICEF. (2017a, September 29). *Bangladesh Humanitarian Situation Report (Rohingya influx)*. Retrieved 23 September 2021 from https://reliefweb.int/report/bangladesh/bangladesh-humanitarian-situation-report-no-4-rohingya-influx-29-september-2017

UNICEF. (2017b, September 25). *Bangladesh: Humanitarian Situation Report No. 3 (Rohingya Influx)*. Retrieved 20 August 2021 from https://reliefweb.int/report/bangladesh/bangladesh-humanitarian-situation-report-no-3-rohingya-influx-25-september-2017

UNICEF. (2017c, October 15). *Bangladesh: Humanitarian Situation Report No. 6 (Rohingya Influx), 6 – 12 October 2017*. Retrieved 20 August 2021 from https://reliefweb.int/report/bangladesh/bangladesh-humanitarian-situation-report-no-6-rohingya-influx-6-12-october-2017

UNICEF. (2017d, September 12). *Bangladesh Humanitarian Situation report (Rohingya influx)*. Retrieved 12 August 2021 from https://reliefweb.int/report/bangladesh/bangladesh-humanitarian-situation-report-no-2-rohingya-influx-12-september-2017

UNICEF. (2020, February 25). *UNICEF, Bangladesh Humanitarian Situation Report No. 53*. Retrieved 20 August 2021 from https://reliefweb.int/report/bangladesh/bangladesh-humanitarian-situation-report-no-53-1-january-31-december-2019

UNICEF. (2021, March). *Bangladesh: Assessment of the Education Sector Response to the Rohingya Crisis – Rohingya Refugee Response*. Retrieved 20 August 2021 from https://reliefweb.int/report/bangladesh/bangladesh-assessment-education-sector-response-rohingya-crisis-rohingya-refugee

WFP. (2017, December). *Refuge influx Emergency Vulnerability Assessment Cox's Bazar, Bangladesh*. Retrieved 20 August 2021 from https://www.wfp.org/news/ news-release/new-rohingya-arrivals-bangladesh-risk-poor-diets

WHO. (2011). *Physical Status: The Use and Interpretation of Anthropometry Report of a WHO Expert Committee*. Technical Report Series No. 854. Retrieved 20 August 2021 from http://www.who.int/childgrowth/publications/physical_ status/en/

WHO. (2017). *World Health Organization, Bangladesh Steps Up Vaccination for New Rohingya Arrivals as Measles Cases Rise*. Retrieved 21 September 2021 from http://www.searo.who.int/mediacentre/releases/2017/1671/en/

WHO. (2021a, July 22). *Rohingya Crisis Situation Report #14*. Retrieved 23 September 2021 from https://reliefweb.int/report/bangladesh/rohingya-crisis-who-bangladesh-situation-report-14-22-july-2021-period-covered

WHO. (2021b, August 5). *Rohingya Crisis Situation Report #15*. Retrieved 23 September 2021 from https://reliefweb.int/report/bangladesh/rohingya-crisis-who-bangladesh-situation-report-15-5-august-2021-period-covered

WHO. (2021c, August 19). *Rohingya Crisis Situation Report #16*. Retrieved 23 September 2021 from https://reliefweb.int/report/bangladesh/rohingya-influx-overview-key-changes-during-2018-monsoon-season

WHO. (2021d). *Cox's Bazar Rohinyga Camps Epidemiological Summary*. Retrieved 12 July 2021 from https://cxb-epi.netlify.app/?fbclid=IwAR3o9bRz7871J8A95a0ZgzS5rEBSNrC2FFQqAxRh3kp-aLUYmfgY9Fv9t4o

KEY TERMS AND DEFINITIONS

Action Against Hunger (ACF): A community-based organization, originated in France and aiming in ending of hunger around the world.

Demographic Survey: Surveys that actually collect information about characteristics of population about a cause and effect of a certain change.

Madrasa/Maktab: Elementary educational institution for teaching children Islamic subjects.

Malnutrition: A condition that happens as a consequence of poor nutrient containing diets.

Refugee Relief and Repatriation Commissioner (RRRC): A government agency of Bangladesh for providing relief to Rohingya refugees and actually aiming for the successful repatriation of the refugees to Myanmar.

Refugees: These are the population who have experienced war, violence, loss of loved ones and have crossed the international border to save their lives.

United Nations Development Programme (UNDP): An organization working towards the aim of eradication of poverty and achieving economic growth and human development.

United Nations High Commissioner for Refugees (UNHCR): A United Nations agency for protecting refugees and providing aid to these refugees.

World Health Organization (WHO): An agency of United Nations aiming in connecting nations and partners to expand health related response.

Chapter 9
Refugee Camps:
Reconsiderations for a New Age

Zeba Zaidi
University of Warwick, UK

Reyes Garcia
University of Warwick, UK

ABSTRACT

This chapter seeks to explore the key considerations that should be employed by camp coordinators and designers to mitigate the fundamental issues being faced by camps today and provide a safer and more durable living environment for refugees. It also aims to provide alternative camp design and material solutions and provide insight into the use of new technologies to aid these considerations. Specifically, this chapter will be exploring GIS technologies and their use in field operations. The need to rethink refugee camps is vital given the continuously increasing influx of refugees worldwide. However, within the context of the COVID-19 pandemic, this study becomes even more critical for the safety of refugees, camp workers, and host countries alike. This chapter will delve deeper into the topics of camp design and layout, camp shelter and materials, and employing technologies such as GIS in the process of re-designing camps.

INTRODUCTION

Global conflict and climate change are causing the displacement of millions of people every year. "As of 2020, there are roughly 80 million people displaced from their homes" (UNHCR, 2020). Given recent political, health and environmental trends, these figures will see an inevitable rise. Coastal populations in the Global South have started to feel the effects of climate change and have had to leave their homes as a result. Climate refugees can already be seen in Bangladesh and India (Chowdhary, 2020). These numbers are predicted to rise exponentially with increases in severe weather conditions (Kaczan & Orgill-Meyer, 2019). Similarly, rising right-wing politics and growing extremism will continue to displace millions of people, as has been seen most recently in Iran and Myanmar (NRC, 2021; Regen-

DOI: 10.4018/978-1-7998-9190-1.ch009

Figure 1. Syrian refugees pictured in a Jordanian camp in 2016 (Al Jazeera, 2016).

cia, 2021). Our current solutions for migrants and internally displaced persons (IDPs) are 'temporary' refugee camps set up in various host countries across the world that are supposed to be a safe haven for refugees until they can seek asylum elsewhere. While they attempt to help refugees and asylum seekers, today's refugee camps are the site of heinous human rights violations. With many camps unable to provide residents with basic human necessities, such as food and water, refugee camps are failing to act as a safe space for migrants. Instead, they have become an extension of the crisis that many refugees are escaping from. Furthermore, although camps are envisaged as temporary accommodation, the nature of modern-day conflicts and natural crises often means that these camps continue to grow for years, with some lasting as long as 67 years, like the Palestinian camp in Zarqa in Jordan (Chamma & Mendoza, 2016). However, even while adopting the permanence of a modern-day city, the sentiments within and around these camps remain transitional. This is reflected in the quality of the camp. A vicious cycle is formed whereby the camp lives on under the guise of a temporary settlement and fails to meet the basic needs of those it is supposed to protect.

The poor conditions of refugee camps have had a profoundly deep and negative impact on the mental and physical well-being of those that reside within them (Byler, Gelaw & Khoshnood, 2015). These conditions have been exacerbated by the current COVID-19 global pandemic (IRC, 2020). With little access to proper sanitation facilities and overcrowded facilities, refugee camps risk becoming an incubation site for COVID-19 outbreaks. The poor conditions of today's camps and their impact on the lives of refugees are perhaps best exemplified by the Moria Refugee Camp on the island of Lesvos in Greece. A camp that has been likened to hell by its residents, Moria has earned a reputation as the worst camp in the world (Grant, 2020). The poor camp conditions were confirmed after the 9 September 2020 fire in the camp. At the time of the fire, the camp was home to roughly 19,000 migrants, even though it was originally designed for a total of 3,000. Since the fire, 13,000 people have been further displaced from this campsite and are living in worse conditions. The areas designed and designated as safe spaces for vulnerable people are becoming sites that they fear and resent as much as the places they are fleeing (Kingsley, 2020).

As engineers and humanitarians, we must do better to support some of the most defenceless people in the world. While we may not be able to solve the crises that they are fleeing, we may be able to provide them with the safe space they need during this transition. A carefully considered, efficient and sustainable design can cater to many of the problems faced by refugees in refugee camps today.

This chapter proposes modular spatial and structural alternatives to current camp design standards as set in the United Nations High Commission for Refugees (UNHCR) camp planning guidelines to improve the current living conditions of refugees, specifically within the context of a pandemic and the growing need for social distancing. This chapter will use the Greek Moria camp as a case study to highlight the importance of camp design as a preventative tool against death and disease within the context of a global pandemic, and propose new standards to meet the requirements of a worsening refugee crisis. Moria was selected because it was identified as "the world's worst refugee camp" (Grant, 2020). This was corroborated during the study when a large fire destroyed the camp on 9 September 2020, leaving thousands dispossessed until today. The study of the Moria camp utilises drone and satellite imagery and ArcGIS to conduct a geographical analysis of the area surrounding the Moria camp. Thus, this chapter will also explore the efficacy of geographical information systems (GIS) and suggest ways in which they can be used for such purposes.

Case Study Context

This chapter focuses on one particular refugee camp, namely the Moria camp, on the island of Lesvos in Greece. The camp was started in 2015 with a capacity for 3,000 refugees, but has since become home to roughly 19,000 refugees, without much spatial expansion (Kingsley, 2020; Cavedon, 2019). Moria's location was chosen specifically by the Greek government as the ideal location for temporary asylum seekers who could be sent back by sea (Kingsley, 2020). However, although many European countries lent money to Greece to care for these refugees, most European leaders refused to allow the refugees to leave Greek camps for sanctuary in their countries, thus leaving an ever-increasing population in temporary accommodation with nowhere to go (Kingsley, 2020). Moria has garnered a reputation as the worst refugee camp in the world and has "been considered an emblem of Europe's hardening approach to migrants" (Kingsley, 2020). The sheer number of people, in combination with a small space, lack of access to basic necessities, and hostile authorities, has created an extremely unsafe and inhospitable environment. Migrants are "crammed into small, inhospitable spaces and forced to wait more than 12 hours in line for food that is often spoiled. There is one shower and one toilet for every 70 to 80 people" (Cavedon, 2019). The safety and mental health of residents is also in poor condition. "Numerous accusations of sexual assault and knife attacks have been made by residents, but no investigations or disciplinary actions have been pursued. Over one-quarter of the children living in the camp have contemplated suicide. Moria has become a death trap instead of a safe haven from war, starvation and abuse" (Cavedon, 2019).

A further challenge has been faced by both migrants and host governments with the rapid spread of COVID-19. With thousands packed tightly into small spaces with a lack of access to basic hygiene facilities, refugee camps, and Moria specifically, are a potential incubator for the disease (Schmitz, 2020; Schülke-Gill, 2020). Although the camp had been in lockdown since 23 March 2020 (Godin, 2020), there were over 240 confirmed cases in the camp that could have spread rapidly. However, only 1,600 tests were conducted in the period of four months since the start of the lockdown, so it is highly likely that there were many unreported cases, as the camp had a population of 19,000 at the time, meaning not even 10% of the population was tested (Bathke, 2020). Although this study had initially aimed to

Figure 2. An aerial view of the recent fire at Moria camp (Kingsley, 2020)

provide solutions to the growing number of social problems in the Moria camp due to overcrowding, during the course of this study, the camp experienced a large fire that left over 13,000 residents without shelter (Mackenzie, 2020). The fire destroyed the majority of the camp, and the resulting smoke and debris severely impacted residents in the other parts of the camp. Since then, the camp residents have been moved to various locations, such as the Kara Tepe camp, which was also shut down in early 2021. There has been no stable or safe solution provided to the residents of the Moria camp.

OVERVIEW

Camp Space

According to Byler, Gelaw, and Khoshnood, "two-thirds of all the refugee camps in the world are either at or over their intended capacities" (2015). The capacity of a camp refers to the maximum number of people who can be accommodated relative to the space available. This ratio is outlined by the UNHCR. "Most camps exceed the UNHCR guidelines on how many inhabitants they can have" (Chamma & Mendoza, 2016), but this is an inescapable result of the nature of conflicts and how they unfold. The overcrowding of camps cannot solely be blamed on the additional influx of refugees, but is "a consequence of flaws within the guidelines" (Kennedy, 2005)

This is perhaps best seen in the guidelines for the minimum area per person within a camp. The guidelines provided by the UN as they currently stand are 900,000 m^2 for 20,000 refugees. According to these guidelines, 20,000 is the limit of a camp (UNHCR, 2020). According to these figures, a camp must have a minimum of 45 m^2 of space per person. While this figure may seem high, this is not the amount of space available to each refugee. Indeed, when space for "fire breaks, non-residential buildings, and buffer zones between shelters" (Kennedy, 2005) is accounted for, the space left available to refugees is significantly less.

Figure 3. Aftermath at the campsite (Kingsley, 2020)

The lack of appropriate space available to refugees has resulted in the rise of makeshift settlements. These makeshift settlements are a cluster of improvised tents built by people who have either been rejected from the camp or taken into a camp with no space. These shelters are usually constructed of flimsy "simple materials available on site such as cardboard sheets, blankets and sleeping bags, and/or nylon and tarpaulin sheets stretched over a frame made of timber or branches collected locally" (Martin, Mica, & Katz, 2019). Figure 4 shows the use of discarded metal in an attempt to build a shelter-like structure.

A literature search found no information regarding the actual size of the camp in 2020. The UNHCR was contacted several times to get access to information regarding the camp; however, no response was received. Therefore, GIS mapping and drone imagery were used to map a rough estimation of how much space the camp covered with informal settlements. The camp currently occupies 0.2 km^2 or 200,000 m^2. Given that the estimated population of the camp in 2020 was 19,000, this would mean that there is roughly 10.5 m^2 of space for each person. This is far below the UNHCR guidelines, which state that a camp should have a minimum of 45 m^2 space per person. The UNHCR guidelines provide the basic guidelines a camp must meet to fulfil the rights of its inhabitants. The Moria camp as it stood was in clear violation of these human rights (Zander & Essife, 2019).

Camp Design

Byler, Gelaw and Khoshnood (2015) describe refugee camps as "temporary settlements built to house displaced persons from war, conflict, or persecution" (2015). Due to the urgent need for development and execution, refugee camps are built in a rush without following strict guidelines, with the sole aim of delivering immediate humanitarian relief in the form of food water and shelter (2015).

This need for immediate response to critical necessities often means that camps are not planned to factor in non-urgent, but essential, aspects of human life such as liveability, economic independence and family life. This means that camps are poorly designed, catering to only the bare essentials for survival, and take away the dignity of those that reside within them (Byler, Gelaw & Khoshnood, 2015; Chamma

Figure 4. A makeshift shelter constructed from discarded metal in Italy, 2018 (Emergency UK, 2020)

& Mendoza, 2016). This is usually known as the **emergency phase** of a camp. This phase is immediately after the crisis, when there is an immediate need for lifesaving services. After the emergency phase comes the **transitional phase**. This is where the camp starts to develop more facilities that extend beyond the urgent initial response. Finally, the camp enters the final stage, which is known as **closure**.

Figure 5. An aerial photograph taken of the camp in 2019 (Kingsley, 2020)

Figure 6. An ArcGIS estimation of the camp area in 2020

This happens towards the end of the camp's life when it is about to be decommissioned (UNHCR 2020). Unfortunately, despite many guidelines in place, due to the nature of the crisis, many camps are still built inefficiently in the emergency phase. By the time they reach the transitional phase, they have too poor an infrastructure to build upon (Chamma & Mendoza, 2016). "From the offset, camps lack basic infrastructure networks for private use such as electricity, drainage, and sewage" (2016).

Envisaged as temporary accommodation, camps lack plans for further growth. They are often unable to cater to the constant influx of refugees they receive and remain in a constant state of emergency, with some failing to provide even the bare minimum necessities to their residents (Chamma & Mendoza, 2016).

This means that camps are often so overburdened with incoming refugees that they cannot improve the camp layout. This is in part due to the nature of the modern camp design. The standard design for today's refugee camps still uses the UNHCR design, which was developed in 1979 (Byler, Gelaw & Khoshnood, 2015). This outdated design greatly hinders the camps' ability to scale up, which in turn affects the delivery of aid services, the incidence of disease, the safety and the sense of identity and dignity of refugees. Furthermore, due to a lack of emphasis on liveability and other cultural factors, camps develop into a "warehouse-type" (Jahre, Kembro, Adjahossou & Ataly, 2018), which reduces dignity for refugees and makes it more difficult for them to adapt to their new environment (Byler, Gelaw & Khoshnood, 2015; Jahre, Kembro, Adjahossou & Ataly, 2018).

Further, even within a well-planned camp, there emerge what are called 'hierarchies of space' (Kennedy, 2005). This refers to the advantages or disadvantages that may come with the given location of accommodation within a camp. For example, when accommodation is farther away from the camp centre, it may feel isolated or as though facilities are inaccessible. Conversely, those who reside closer to the camps' communal areas may feel as though they have a reduced sense of privacy and security, even though they may derive some benefit from their proximity to critical facilities. To avoid this, there should be a clear but subtle distinction between public and private spaces and "intermediary spaces between the two" (Kennedy, 2005). Further, there should be an understanding that while some spaces may be exclusively private or exclusively public, there will be spaces that are a combination of the two. Additionally, rather than one large community, there should be a division of the main camp into small communities that will

Figure 7. A U-shaped compound design used in the Nyarugusu Refugee Camp, Tanzania (Adjahossou et al., 2018)

include smaller public spaces available only to these communities. These public spaces are an example of spaces that are both private and public, as they are only open to a certain part of the camp. This type of design also reduces the hierarchy of spaces. This community-centric approach to designing refugee camps will not only allow for more efficient delivery of services, but will also provide a greater sense of privacy and security for those living within these communities. People within these communities will develop a greater sense of bonding and familiarity and thus will have an enhanced sense of community and security (Kennedy, 2005; Bulley, 2014). Adjahossou et al. (2018) presented a U-shaped design for compounds that could provide larger communal spaces for residents to interact and socialise.

Figure 8 shows the initial plan created for the Moria camp. Although the camp faced many challenges due to overcrowding, it also had many flaws in its inception, as can be seen in the image below. Some of the key issues that can be seen from this image are the following: long/inadequate distances from critical facilities such as toilets and water taps, a lack of specially designated areas for men and women to use private facilities like showers and toilets, and a disproportionate distribution of facilities.

Shelter Design

Camp planning should always start from the smallest unit and then work upwards (Kennedy, 2005). The smallest unit, in this case, would be the housing unit. In current camp planning standards, the housing unit is used as the method of expansion for the camp, such that when the camp increases in population, the camp will not increase the number of housing units but enlarge the size of the already given housing units (Byler, Gelaw, & Khoshnood, 2015). This solution has two key problems. The first is that it only caters to internal population growth (i.e., births) within the existing camp population. It fails to account for incoming refugees who may not be able to reside in the existing accommodations. The second problem with this approach is that housing units are not spaced out in a way that allows for this expansion to

Figure 8. The Moria camp's original camp layout (UNHCR, 2016)

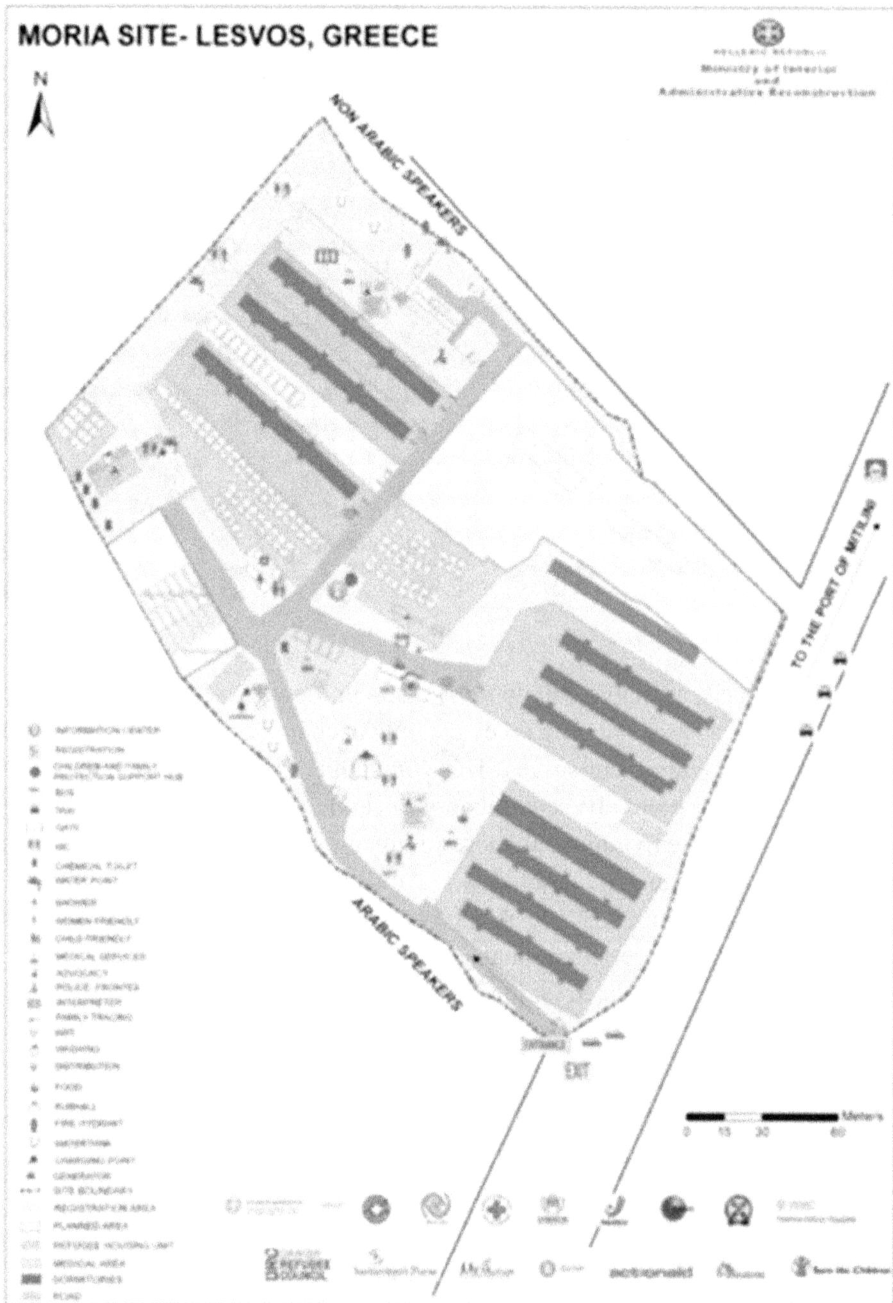

take place efficiently. The expansion of housing units is rarely planned with consideration of the proximity of surrounding buildings and access to key facilities (Byler, Gelaw & Khoshnood, 2015). Camp accommodation currently utilises tents. These tents are only available in two sizes, medium and large, and cannot be changed or expanded further if the situation changes. Furthermore, these tents lack any sort of basic security measures. They are made of a combination of fabric and plastic, and provide only basic shelter for residents. The lack of proper infrastructure has a direct effect on gender-based violence, mortality and general safety (Byler, Gelaw & Khoshnood, 2015).

Poorly designed and laid out shelter designs also inhibit the installation of necessary individual facilities such as personal toilets or showers. In the emergency phase, communal latrines and showers are put in place (UNHCR, 2020). These are replaced as the camp develops, and personal facilities are installed with housing units. If housing units do not have space to expand, refugees will need to rely on overcrowded communal latrines or open defecation. Both are highly dangerous and cause preventable deaths and diseases (Byler, Gelaw, & Khoshnood, 2015). Thus, properly designed housing units are a barrier to disease and mortality (Byler, Gelaw & Khoshnood, 2015).

To overcome the mistakes of the current shelter design, there needs to be a recalculation of the space needed per person in a camp and the space needed for one housing unit. "Once the space for the pathways and firebreaks is added in, and a non-residential block added for each eight residential blocks, the final area per person on a camp-wide basis would be 61 m^2 at the end of the ninth year. This would necessitate an initial calculation for the first year, before any internal population expansion, of 89 m^2 per person—almost double the UNHCR recommendation" (Kennedy, 2005).

Materials

The materials used within a camp are critical. Currently, camp tents utilise plastic and cloth, both of which are highly flammable materials. Tightly packed structures built with highly flammable materials around "open fires, kerosene stoves, lamps, candles, smokers' materials and electrical equipment create an environment that is at high risk for uncontained fires" (Atiyeh & Gunn, 2017). Additionally, "most camps completely lack water or firefighting capabilities" (2017); thus, it is imperative for the safety of the camp residents that materials used to build the camp are non-flammable and durable. Figure 9 depicts the highly risky scenarios that take place in refugee camps. The picture shows open fires and (what can be presumed to be) kerosene stoves close to tightly packed cloth tents.

The materials used by residents of the Moria camp since its construction were predominantly sheet metal and plastic. The extensive use of plastic and cloth to build shelters, as can be seen in Figure 10, is what allowed the fire at Moria to grow to such a large size.

Material and Shelter Solutions

Fixing the shelter design and materials used in a camp could single-handedly reduce the risk of fires and other disasters while also increasing security and liveability. This section seeks to provide alternative shelter and material solutions to those currently being used by the UNHCR.

Wall/ Roofing Building Materials

Table 2 provides a broad overview of the advantages and disadvantages of using different material solutions. However, as it is a broad generalisation, it overlooks some key factors to address within all categories of material solutions. The material solutions used in the Moria camp will be explored in greater detail.

Figure 9. Syrian refugees pictured in their accommodations in a camp in Lebanon in 2011 (Bowman, 2020)

Prefabricated Shelter Solutions—Better Shelter

The Better Shelters 'Relief Housing Unit' is an example of modular prefabricated metal shelters that are currently used by camps under the UN and were initially brought to the Moria camp (Murphy, 2015). This shelter solution has been promoted as a modular, durable shelter design that factors in the basic needs of refugees in terms of privacy, security and familiarity. The Better Shelter is a good example of how the assumptions made in the above table regarding materials and their strengths can be misleading. Although the Better Shelter is made from galvanised steel, it is neither flame retardant nor non-combustible. The Better Shelter guidelines recommend that there should be 5 m gaps between structures

Figure 10. An image of the Moria camp housing units in 2019 (Schulke-Gill, 2020)

Table 1. A list of the different types of shelters provided by the UN and where they are being used (UNHCR, 2020)

Shelter solution	Pros	Cons	Applications Examples
Family tents	Traditional relief tent; lightweight; proven design; good headroom; can be winterised; large production capacities.	Inflexible; may be unstable in high winds or heavy snow, difficult to heat. Where tents are used for long durations, provisions for repair materials should be considered.	Beddawi camp, Lebanon
Plastic sheeting	The most important component in many relief operations; UV resistant; heavy-duty; lightweight, flexible; large production capacities.	No frame is provided, migrants must collect/ create their own frames. Collecting wood for support frames or stick skeletons could considerably harm the environment if collected from surrounding forests. It is important to consider sustainable sources of framing.	Moria camp, Greece
Materials and tools for construction	Suitable local materials are best, if available, and must be suitable for variance in the seasons, culturally and socially appropriate and familiar.	Require time and training.	
Prefabricated shelter and containers	Permanent or semi- permanent structures; long-lasting.	High unit cost; long shipping time; long production time; transport challenges; transport emissions; assembly challenges; inflexibility; disregard cultural and social norms.	Azraq camp, Jordan
Rental subsidies	Greater sense of independence; greater integration in a community; an influx of income to host community.	Difficult to monitor that shelter meets standards; competitive market may result in exploitation and abuse; inflation and speculation may occur; upgrades or repairs may be needed.	Examples seen in Turkey

to avoid the spread of flames, as well as no smoking or cooking inside (Better Shelter, 2020). While it is thought to be a safe solution in terms of fire safety, the Better Shelter has failed to meet Swedish fire standards (Murphy, 2015).

Temporary Solution vs. Permanent Settlement?

Sustainable and durable infrastructure is key when designing a new camp. One of the biggest questions plaguing camp planners today is whether they should design camps to be temporary safe havens for transitory refugees, or whether they should be designed to resemble cities and permanent settlements to provide a more stable environment for long-term refugees. The answer, as it is dependent on the nature of the crisis, is often unclear. The average lifespan of a refugee camp is 17 years, while the oldest refugee camp today was established 67 years ago (Chamma & Mendoza, 2016). Designing a camp that is too temporary can result in overcrowding, lack of sufficient resources, and deterioration of camp facilities over time. More permanent housing solutions have been linked to a better quality of life for camp residents (Byler, Gelaw, & Khoshnood, 2015). So why not build all camps to be permanent? This can have negative repercussions for the area. While interviewing refugees at the Azraq and Zaatari camps in Jordan, refugees were asked what their plans and hopes were for the future. All the refugees responded by saying, "We want the conflict to end tomorrow and we want to go home". This question was followed by asking residents if they would feel differently had the camp been a five-star settlement with all the

Table 2. A table of the different materials that can be used on refugee camps, their advantages and disadvantages and where they are being employed (Slater, 2014)

Material	Pros	Cons	Applications
Mud	Available locally; acts as a conductor during the day and insulator at night; adapts to the local environment; readily available in the area.	Not possible in areas with scarce resources.	Za'atari camp, Jordan
Metal Sheets	Heavy duty; durable; water-resistant; non-flammable.	Expensive (where expensive materials are used it should be noted that migrants may sell these resources to purchase other items); inflexible; transportation costs and emissions.	Azraq camp, Jordan
Plastic Sheets	Waterproof; UV resistant; heavy-duty; lightweight, flexible; large production capacities.	Highly flammable; may not withstand in areas with high winds; may not provide adequate protection for external migrants.	Moria camp, Greece
Canvas Tarp	Waterproof; UV resistant; heavy-duty; lightweight; flexible; large production capacities.	Highly flammable; may not withstand high winds; may not provide adequate protection for external migrants.	Moria camp, Greece
Wood	Sturdy; can provide better ventilation; suitable for humid rainy environments as they keep out rain but do not trap heat (Slater, 2014).	Collecting wood for support frames or stick skeletons could considerably harm the environment[1] if collected from surrounding forests. This may also lead to problems with the host country.	Karen Refugee Camp, Thailand
Recycled Materials	Reduces waste; cheaper cost of building.	May not be readily available.	Plastic bottle houses in Sahrawi Camp, Algeria (World Habitat Org, 2018)

facilities of a normal city (Chamma & Mendoza, 2016). The response was the same: refugees wanted to return to their homes as soon as possible. This hinders camp planners from designing permanent formal settlements for refugees, as there is the question of what would happen to camp infrastructure after the closure of the camp (Chamma & Mendoza, 2016). This dilemma faced by planners often results in inadequately designed camps that are unable to cater to the needs of their residents. A potential solution

Table 3. The features and materials used in the Better Shelter (2020)

Features [18]	Materials [18]
17.5 m^2	Roof and wall: polyolefin foam panels treated with UV protection
160 kg	Frame and foundation: galvanised steel
Lockable door	Windows, ventilation frames, hatches: lightweight galvanised plastic
4 windows and 4 ventilations	Door and door frame: UV stabilised polymer plastic
Solar-powered	Smaller components: polymer plastic
36-month lifespan	Floor sheet: woven high-density polyethylene fibres
Price: $1,250 (Wainwright, 2017)	Assembly tools: steel

Figure 11. An image of the inside of a Better Shelter Unit (2020)

for this would be a set of modular units and camp designs that are sustainable and can be changed based on the needs of the people.

Politics

Although the role of politics surrounding refugee camps is too broad to discuss in detail, its relevance in the planning and management of a camp means it is too critical a factor to overlook[2]. This can be seen through the difference in camp planning and management between a camp where the host government is supportive of refugees and a camp where the host government is hostile towards refugees. Two camps that are often praised for their planning and maintenance are the Azraq camp and the Za'tari camp, both in Jordan. The migrants that these camps cater to are Syrian refugees escaping violence within their own countries. The Jordanian government is much more amenable to the causes of the Syrian refugees as they are culturally similar to the Jordanian people, and the Jordanian government has not had a long-standing history with the Syrian government (Chatham House, 2015).

Many of the refugees arriving in Greece are Syrian refugees who have been re-routed from Turkey. Turkey is one of the main transit countries for refugees arriving in Europe due to its location. The Greek

Figure 12. The BetterShelter Housing Unit from the Outside (2020)

government has a long-standing history of hostility with the Turkish government as well as ongoing border disputes (Enria & Gerwins, 2020). This has caused severe mistrust between both the Greek authorities and incoming migrants, which was exacerbated by the COVID-19 pandemic and the general fear surrounding it (Enria & Gerwins, 2020; Donadio, 2019). Cultural dissimilarities and Islamophobia also create a great deal of tension between the two countries (Osiewicz, 2017). This hostility can be seen not only through the poor construction and maintenance of the camp but also through the extremely poor treatment of migrants arriving in Greece. When governments are hostile towards migrants arriving in their countries, they are less likely to authorise permanent settlements to be created to cater to them. An understanding of political sentiment towards refugees will help camp organisers immensely when planning the materials to use and the permanence of structures.

Social Distancing and COVID-19

Aside from the already precarious situation of a refugee camp, another vulnerability faced by refugees and migrants is the risk of COVID-19. Refugees and migrants remain some of the most exposed and defenceless people to COVID-19, as they have limited access to sanitation facilities and are often packed together in large numbers in tight spaces (Subbaraman, 2020). To tackle this vulnerability effectively, it is necessary to understand that host countries may often be vulnerable themselves and unable to provide adequate measures. Approximately 80% of the world's refugee camps are located in low- to middle-income nations that have faced their own challenges in dealing with the disease (UNHCR, 2020). Although aid agencies are doing their best to provide refugee camps with adequate equipment to combat the virus, the current design of refugee camps greatly increases the spread of disease. "Even if one infected person enters a camp, it could threaten the lives of all who reside within it" (Subbaraman, 2020). Some camps have been able to handle the COVID-19 pandemic well because most camps operate in a controlled environment and undergo constant monitoring of who is entering and exiting the camp (Egeland, 2020). However, this was not the case at the Moria camp. "Men, women, and children were walking in and out of the main gate of hotspot zones without any restriction" (Tazzoli, 2020). The mobility of camp residents was not restricted until after the first COVID-19 case had been confirmed within the camp (Tazzoli, 2020). This facilitates new sources of infection and highlights the need for adequate social distancing measures from the offset.

GIS and Drone Technology

With the changing nature of conflict and the increasing need for rapid response for millions of people at a time, the ability for old methods to efficiently cater to the needs of vulnerable people weakens. Refugee camp planning must start utilising all the tools available to it to most effectively help those in need. Two tools that would greatly benefit camp planners are drone technology and the GIS. The use of drone technologies is very useful in relief work, as it provides an easy and efficient tool for mapping that helps aid workers contextualise and analyse affected areas. Today, drone technology is being used in a wide variety of situations. Using images procured from drones, UN refugee agencies have been able to visualise the locations of camps more accurately. This technology also allows agencies to conduct easier needs assessments and identify possible solutions, for example, the closest water bodies for water, sanitation and hygiene (WASH) facilities. This technology can go beyond the needs of a camp's emergency phase. In 2013, small drones were used to study the environmental damage caused by refugees creating informal settlements near camps in Niger (UNHCR, 2020).

GIS is a "software that allows you to conceptualise and analyse spatial and geographic data" (GIS, 2020). It can be used in combination with drone imagery to strengthen the spatial understanding of an area. The physical space available to a camp is sometimes limited due to its spatial surroundings; therefore, these technologies should be a vital part of camp planning and area assessment. GIS is more limited in its ability to provide accurate data as it is on a global database that is monitored; thus, important information, especially concerning refugee camps, can be easily hidden. In this study, GIS will be used to understand spatial restrictions for future camp site designs. Some of the ways that GIS can be used are as follows (Ortiz, 2020):

- Cartography and humanitarian intelligence: rapid assessment, presentation of information, visualisation, land use, infrastructure and facility mapping, remote sensing for monitoring and evaluation.
- Crisis simulation and impact models: analysing 'what-if' scenarios and consequences of a disaster, incident modelling, forecasting.
- Risk assessment: identification of 'hotspots', a combination of vulnerability and hazard assessment to evaluate at-risk populations.
- Vulnerability assessment: the integration of socioeconomic and environmental data to serve as an early warning alert.
- Decision-making support: network and location analysis for resource allocation and optimisation.
- Surveying and communications: citizen participation and crowd-sourcing, status reporting, programme assessment.

Table 4. The original UNHCR guidelines regarding camp structure (UNHCR, 2020)

Module	Structure	Approximate Number
Family	1x family	4-6 persons
Community	16 x families	80 persons
Block	16 x communities	1,250 persons
Sector	4 x blocks	5,000 persons
Settlement	4 x sectors	20,000 persons

RETHINKING CURRENT STANDARDS

The UN currently suggests that the area per person for a family shelter should be 4.5 m^2. However, this leaves very little space for growth. If we assume even a minimum of 4% growth per year, the area per person is reduced to 3.2 m^2. Allowing just one square metre of this space for storage, materials, storage, and home enterprise, the size of the shelter is reduced almost to the point where the refugee lacks sufficient space to lie down and sleep. Setting aside these reservations regarding the adequacy of the UNHCR guidelines and assuming that 4.5 m^2 per person is sufficient, then the necessary shelter space for a family of five would be 22.5 m^2. In reality, this figure should be 31.5 m^2, allowing the family to expand to seven members over time. Accordingly, only 11 families should occupy a community block rather than the UNHCR-mandated 16 (Kennedy, 2005).

While this highlights the general problem with the current UNHCR guidelines, these calculations do not take into account the extra space needed for precautions to be taken against pandemics such as COVID-19. Such small spaces with a large density of people tend to be a breeding ground for disease. Not providing adequate space within refugee camps for social distancing and safe health practices could prove catastrophic. For this reason, only six families should occupy each community block. Further, there should be an increase in the number of communal latrines, water taps, showers, waste bins, etc., as these communal spaces are at high risk for spreading diseases. The number of wash facilities should be increased by 50%. Furthermore, the total area per person should be doubled to allow for social distancing to take place easily. This means the minimum area per person should be 90 m^2.

Table 5 shows the original UNHCR camp structure. Table 6 shows the proposed revised standards factoring in a general need for size increase and COVID-19 precautions. The biggest difference between the two standards is at the community and block levels. Where the UNHCR originally suggested 16 families in one community, the new proposed design only allows for six families in a community. Similarly, while the original standards allowed for 16 communities in one block, the new proposed standards only allow for six communities per block. This will be further explained in the 'New community layout design' section.

Table 6 presents a side-by-side comparison of the original UNHCR standards for facilities along with the new proposal for facilities factoring in COVID-19 precautions. The differences to note are mainly in highly contagious zones such as health centres, which have been reduced from 1 per 20,000 to one per 2,000, and feeding centres, which have been reduced from 1 per 20,000 people to one per 1,000 people.

Table 5. Proposed revised standards for camp structure

Module	Structure	Approximate number
Family	1x family	3-7
Community	6 x families	42
Block	10x communities	420
Sector	4 x blocks	1,680
Settlement	8 Sectors	13,440
	10 Sectors	16,800

Table 6. A comparison of the original UNHCR camp standards for facilities with the newly proposed standards (UNHCR, 2020)

Description	Original Standards	Proposed Revised Standard	Further Considerations
Communal Latrine	1 per 20 persons	1 per 10 persons	Separate latrine areas for men and women.
Latrine Distance	Not more than 50 m from a shelter and not closer than 6 m	Not more than 25 m from a shelter and not closer than 6 m	
Shower	1 per 50 persons	1 per 20 persons	Well-drained shower areas, separate for men and women.
Water Supply	20 litres per person per day	20 litres per person per day	
Water Tap Stand	1 per 80 persons	1 per 20 persons	
Water Distance	Max.200m from household	Max.50 m from household	
Refuse Pit	1 per 500 persons	1 per 50 persons	
Health Centre	1 per 20,000	1 per 2,000 people	
Referral Hospital	1 per 200,000 persons	1 per 10,000 people	
Distribution Centre	1 per 5,000 people	1 per 1,000 people	
Feeding Centre	1 per 20,000 people	1 per 1,000 people	
School	1 per 5000 people	1 per 2,000 people	
Market Place	1 per 20,000people	1 per 10,000 people	

A New Community Layout Design

Figure 13 depicts the proposed layout design for one community, which will host six houses rather than the recommended 16. Each community will be roughly 54 metres wide and 48 metres long. This will give it an area of 2,592 m^2. At maximum capacity, each community will have 61.7 m^2 of space per resident. The communities will not be closed off by hard borders; instead, they will be divided by soft dirt roads and defined by green areas and vegetation planted along the borders by the residents. This way, the residents can choose where the entrance to their community can be, depending on how much privacy they want. The vegetation and soft borders will also improve liveability and not make the divi-

Figure 13. Proposed layout for one community

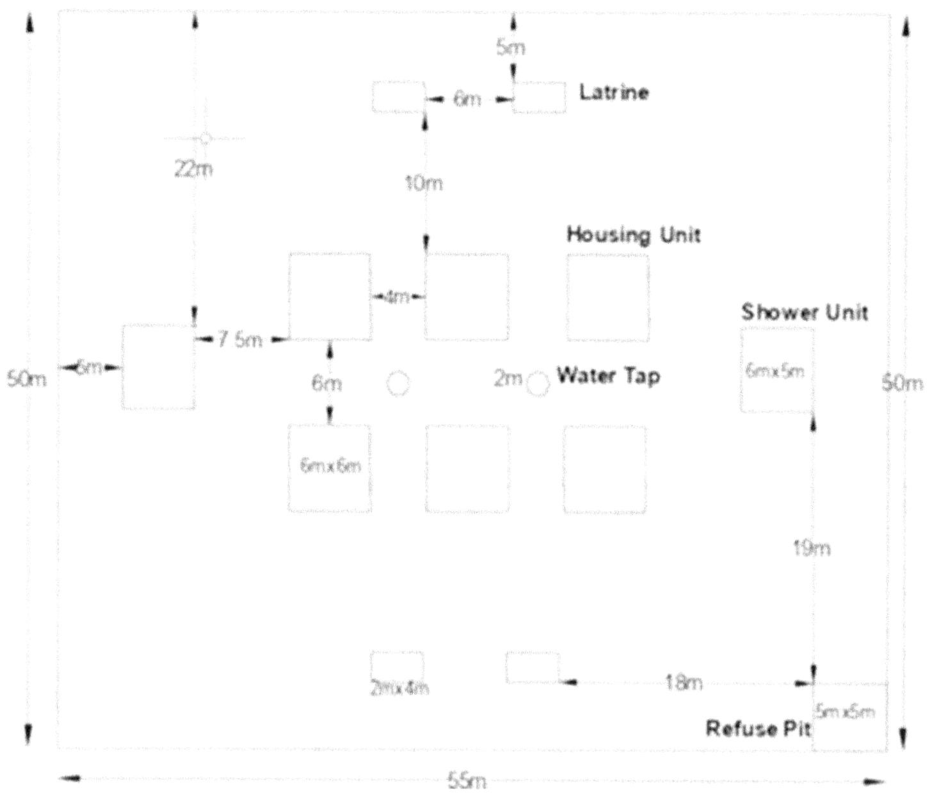

sions between communities feel rigid and camp-like. This is to cater for the space that will be required in the coming years, as well as social distancing practices.

Housing Units

Each house will have a capacity of holding up to seven people; thus, one community can host a maximum of 42 people. Each housing unit will be 6 m by 6 m. This will provide ample space for growth over the years. The ratio of covered living space to open living space per each housing unit will depend on the refugees, how big their family is, how much space they need, etc.

Outdoor Space

A large amount of outdoor space per housing unit will be allocated to provide refugees with the ability to grow some food within their home gardens. This has been done to increase independence amongst refugees as well as limit the need for refugees to visit food distribution centres and feeding centres (public spaces) and facilitate social distancing.

Latrines

Each community will have four communal latrines. This means that at maximum capacity, each community will have one latrine for every 10.5 people. This is in accordance with the UNHCR guidelines. The latrines will be a maximum of 10 m away from the start of the housing block that they are in line with. This is so that they are far enough that no smell is carried back to the housing facilities, but close enough to all housing units in the community for users to not feel unsafe. Two out of four of the communal latrines will be solely for women and two will be solely for men. Each side of the community will have one latrine for men and one for women. This is to ensure that all houses have the same access to toilets for both genders and that any latrine is no more than 22 m from any housing unit.

Shower Units

There will be two communal showers, one on each side of the community. One will be solely for women and one will be for men. At maximum capacity, there will be one shower for every 21 people. The showers will be located 7.5 m from the first housing unit on their respective side. This distance was chosen to provide enough of a sense of privacy when taking a shower, but not isolate the showers so much that they feel unsafe. A shower unit will be no greater than a 22 m distance from any house.

Water Taps

Each community will also have two water taps. This means that at maximum capacity, each community will have one water tap per 20 people. The water taps will be located centrally between the houses to provide residents with easy access.

REDESIGN OF REFUGEE CAMP AND DISCUSSION OF IMPLICATIONS

It is proposed that ten communities together will form one **modular block**. One block will have a maximum capacity of 420. According to these numbers, each block should have 42 toilets, 21 showers, and 21 water taps. However, the facilities described above only total 40 toilets, 20 showers and 20 water taps within each community. This means there should be two more toilets, one more shower and one more water tap to fit the UNHCR standard of facilities per person. To accomplish this, the extra toilets, water tap, and shower can be placed in the communities that are farthest from the block centre, to diminish the hierarchy of spaces and provide benefits for those who are farther away from common spaces.

The **dirt roads** that separate the communities will be 4 m wide to create enough distinction between each community. This distinction can deliver a greater sense of security and privacy amongst the residents. These small communities will help build a sense of belonging and community at a time when social distancing may be in place. The sense of community and belonging will help create a sense of security, as residents will know more easily who their neighbours are and who belongs to their areas. Furthermore, this sense of community will make it possible for camp administrators to put in place policies such as social distancing, as people will have more of a sense of responsibility towards their community as well as a greater sense of accountability. The 4 m roads will also allow for large transport vehicles to go to each community individually to deliver goods or services such as emergency ambulances or water sanitation trucks. **Streetlights** should be placed at every intersection of the block and at the end of each road. This will increase both camp security and liveability.

An **L-shape** design is proposed to fit ten communities into one block and maximise space. Four blocks will be put together to form one sector with a maximum capacity of 1,680 people. The four blocks will be laid out in such a way that there is a large space in the middle for communal facilities. These facilities will be a health centre, a school, two food distribution centres, two feeding centres and one large open space for leisure activities, as shown in Figure 10. According to the UNHCR guidelines, there should be one health centre for every 20,000 people. However, due to social distancing needs, our revised standards provide one health centre for every 1,680 persons. This is to limit the number of people who come in contact with communal spaces like health centres that would also be hotspots for disease. These considerations have also been applied to the feeding centres, schools, and distribution centres. The original design of the camp included one school and distribution centre for every 5,000 persons and one feeding centre for every 20,000 persons. This design has two distribution and feeding centres for 1,680 people; that is, one feeding and distribution centre for every 840 people and one school for every 1,680 persons. These new standards ensure limited contact between residents. Furthermore, the layout of the four blocks to create an internal space is intended to ensure the **privacy and security** of that space and also try to limit it to those who reside within this sector. This, again, is a precaution against the spread of disease.

A common problem faced by camp residents is the **hierarchy of spaces**. While proximity to communal spaces may be convenient, it can also hinder privacy and security. To curtail this with communities close to communal spaces in the camp, a road with a 2 m width will go around the exterior of the communal space, creating a barrier between the community and the communal space. A further tall hedge will be placed on the exterior of the community and a small hedge on the interior of the communal space to create a pleasant and soft division. This is intended to create a greater sense of privacy for the residents.

Proposed Settlement Layout

The proposed camp structure is modular and can be adjusted to different camps' needs. After the sector-level design is completed, the layout of a settlement is to be designed. These layouts would keep in mind the size of the population and the space available. For the sake of brevity, as well as a clear understanding of the processes undertaken during the study, only one camp layout design was explored above. This layout option was chosen because it met the minimum criteria for space and had an even number of sectors that provided a more uniform camp layout and helped avoid the hierarchy of spaces.

The option to be discussed further in this chapter will have eight sectors in one settlement. Each settlement will have a maximum capacity of 13,440 people. The sectors will be divided by roads that will be roughly 5 m in width. The roads are designed to be wide, as they will be used by large delivery vehicles to deliver materials throughout the camp.

The space in the middle of these sectors will cater to the larger facilities that are needed. It will include a referral hospital, a marketplace, administrative offices and cultural spaces. Given that each sector has its own area for pertinent facilities, the hierarchy of spaces is reduced as those towards the camp edge will not feel as though they are completely isolated and distanced from the camps centre. Busy facilities in the camp centre have been placed at a far enough distance so as to not disrupt the privacy of those living close to the camp centre.

Figure 14. Proposed layout for the common spaces in an 8-sector settlement

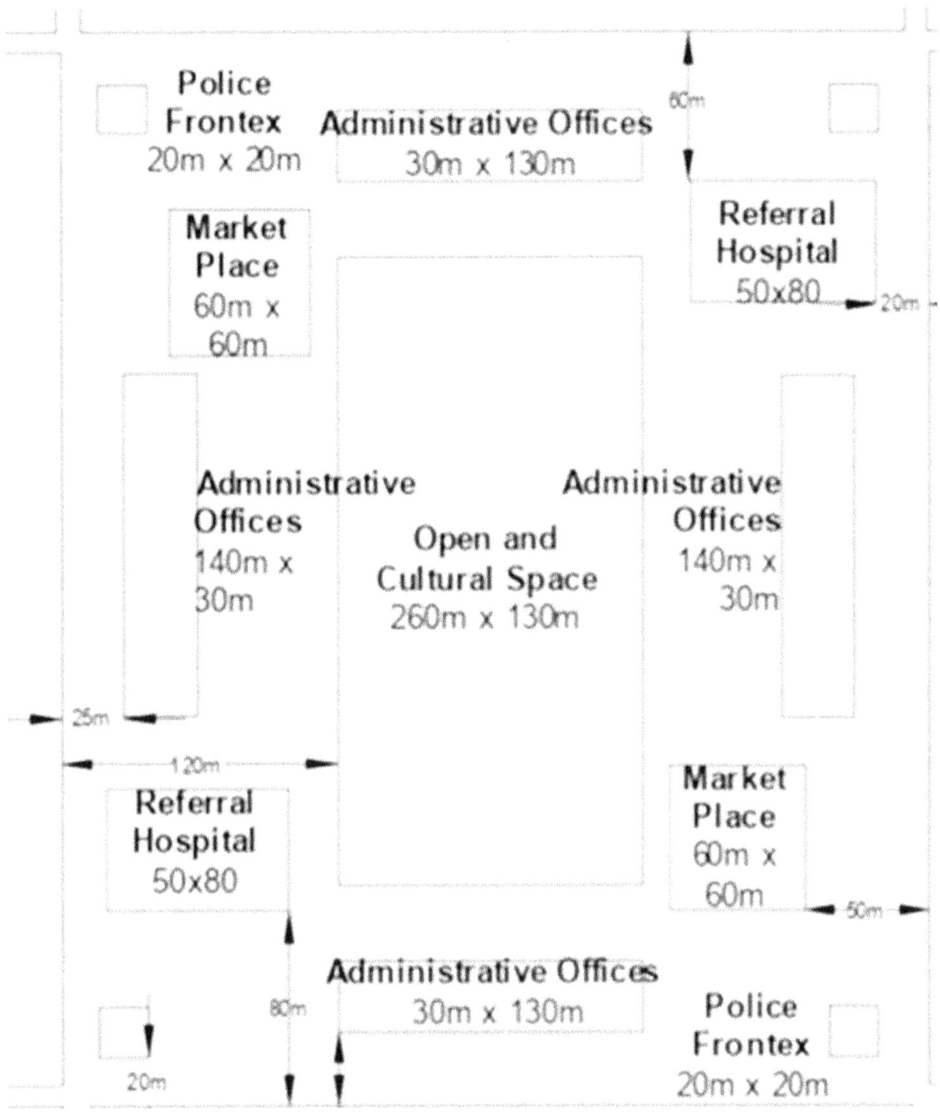

Common Spaces

At the settlement level, the common spaces have four separate areas for administrative offices. These **administrative offices** can be offices of the NGOs working within the camp, information centres, advocacy helpers, interpreters, family tracing, Wi-Fi and charging services, children's and women's organisations, and other such service offices. They have been placed in every direction of the camp so as to allow for easy access, no matter where in the camp a resident may live.

While in the previous layout of the Moria camp, there were only two **police stations** on opposite ends of the camp, the new design proposes four police frontexes at each corner of the common space. The police frontexes will be 20 m away from the boundary of the surrounding sectors. This distance was chosen to provide the residents with ample distance to have privacy, but close enough so that they feel safe as the police are within reach.

Spaces for communal activities and facilities such as **hospitals,** which could be highly contagious zones, have been given large areas to allow for social distancing practices within them. There are also two of each facility within a settlement to ensure that fewer people are using the same facilities. Buildings such as hospitals have the highest height; this is again to allow more space in the facility.

One feature that this plan emphasises is the need for **open and cultural spaces**. Most refugee camps lack space for cultural or social activities. This design lays out large areas solely to encourage social activities. The cultural space can also be used to develop religious spaces, such as mosques. These features will have a significant impact on improving the mental and emotional lives of refugees.

The presence of these many common spaces may seem like an inefficient use of land and resources, as the world transitions back to normal. It is worth mentioning that this layout was designed keeping specifically in mind highly infectious diseases and the need for social distancing. In times without prevalent contagious diseases such as COVID-19, this design can be adjusted to require fewer more centralised facilities, which would occupy less land and use fewer resources, keeping in mind the hierarchy of spaces and other key principles.

The settlements in the camp can be divided by language. The Moria camp currently has been geographically divided into Arabic and non-Arabic speakers. However, where previously the facilities of the camp were disproportionality distributed amongst the Arabic and non-Arabic speakers, this study proposes a more equal distribution. This will help in fostering a greater sense of community in the residents, as they will be living amongst familiar and similar people with whom they can communicate easily. Furthermore, it will be easier to provide services to communities if all residents speak a common language.

GIS WITHIN THE CONTEXT OF MORIA

The original study of the Moria camp used several variations of the camp design and studied them against GIS maps available. However, for the sake of brevity and clarity, this chapter will explore only one of the proposed designs. This is so that the reader gets a basic understanding of the data and analysis available through ArcGIS.

In the case of the Moria camp, ArcGIS was used to select the appropriate site for the camp to be rebuilt. To do this, the population and camp size projections were used to calculate the minimum area required. This minimum area was then mapped onto ArcGIS to see how much of the camp's surrounding area would be needed. Figure 15 shows a satellite image of the surrounding area of the camp. By studying this space against a terrain map, as seen in Figure 16, it can be seen that it is in a highly mountainous region with mountains as high as 450 m. This means it would be extremely difficult to build in this area (Corsellis & Vitale, 2005).

Given this factor, the camp can more conveniently grow eastwardly from its current location. Although this space would also have slopes, the slopes in this design may actually work in its favour as they are not as high as those in the surrounding area and still provide a good gradient for natural drainage of rainwater and sewage (UNHCR, 2020).

The following figures use the previously predicted camp size requirements to envisage the estimated amount of space needed in each population growth scenario. In these scenarios, the area was selected in this manner, as below the selected area there is a main road that connects existing towns. This road

Figure 15. A satellite image of the area taken on ArcGIS

cannot be built upon; hence, the growth is depicted in a northern direction. South of this road is another settlement; hence, no construction can take place there. West of the camp there are high slopes, as can be seen on the topographical map above; therefore, building in that direction may also be limited but preferable to the east of the camp where there is already another existing settlement. While there are smaller roads in the highlighted areas, these roads may be allowed to build upon, as they do not lead to a significant settlement and are not connecting two different parts of Lesbos island. Based on these scenarios, the proposed area of the camp was defined as shown in Figure 17. Note that below the selected area is the main road that connects existing towns. This road cannot be built upon; hence, the growth of the camp is depicted in a northern direction. South of this road, there is another settlement; hence, no construction can take place there. West of the camp, there are high slopes, as can be seen on

Figure 16. A topographical image of the region surrounding the camp

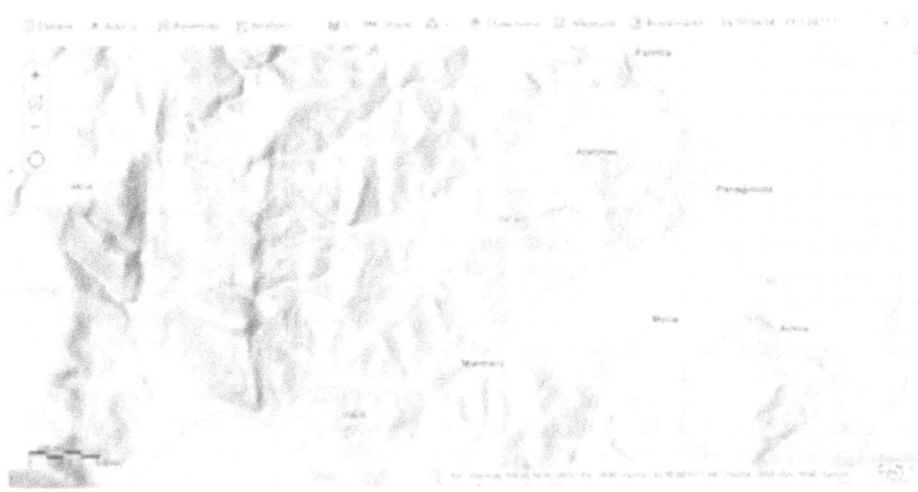

Figure 17. An image of the proposed 8-sector camp layout imposed onto the topographical map of the area

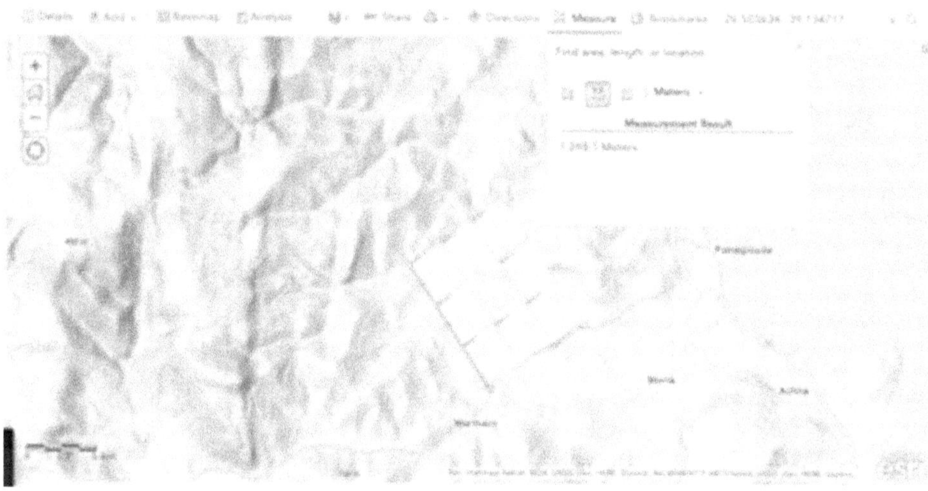

the topographical map in Figure 17; thus, building in that direction may also be limited but preferable to the east of the camp, where there is already another existing settlement.

Figure 17 is an ArcGIS conceptualisation of how much space one settlement will take in relation to the area near Moria. Should the settlement be placed here, it would have a height of 137 m above sea level at its highest. The slopes in this region are still relatively flat; thus, they would not be too difficult to build on. Only a settlement of this size can be placed here, as a larger settlement would be too close to an already existing settlement nearby.

Figure 16 shows a preexisting settlement by the name of Afalonas. Given the hostile attitudes of the locals towards the refugees and the camp, it may also not be viable to build the camp close to those existing settlements.

The only other space left for building would be uninhabited hilly areas west of the camp, which have dense vegetation. To build the camp in this direction, the planners may have to deforest large areas of lands, as the trees will not only get in the way of building the camp site but may also be a fire hazard. However, not only is this a highly unsustainable practice, it also increases the risk of landslides in the region. Deforestation enables soil erosion, which in turn can lead to landslides. Accordingly, this option is deemed unfeasible.

Potential of GIS

Although the use of GIS was limited to mapping and design within this particular study, other studies have explored its potential uses in camp maintenance, registration, demographic data studies, and more.

An example of ArcGIS being used for camp maintenance can be seen in the Al Za'atari camp in Jordan. The Al Za'atari camp is the largest refugee camp in the world. The camp was originally designed to host roughly 20,000 refugees, but as of 2021, it hosts 78,908 refugees. The rapid expansion of the camp made the management of camp facilities ever more difficult, as "Management includes distributing facilities throughout the camp, monitoring the condition and efficient operation of facilities, and determining the need for additional facilities" (Ghattasheh, 2019). Camp organisers needed to find a way to navigate through all the facilities quickly and efficiently while understanding accessibility in real time. Using GIS, camp organisers were able to create an interactive map showing the entire camp that contained all the facilities, landmarks, street names, caravans, and tents within the camp, as well as a map of the wastewater network (Ghattasheh, 2019). "REACH GIS experts created thematic maps to show statistics for the demographic data on Syrian refugees. ArcGIS for Desktop is used to map all camp facilities and analyse the accessibility of these facilities. As part of this proximity analysis, Theissen polygons were created to calculate the number of tents served by a water tank, kitchen, restroom, or other specific facility. After calculating these statistics, facilities could be rearranged to provide equitable access to them" (Ghattasheh, 2019). As previously highlighted in the chapter, the ability to monitor camp demographics and access to facilities can make the difference between effective and ineffective camps. When used effectively, GIS can play a key role in the successful design and operation of a camp. It is also a much less resource-intensive process, as it can carry out tasks that usually require multiple people and large investments, from just one computer and internet connection. This also makes it a more cost-effective option to employ.

CONCLUSION

This chapter proposes modular spatial and structural alternatives to current camp design standards as set in the UNHCR camp planning guidelines to improve the current living conditions of refugees. It identifies the key oversights and failures that current camp designs make concerning material and spatial considerations during the emergency phase of a camp. It goes on to highlight key technologies, specifically GIS and drone technologies that are instrumental and should become standard practices for future camp analysis and design. A thorough revision of the current standards is conducted within the context of the global pandemic. Finally, this chapter proposes a spatial redesign of the camp to go beyond providing a safe environment for camp residents and promoting the mental and physical well-being of the camp residents. Using the suggested layout plan at the original Moria campsite could have avoided some of the key errors that were made during the camp's lifespan, some of which inevitably led to the dangerous fire. The suggested spacing would not have allowed the fire to grow to the size it did. A more critical look at the materials used could have prevented the use of flammable materials. As engineers and humanitarians, we must do better to support some of the most defenceless people in the world. While we may not be able to solve the crises that they are fleeing, we may be able to provide them with the safe space they need during this transition. A carefully considered, efficient, and sustainable design can prevent many of the problems faced by refugees in refugee camps today.

REFERENCES

Atiyeh, B., & Gunn, S. (2017). *Refugee camps, fire disasters and burn injuries. Ann Burns Fire Disaster.* https://www.ncbi.nlm.nih.gov/pmc/articles/PMC5946753/

BBC. (2020). *Lesbos: Hundreds test positive for Covid-19 after migrant camp fire.* https://www.bbc.co.uk/news/world-europe-54239446

Better Shelter. (2020). *Designed with and for Refugees.* Product: Better Shelter. https://bettershelter.org/product/

Block, I. (2019). *Refugee Shelters Could Be Built from Concrete Fabric in 24 Hours.* Dezeen. Available at: https://www.dezeen.com/2019/09/17/cortex-shelter-cutwork-refugee-architecture/

Bowman, B. (2020). *As Coronavirus Spreads, Experts Warn Refugees Are Particularly Vulnerable.* The Globe Post. Available at: https://theglobepost.com/2020/02/28/coronavirus-outbreak-refugees

Bulley, D. (2014). Inside the Tent: Community and Government in Refugee Camps. *Security Dialogue, 45*(1), 63–80. doi:10.1177/0967010613514788

Byler, R., Gelaw, F., & Khoshnood, K. (2015). G: Considerations for Altering the Standard Refugee Camp Design for Improved Health Outcomes. *2015 IEEE Global Humanitarian Technology Conference (GHTC).* 10.1109/GHTC.2015.7343964

Cavedon, S. (2019). *The Failures of Refugee Camps in Europe.* Columbia Political Review. http://www.cpreview.org/blog/2019/11/the-failures-of-refugee-camps-in-europe

Chamma, N. & Mendoza, C.. (2016). *Rethinking Refugee Camp Design: From 'Temporary' Camps To Sustainable Settlements.* https://www.researchgate.net/publication/317694602_RETHINKING_REFUGEE_CAMP_DESIGN_FROM_'TEMPORARY'_CAMPS_TO_SUSTAINABLE_SETTLEMENTS/stats

Chatham House. (2015). *The role of Local Government in Addressing the Impact of Syrian Refugees: Jordan Case Study.* https://www.chathamhouse.org/sites/default/files/events/02-06-15-jordan-workshop-summary-final2.pdf

Chowdhary, N. (2020). Environmental Refugees: A Humanitarian Crisis in India and Bangladesh. *Refugee Crises and Third-World Economies,* 37-43. doi:10.1108/978-1-83982-190-520201007

Corsellis, T., & Vitale, A. (2005). *Transitional Settlement: Displaced Populations.* Oxfam GB.

Dabaieh, M., Emami, N., Heinonen, J., & Marteinsson, B. (1970). *A life cycle assessment of a 'minus carbon' refugee house: Global warming potential and sensitivity analysis: Semantic Scholar.* https://www.semanticscholar.org/paper/A-life-cycle-assessment-of-a-'minus-carbon'-refugee-Dabaieh-Emami/7e7136057c4617b313a560d7c76b0845ed6835a4

Denselow, J. (2016). *The end of the refugee camp?* https://www.aljazeera.com/opinions/2016/5/20/the-end-of-the-refugee-camp

Donadio, R. (2019). *Welcome to Europe. Now Go Home.* https://www.theatlantic.com/international/archive/2019/11/greeces-moria-refugee-camp-a-european-failure/601132/

EMERGENCY UK. (2020). *Italy: The Makeshift Camps of San Ferdinando.* EMERGENCY UK. https://emergencyuk.org/2018/03/italy-the-makeshift-camps-of-san-ferdinando/

Enria, N., & Gerwins, S. (2020). *Greek-Turkish border crisis: Refugees are paying the price for the EU's failure to reform its asylum system.* https://blogs.lse.ac.uk/europpblog/2020/03/25/greek-turkish-border-crisis-refugees-are-paying-the-price-for-the-eus-failure-to-reform-its-asylum-system/

Fonseca, C. (2015). *Reality check for the life-cycle cost approach in the context of refugee camps.* https://www.ircwash.org/blog/reality-check-life-cycle-cost-approach-context-refugee-camps

Ghattasheh, E. (2019). *Managing Syrian Refugee Camps Using ArcGIS.* https://www.esri.com/about/newsroom/arcuser/managing-syrian-refugee-camps-using-arcgis/

GIS. (2020). *What is GIS?* Geographic Information System Mapping Technology. https://www.esri.com/en-us/what-is-gis/overview

Grant, H. (2020). *'Moria is a Hell': New Arrivals Describe Life in a Greek Refugee Camp.* The Guardian. https://www.theguardian.com/global-development/2020/jan/17/moria-is-a-hell-new-arrivals-describe-life-in-a-greek-refugee-camp

IRC. (2020). *COVID-19 Threatens Refugee Camps.* International Rescue Committee (IRC). https://www.rescue-uk.org/coronavirus?utm_source=www.rescue.org&utm_medium=georedirect

Jahre, M., Kembro, J., Adjahossou, A., & Altay, N. (2018). Approaches to the design of Refugee Camps. *Journal of Humanitarian Logistics and Supply Chain Management*, *8*(3), 323–345. doi:10.1108/JHLSCM-07-2017-0034

Kaczan, D. J., & Orgill-Meyer, J. (2019). The impact of climate change on migration: A synthesis of recent empirical insights. *Climatic Change*, *158*(3-4), 281–300. doi:10.100710584-019-02560-0

Kennedy, J. (2005). Challenging Camp Guidelines. *Forced Migration Review*, (23), 46–47.

Kingsley, P. (2020). Fire Destroys Most of Europe's Largest Refugee Camp, on Greek Island of Lesbos. *The New York Times.* https://www.nytimes.com/2020/09/09/world/europe/fire-refugee-camp-lesbos-moria.html

Martin, D., Mica, C., & Katz, I. (2019). Rethinking the Camp: On Spatial Technologies of Power and Resistance. *Progress in Human Geography*, *44*(4), 743–768. doi:10.1177/0309132519856702

MSF. (2015). *Greece: Testimonies from migrants and asylum seekers in Lesbos.* MSF. https://www.msf.org/greece-testimonies-migrants-and-asylum-seekers-lesbos

Murphy, T. (2015). *Swiss Ditch Ikea Refugee Shelters Over Fire Concerns.* https://www.humanosphere.org/basics/2015/12/ikea-shelters-fail/

NRC. (2021). *Humanitarian needs in Iran rise as 300,000 Afghans arrive since Taliban takeover.* https://www.nrc.no/news/2021/november/humanitarian-needs-in-iran-rise-as-300000-afghans-arrive-since-taliban-takeover/

Ortiz, D. (2020). *Geographic Information Systems (GIS) in Humanitarian Assistance: A Meta-Analysis.* https://repository.upenn.edu/cgi/viewcontent.cgi?article=1023&context=pathways_journal

Osiewicz, P. (2017). *Europe's Islamophobia and the Refugee Crisis.* https://www.mei.edu/publications/europes-islamophobia-and-refugee-crisis

Papantonis, A., & Becatoros, E. (2020). *Greek Minister Says 5,000 Move into New Refugee Camp on Lesbos After Fire.* https://www.stripes.com/news/europe/greek-minister-says-5-000-move-into-new-refugee-camp-on-lesbos-after-fire-1.645470

Regencia, T. (2021). *Myanmar Coup Displaces Thousands as Global Refugee Numbers Rise.* https://www.aljazeera.com/news/2021/6/18/unhcr-urges-action-as-refugees-hit-record-high-of-82-4-million

Schmitz, F. (2020). *Europe's Largest Refugee Camp Braces for COVID-19 Outbreak.* https://www.dw.com/en/europes-largest-refugee-camp-braces-for-covid-19-outbreak/a-54640747

Schülke-Gill, B. (2020). *MSF: Moria Refugee Camp'An ideal Breeding Ground for a Rapid Spread' of Corona Virus.* https://www.dw.com/en/msf-moria-refugee-camp-an-ideal-breeding-ground-for-a-rapid-spread-of-coronavirus/a-52824830

Slater, J. (2014). *Urban Systems of the Refugee Camp.* https://surface.syr.edu/cgi/viewcontent.cgi?article=1279&context=architecture_tpreps

Subbaraman, N. (2020). *'Distancing is Impossible': Refugee Camps Race to Avert Corona Virus Catastrophe.* https://www.nature.com/articles/d41586-020-01219-6

Tazzioli, M. (2020). *Confine to Protect: Greek Hotspots and the Hygienic-Sanitary Borders of Covid-19.* https://www.law.ox.ac.uk/research-subject-groups/centre-criminology/centreborder-criminologies/blog/2020/09/confine-protect

UNHCR. (2016). *Lesvos Moria Campsite.* United Nations High Commissioner for Refugees. https://data2.unhcr.org/en/documents/details/46995

UNHCR. (2020). *Figures at a Glance.* United Nations High Commissioner for Refugees. https://www.unhcr.org/uk/figures-at-a-glance.html

UNHCR. (2020). *Corona Virus Outbreak.* United Nations High Commissioner for Refugees. https://www.unhcr.org/uk/coronavirus-covid-19.html

UNHCR. (2020). *Emergency Handbook.* United Nations High Commissioner for Refugees. https://emergency.unhcr.org/entry/45581/camp-planning-standards-planned-settlements

World Habitat Org. (2018). *Plastic Bottle Houses for Sahrawi Refugees.* https://world-habitat.org/world-habitat-awards/winners-and-finalists/plastic-bottle-houses-sahrawi-refugees/#award-content

Zander, M., & Essife, A. (2019). *Refugees Endure Living Hell of Greece's Moria Camp on Lesbos.* https://www.dw.com/en/refugees-endure-living-hell-of-greeces-moria-camp-on-lesbos/a-51305549

KEY TERMS AND DEFINITIONS

Cartography: Referring to the activity of mapping.

GIS (Geographical Information System): A software that combines mapping data with other informational databases to provide robust spatial and geographical analysis.

Hierarchy of Spaces: Hierarchy of spaces refers to the disparity between different areas of a camp that arises due to the disproportionate distribution of resources.

IDPs (Internally Displaced Persons): An IDP is someone who has been forced to leave their home due to political conflict or persecution and is seeking immediate shelter in another country or region.

Refugee: A refugee is defined as someone who has been forced to leave their home due to political conflict or persecution and is seeking immediate shelter in another country or region.

Topography: Referring to all physical features (natural and artificial) present in any area and their arrangement.

ENDNOTES

1. While this chapter does not explore the environmental impacts of camp design, it is something that is becoming of increasing importance in modern camps. Camp planners should conduct life cycle assessments for the materials used within a camp.
2. Another key factor for government attitudes is economics. At the start of the migration crisis in 2015, Greece was—and still is—recovering from a crippling economic crisis which made the government more hesitant to opening its doors to refugees (Donadio, 2019). However, this topic is again too broad to cover here.

Chapter 10
Application of Artificial Intelligence Towards Successful Ageing:
A Holistic Approach

Stavros-Theofanis Miloulis
https://orcid.org/0000-0001-5091-5530
National Technical University of Athens, Greece

Ioannis Kakkos
National Technical University of Athens, Greece

Athanasios Anastasiou
National Technical University of Athens, Greece

George K. Matsopoulos
National Technical University of Athens, Greece

Dimitris Koutsouris
National Technical University of Athens, Greece

ABSTRACT

Global population ageing is a pressing issue with significant humanitarian implications associated with compromised independence, social isolation, frailty, and diminished quality of life affecting millions of older individuals and their families worldwide as well as healthcare and social infrastructures. In this context, this chapter presents a personalized smart-living framework for promoting healthy and successful ageing, combining interdisciplinary knowledge and technologies including internet of things (IoT), artificial intelligence (AI), and affective computing (AC). The proposed intervention tackles the health-related, societal, and emotional impacts of ageing within a society with a rising mean global age aiming to mitigate the effects of decline and reduce the associated risks. The outlined architecture is based on sensor-derived information blended with non-sensor-based computerized modules and voice-based inputs for establishing a smart ecosystem that enables older people to live independently inside and outside their home, maintaining a high cognitive and physical status.

DOI: 10.4018/978-1-7998-9190-1.ch010

INTRODUCTION

Over the last decades, a dramatic demographic shift has been taking place regarding worldwide population, caused by an extended life expectancy for residents of modern societies (Lunenfeld & Stratton, 2013). This fact has translated into an increase in the number of older adults, who in 2018 surpassed children and in the next 30 years are expected to reach 2 billion globally, the majority being over 80 years old (Kastner et al., 2019). In this respect, it is a well-known fact that mental and physical status become increasingly vulnerable over the course of ageing, associated with a number of adverse outcomes that affect quality of life. Health-related impacts of ageing are accompanied by further negative consequences, since compromised independence and lack of self-confidence may cause stress, lack of motivation, social isolation and overall frailty, all acceding to a vicious cycle besetting older individuals. Conjointly to the normal decline process, the accumulative effects of dietary and lifestyle habits of an individual pose as significant risk factors for the incidence of chronic conditions such as hypertension and diabetes (Walther et al., 2017), accounting for additional disease and treatment burdens inflicted on individuals and their families, as well as healthcare and social systems (Wallace et al., 2015).

These facts call for effective solutions for the overall welfare of an ageing population, implicating health, social and emotional domains. In this view, the notions of healthy and successful ageing (Longobucco et al., 2019) are currently being promoted in order to mitigate the effects of normal decline and reduce the associated risks by contributing in keeping the older individuals healthy and active, as well as in maintaining high cognitive and physical status. In cultivating these goals, the effectiveness of such interventions entails analysis of the multifactorial determinants of healthy living by considering behavioral, social and medical aspects for establishing safe (i.e. low-risk) environments, assisting older adults in adopting a healthier lifestyle related to aspects involving nutrition, physical/cognitive activity, leisure and social engagement.

Towards addressing these challenges, the combination of novel technologies with acquired knowledge from clinical and social fields bears the potential to cultivate the development of holistic protocols that employ Artificial Intelligence (AI) for unobtrusive monitoring and comprehensive assessment, promoting new best practices and collective guidelines for smooth penetration into societies. From this perspective, this chapter proposes a generalized framework that incorporates concepts such as Internet-of-Things (IoT), AI and Affective Computing (AC), for establishing a healthy and successful ageing ecosystem that enables older people to live independently inside and outside their home. The outlined framework is based on sensor-derived information blended with additional non-sensor-based computerized modules and voice-based user inputs, backed by an interoperable ICT (Information and Communication Technology) framework utilizing cloud computing.

Initially, the chapter focuses on the specific impacts of ageing on individual, regional and global scales, delineating the humanitarian context and challenges posed with reference to health, emotional, social and financial aspects. Next, the detailed methodological components of the proposed smart-living environment are presented, along with the projected tangible influence and real-world scenarios. Additionally, some key barriers and potential solutions are identified, followed by a description of the current state-of-the-art and future prospects associated with the full implementation of the envisioned architecture.

Table 1. Notions currently employed on assessing population health

Notion (Abbreviation)	Description
Heath-Adjusted Life Expectancy (HALE)	Life expectancy adjusted for years lived under health burden (disease, disability, etc.)
Quality-Adjusted Life Years (QALYs)	Represents the years lived while maintaining functional capacity
Disability-Adjusted Life Years (DALYs)	Reflects the years lost due to disability, poor health or premature mortality
Potential Years of Life Lost (PYLLs)	Comparative measure between average life expectancy and premature death due to health conditions

Ageing Population: A Humanitarian Crisis in The Developed World

Out of the total disease burden on a global scale, more than 20% appertains to population ageing (Prince et al., 2015), mainly concerning individuals over 60 years old. The most prevalent conditions (ranging from 6-30%) within this age group include cardiovascular diseases, diabetes, pulmonary disease, visual impairment, musculoskeletal disorders, cancer, stroke and dementia (Prince et al., 2015). From 1990 to 2017, nearly 28% of deaths were attributed to ageing, representing a 10% to 20% increase on death burden for more than 20 countries worldwide (Cheng et al., 2020). This mortality crisis has been identified since over a decade ago, with researchers having predicted age-related increases up to 50% in deaths caused by cardiovascular conditions by 2030 (Moran et al., 2010). Real-world data indicate that although humanity seems to constantly progress on the disease-fighting field, we are not gaining much ground on overall deaths, with age-related mortality outbalancing pure disease-related mortality. Ultimately, population growth that comes with increased life expectancy also bears an association with increased mortality. In this regard, statistics provide further insight regarding its constituents, inevitability and available intervention margin. Indicatively, despite the universal increase in pure numbers for both males and females, the corresponding link appeared to be much weaker for low-income countries (Cheng et al., 2020). It may be assumed that wealthier countries are generally in a better position to provide quality health services, dealing more effectively with non-ageing-related threats and thus attaining a higher life expectancy for their citizens. However, alongside the obvious beneficial impact (i.e. people live more), indices also highlight the emergence of challenges with reference both to the quality of life related to this increased expectancy and to the sustainability of long-term health service provision.

Within this framework, life expectancy has long ceased to constitute the lone primary metric for assessing population health. Quality of Life (QoL) is the principal focus, with Health-Adjusted Life Expectancy (HALE), Quality-Adjusted Life Years (QALYs), Disability-Adjusted Life Years (DALYs) and Potential Years of Life Lost (PYLLs) being some key terms employed by regulatory bodies, health researchers and professionals (Gold et al., 2002; Labbe, 2010; Wiesner & Bittner, 2004). Short descriptions of these terms are provided in Table 1:

Reports have established a clear link between DALYs and ageing (Li et al., 2021), with about 65% corresponding to YLLs. Exploring potential risk factors, researchers have conducted numerous surveys and employed mathematical models seeking to identify explanatory variables affecting Quality of Life (Ghosh & Dinda, 2020; Martinez-Martin et al., 2012; Raggi et al., 2016; Talarska et al., 2018; Vo et al., 2019). Significant factors include age, education level and learning difficulties, residence (i.e. country), a non-risky home environment, chronic conditions, disabling pain, vision/hearing problems, physical

/ mental functioning, emotional status, social networking, alcohol consumption, smoking, etc. Within this set, age stands out as a major interaction factor, since the risk and impact of all the other variables is augmented for the older population (Prince et al., 2015). From a macroscopic point of view, the notion of QoL comes down to self-independency and self-perceived status, paired with existing disease burden. Remarkably, the majority of the identified points are not irrevocable (Alberro et al., 2021). In spite of age being a decisive factor, its correlation with elements such as emotional status, social networking, physical activity and home environment greatly varies among individuals. Nonetheless, their impact can be mitigated via medical and behavioral interventions targeting an increase in QALYs and HALE.

Besides, such an effect extends well beyond an improved QoL. Pairing the burden of ageing on individuals and their families with the constantly growing older population, the scale of the translated impact on health systems becomes evident. Within everyday practice, there emerges a growing demand for medical resources, while also limited by financial constraints threatening the sustainability of healthcare services.

All the aforementioned consequences of ageing bear a considerable cost for public infrastructures (*138 - The Economic Consequences of Ageing Populations. - K. Mc Morrow and W. Roeger - European Commission*, n.d.). More specifically, resource demands are constantly rising due to population growth (i.e. more people to treat), poorly balanced age ratios and elevated per-person costs, stemming from an increased life expectancy and the corresponding age-related issues. With respect to population growth and age ratios, the financially active part of the population is required to provide the primary support to health systems while at the same time shrinking in comparison to the older population that needs increasingly more resources from healthcare (Cristea et al., 2020). As a result, social security and healthcare face a pressing sustainability challenge, one that cannot be resolved simply by increasing retirement age limits or devising ways to increase available resources for meeting demands. Without a tangible impact on quality of life, escalated costs will account for no result, caught in a loop fed by a higher absolute life expectancy.

On that premise, the next big challenge for healthcare systems, governments and policy makers in developed countries is to proactively work towards the following three axes:

- Improving resource allocation
- Expanding reachability
- Alleviating ageing impacts before they escalate with respect to healthcare resource demand

The subject of resource allocation comprises a purely organizational issue, where health systems and policy makers need to establish operational models that provide optimal resource consumption per case. In the matter of reachability, there exists an undeniable under-exploitation of resources with reference to novel methodological and technical tools poorly (or not at all) adopted in clinical practice. This equates to an accessibility issue, for capabilities regarding remote patient monitoring, smart sensing, patient stratification, risk prevention and support. All these tools may be put into action for promoting successful ageing centered on expanding QALYs, improving independence and self-perceived health (Urtamo et al., 2019), as well as fostering social engagement and favorable mental/emotional status (Annele et al., 2019).

In this regard, it should be emphasized that successful ageing does not necessitate perfect health, being attainable even under disability and multimorbidity. Its definition is mainly centered on activity capacity and participation. More specifically, activity capacity can be measured in a variety of ways, a

Figure 1. Shaped landscape regarding ageing population, impact and pressing challenges

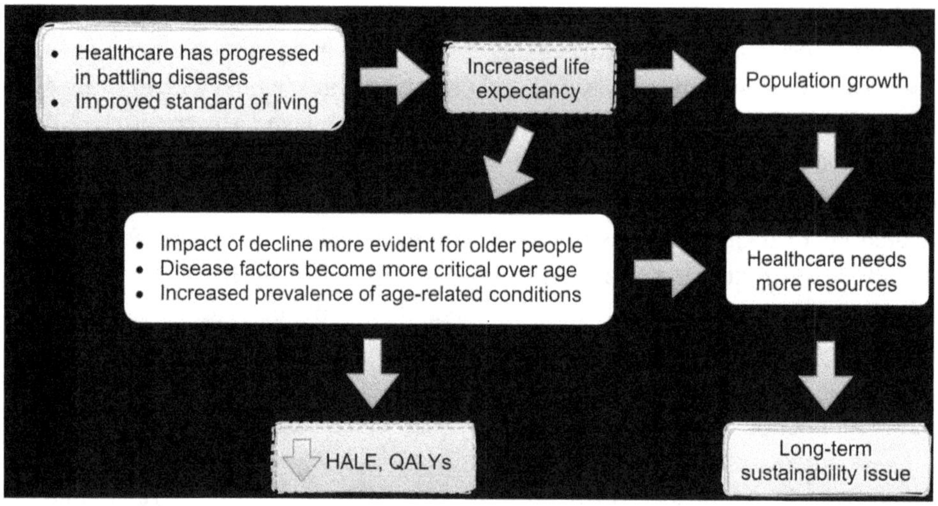

fundamental one being the ability to carry out Instrumental Activities of Daily Living (IADLs) (Guo & Sapra, 2021), which reflect basic everyday actions and enable independency within a society.

Figure 1 summarizes the described landscape regarding the ageing population status and current healthcare positioning. Shaded boxes display the controversies within which an improved healthcare leads to sustainability issues and a sub-optimal quality of life for the population in the receiving end of services:

The environment described in this chapter seeks to tackle these issues from both the individuals' and healthcare perspectives, exploiting novel technologies for monitoring older individuals (from a health status and QoL aspect), empowering them on their everyday lives and aiding in their more effective support by health systems.

In the digital age, more and more older individuals utilize information and communication technologies (ICT) to address aspects of well-being, particularly in regards to social engagement, physical health limitations and engagement in healthy behaviors (Cotten et al., 2014). Since associations between ICT and well-being are well-established, motivating older adults into using technology in their everyday routine can significantly improve their quality of life (Demiris et al., 2013). In that scope, motivations that mediate older population's use of technology include the need to sustain autonomy and independence, to learn about or better manage their health and to connect with others and garner social support (Sims et al., 2017). Acknowledging this landscape, targeted lifestyle interventions can be applied on the behavioral and psychological domains (Baltes & Baltes, 1990), whilst providing the necessary health-related assistance for expanding QALYs and alleviating a considerable portion of the supposedly inevitable ageing impact. Interventions for healthier lifestyle adoption are related to aspects involving nutrition, physical/cognitive activity, leisure and social engagement. Especially for the older adults, lifestyle habits bear a tremendous effect on health status and quality of life (Khalili et al., 2014), while benefits of a healthy lifestyle also transfer into healthcare and social systems that sustain the corresponding burden. Taking into account that the impact on everyday routine heavily depends on parameters within the control of individuals and given that preserving a way of life is associated with lower depression incidence and higher satisfaction rates (Hung & Lu, 2014), a high customization degree is ascertained, based on the personalized needs and preferences of the people involved.

Envisioned Environment for Successful Ageing

Overall Environment: Operating towards fostering successful ageing, the proposed framework (Figure 2) entails unobtrusive monitoring and comprehensive assessment based on non-privacy-sensitive information, using IoT and user-reported inputs in order to define the "Status Level" (SL) of each individual over the course of time. Data recording is conducted via wearable and ambient sensors, establishing a smart-living environment operating at rest, during sleep, during activities and at workplace, both inside and outside home. Wearable sensing, centered around a smart wristband, can be exploited for monitoring physiological parameters comprising electrodermal activity (EDA), electrocardiograpic (ECG) measurements such as heart rate (HR) and heart rate variability (HRV), as well as body temperature measurements. Wearable sensors are also incorporated for motion capture through inertia measurement units (IMUs) and step trackers, while ambient sensors account for environmental settings (temperature and humidity) and safety, convening messages for optimal preferences and triggering alarms in case of emergency (e.g fire alarm). For indoor and outdoor usability, wearable and stationary equipment should be operational both in combination and independently from one another. In addition, voice-based user input provides information with respect to dietary habits, activity planning, social participation and wellness, augmented by pitch analysis used within an affective computing module combined with physiological recordings.

The above can effectively allow intelligent data analytics to infer on overall health status, environmental conditions, lifestyle and risks, thus recognizing key points for health monitoring and dedicated interventions targeting the holistic support of older individuals. This can be achieved by incorporating AI tools for providing an AI-Decision Support System (DSS) core to address all the functionalities proposed. Interventions arising through this analysis promote adoption of "healthy" behaviors through lifestyle adaptations, dealing with loneliness and anxiety issues, gamified cognitive training, as well as support of physical functionality by employing dedicated balance and gait training for reducing fall risk. The latter can be recognized at an early stage and estimated periodically by means of pattern recognition algorithms, based on sensor data and training results. Moreover, blending AI with expert knowledge (derived by specialists and scientific studies) is suggested, for applying behavioral modeling in pursue of creating personalized plans, designed to promote adherence and maintain high motivation. By means of motivational methods and nudge theory, elevated compliance levels may be attained upon applying statistical analysis to identify key contributors, thus defining optimal strategies.

In this regard, establishment of feasible short-term goals linked to status-levels bear the potential to provide the users with tangible and relatable measures, comprising a) stress-related, b) activity-related, c) diet, d) social interaction, e) health-related, f) peer-related and g) self-reported goals. To that end, a continuous feedback-based design could assess the improvement of physical and cognitive aspects, as well as the older adults' adaptation to a healthier lifestyle. More specifically, adjustment of training levels and goal-setting accommodated for personalized support by employing a stepwise approach, tailored to each user. Furthermore, risk minimization and sense-of-safety establishment is reinforced via automated identification of emergencies (e.g. falls or arrhythmias) and contact alerts, granting round-the-clock assistance.

Moreover, encompassing a set of physical, cognitive and balance interventions in the form of gamified training tasks alleviates the decline of lifestyle aspects due to ageing. These serious games can incorporate adjustable difficulty levels and goal-setting based on the individuals' skill level, while by utilizing nudge theory and motivational strategies, higher user-acceptance is achieved ensuring long-term adherence.

Figure 2. Conceptual architecture of the envisioned environment for successful ageing employing novel technologies

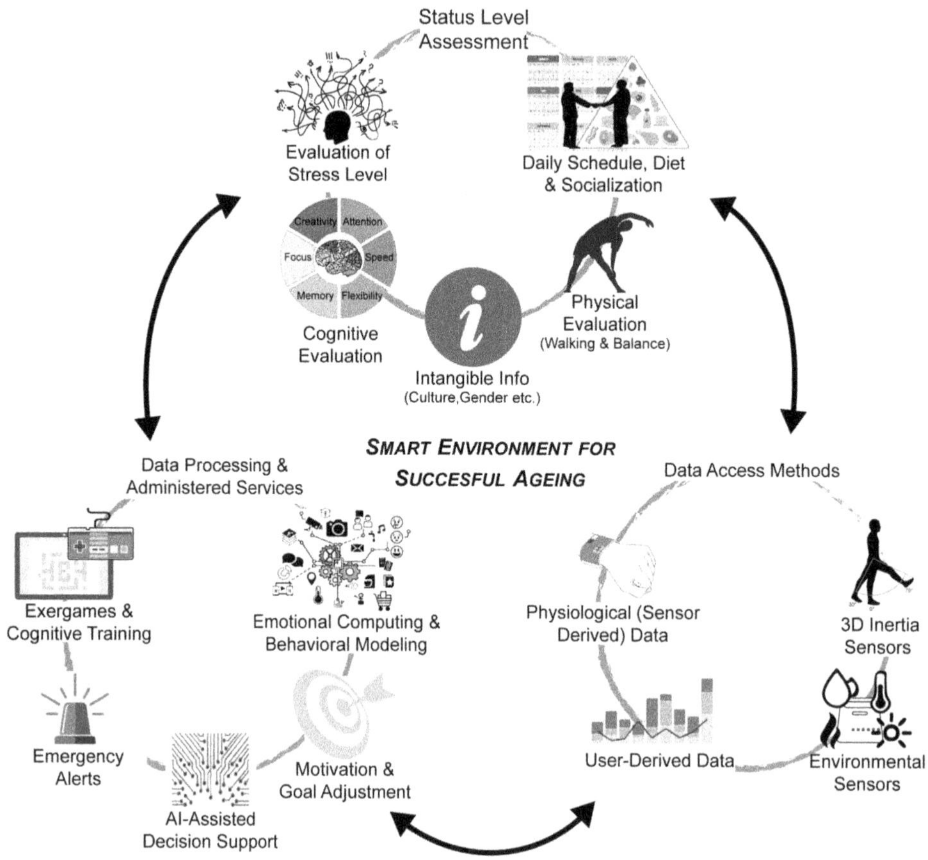

In addition to training, these games can incorporate a social networking design, providing social spaces where older adults can interact, ultimately creating a community of people supporting each other.

Sensing Equipment: Information may be collected, analyzed and monitored either continuously or on an as-needed basis, depending on joint user-clinician decision and profile needs. Within the proposed environment, raw data are collected through i) functional monitoring (measurement of general activities, gestures, motion, falls, resting, sleeping, nutrition), ii) physiological monitoring (heart rate, blood pressure, temperature, electrodermal activity), iii) behavioral monitoring (derived from self-reported aspects such as fatigue and discomfort to stress and emotion measurements from electrodermal activity, heart rate and voice analysis) and iv) user compliance. Overall, implemented services are delivered within a portable unobtrusive compound consisting of environmental sensors, a smart wristband with voice-based user input and a minimum amount of sensing modules, with advanced back-end processing being built upon a cloud-based IoT platform ensuring smooth operation, storage, security and interoperability. The contents of Table 2 comprise a full set of potential components, although the authors encourage case-specific adaptations regarding the specific sensor subset used.

Everyday Monitoring: The overarching concept of the proposed design is to enable the continuous collection of data from the everyday life of the older individuals, for obtaining the evidence needed in

Table 2. Potential hardware components for monitoring successful ageing

Component	Function
Electrocardiography (ECG)	Heart rate (HR) and heart rate variability (HRV) measurement
Photoplethysmography (PPG) sensor	Arterial pulse waveform and blood pressure measurement
Thermistor thermometer	Body temperature measurement
Ambient temperature & humidity Sensors	Temperature & humidity environmental measurements
Ambient fire/smoke sensor	Intelligent fire/smoke detector
Galvanic skin response sensor	Electrodermal activity measurement
GPS receiver	Location transmission in case of emergency only
3D inertia sensor	Motion analysis, fall detection and fall risk estimation
Electronic pedometers	Step tracking
Speaker/microphone	Voice-based interface
Bluetooth	Data exchange and proximal IoT devices (e.g. smart TV)
Wi-Fi adapter	Wireless high-speed connections in indoor environments
5G adapter	Wireless high-speed connections in outdoor environments

order to offer personalized interventions to promote their healthy and independent living (e.g. dietary habits, activity log, socialization level and basic health status). More specifically, the projected architecture features an abundance of available options for the individuals to compose their daily regime accounting for detailed nutritional values, generating statistics that will accurately depict nutrient intake on daily / weekly / monthly basis. Moreover, the incorporation of personalized activity logs, hosting a complete history of physical/cognitive activities, provide the required information for tracking social engagement, outdoor activities, interactions with friends/family/peers on a given time frame and overall independence. Fundamental measurements (weight, hearth rate, blood pressure), self-reported outcomes (emotional status and self-perceived physical state) and emotional/affective computing results representing stress levels and potential depression can all be combined for extracting a personalized health status and assessing risks.

Indicative use cases on the value of continuous monitoring comprise i) detection of abnormal parameter values (i.e. abnormal heart rates), ii) detection of emergency events such as falls, heart attacks or blackouts, or iii) identification of anxiety and depression. When needed (based on AI output) emergency services may be contacted automatically. Particularly regarding stress and depression, their identification is feasible based on data beyond mere user reports. Electrodermal activity, heart rate variability and skin temperature (Mozos et al., 2017), as well as voice indicators and speech patterns (Aguilar et al., 2019), all bear the potential to contribute in interpreting the user's emotional state in order to assess lifestyle parameters and intervention outcomes, while providing emotional support to the ageing population.

Results acquire a significantly enhanced value when examined together with medical history data and evaluated by experts, who - through such a platform - will be able to effectively access and evaluate large data volumes for diverse populations. Anonymized processing of this information allows operation of AI-DSS for providing personalized adjustments, identifying all key parameters linked to healthy ageing and determining their significance.

Physical / Cognitive Assessment and Training Games: It is of utmost importance that any smart environment for successful ageing incorporates physical and cognitive assessment and support modules as part of its comprehensive intervention on preserving functioning capacity. In this domain, muscle weakness, cognitive impairment, depression and overall frailty can be detected in time, while computerized cognitive training and exergames may be implemented via interconnected digital technology for slowing down decline as well as enhancing self-confidence and independence.

With regard to physical functioning, exercise has emerged as a key factor for healthy ageing, since convergent evidence shows that exercise presents a highly beneficial impact regarding the prevention of diseases (e.g. cardiovascular & metabolic), muscle weakness, cognitive impairment, depression and overall frailty. Moreover, physical exercises constitute a key intervention for enhancing self-confidence and independence through body strengthening, an indicative implementation being risk minimization through balance and gait training for fall prevention. For everyday purposes and quantitative evaluation, physical assessment and training can be focused on balance and falls, representing a major factor regarding movement independence and safety. More specifically, falls have been associated with both short-term and long-term physical injuries as well as with usage of health services, presenting an annual incidence among older individuals exceeding 25% (Wang et al., 2016). Moreover, the risk of falling is multiplied in the presence of previous fall history, while also leading to isolation due to reduced self-confidence and inability to safely engage in activities caused by fear of falling (Schoene et al., 2019). Accordingly, the physical status component of the proposed system revolves around balance evaluation and fall risk estimation. The user is equipped with a minimized setup of unobtrusive wearable sensors (operable within both indoor and outdoor settings) that grant a continuous data stream comprising kinematic information (based on sensors placed on key points), analyzed to unveil motion patterns, balance and real-time fall risk and detection. Similarly, exergames targeting motion independence and specialized balance training can be performed at home by each user, regulated based on "physical monitoring" analysis and guided by AI results. Quantification of progress and training effects is achievable based on sensor data, while the difficulty level should be set in the basis of a personalized profile. Balance metrics can be calculated by assessing center of mass or center of gravity (CoM/CoG) in relation to the body's support base (SB), i.e. the area defined by the body points in contact with the support surface (Ismail, 2015). During walking, improper transfer of body weight causes the CoM to move away from the central point of the support base, resulting in loss of balance and eventual falling. Expected outcomes include body strengthening and adoption of kinematic patterns related to minimized risks, thus enhancing independence and encouraging social participation and overall activeness in outdoor settings. As such, gait / balance patterns related to fall risk are captured based on step-tracking and IMU measurements embedded in the smart wristband with an addition of a small number of extra sensors attached in the legs/arms or back (depending on the game module), in order to administer personalized training using exergames (Meekes & Stanmore, 2017) that users will be able to perform both inside and outside home. Namely, one of the main goals is the extraction of personalized "weak points" regarding posture, balance and walking patterns, feeding the data into the corresponding AI model, which in turn will infer on the optimal set of exercises, as well as the overall regime and corresponding goals. Clearly, the latter depend on the parameters of each individual, so that these goals are achievable and the outcomes relatable to the users concerning impact on their everyday life. Such games can be designed to run on a streaming application in a smart TV, tablet or smartphone, providing social networking features that allow other players to communicate and participate in sessions both remotely and on-site. In this regard, interoperability plays a major role rendering the games able to be displayed in different hardware devices. Moreover, through the social networking capabilities, when opting for outdoor training, individuals can be informed of when and where others engage in similar activities and allowing them to socialize and exercise in groups, thus reinforcing the participation element.

Correspondingly, the envisioned environment offers a set of certified cognitive strategies, designed and verified by experts in order to define baseline and validate tangible progress. Prevention of cognitive regression by means of computerized cognitive training has shown promising results regarding its benefits (Giuli et al., 2016), mainly aspiring to achieve transfer of training effects into everyday life by focusing on functions including visuospatial information processing, memory and reaction time. This process should also take into account the current mental status of the user to avoid disappointment for not achieving training goals, resulting in high drop-out rates. Hence, as with physical training, cognitive training should also implement adjustable difficulty levels and heavily depend on the gaming element, enhancing adherence and addressing the often frustrating and repetitive nature of training process.

Motivational Strategies: Sensibly, the emotional status of a person affects many aspects of the daily life to a high degree. Adherence to a healthy diet, socializing and general activity performance and participation, all depend on the contextual emotional status and motivation. Specifically, low motivation and adherence are common issues among older people, who are often reluctant to changes. Motivation is defined as the forces acting on or within a person to initiate a behavior. Understanding how behavior change can be motivated towards lifestyle modifications in older adults is critically important. Primarily, behavioral intentions (aiming to change) are vastly different from actual behavior, since people often do not act in accordance with their intentions (Schwarzer & Luszczynska, 2008). In this context, motivation has been defined as a function of perceived chance of success (CS), goal importance (GI), perceived cost (PC) and inclination to remain sedentary (SI). The problem that individuals usually encounter is that long-term goal achievement often seems to pose high difficulty and require an undetermined amount of time, while the short-term perceived cost is deterring. Therefore, the major aim of a successful motivational strategy is to present the individual with meaningful personalized goals (high GI) paired with a clear and feasible path to success (high CS), while minimizing PC and applying constant monitoring in order to counter SI. PC minimization can be achieved by defining multiple short-term goals that gradually evolve over time, providing accessible checkpoints towards attaining long-term goals. Combined with behavioral modeling utilizing emotional/affective computing, activity monitoring and user reports, individualized goal setting bears considerable potential for assisting users in adopting beneficial lifestyle guidance provided by the AI system. From this standpoint, the AI-DSS should generate meaningful personalized goals based on recorded data and affective computing principles, conducting personality and psychometric assessment towards maintaining high user adherence rates.

In the context of the portrayed architecture, the implementation of composite motivational strategies based on the self-determination theory is suggested (Lee et al., 2016) (SDT). SDT defines three motivation levels: a) amotivation (lack of motivation), b) extrinsic motivation (aiming for a reward) and c) intrinsic motivation (emerging from performing an activity itself). According to SDT, intrinsic motivation is assumed to be the most influential of the three, being closely related to the three basic psychological needs: a) autonomy, b) competence and c) relatedness. Autonomy is defined as the individuals' need to control their lives, competence refers to the acknowledgement of their own abilities and limitations in pursuing goals, while relatedness introduces the social factor, meaning the need to interact and be approved by others. On applying these principles, autonomy is achieved by granting the participants with control over certain intervention parameters, such as preferred exercises (over predefined alternatives), own goals and scheduling. On establishing competence, the AI-DSS takes into account the abilities and baseline performance of the participants in an individualized manner for finer goal-setting and frustration avoidance. In addition, presentation of tangible measures allows the evaluation of outcomes and ascertainment of progress. Finally, the socialization aspect represents the attainment of relatedness.

Through this process, the initial aim is to establish an augmented extrinsic motivation, mainly regarding competence and relatedness, so that the older adults first focus on physical status and independence as well as social participation. On a next level, based on the primary groundwork for developing extrinsic motivation, users are expected to mature, advancing to intrinsic motivation, where the individuals will actively participate in shaping intervention parameters (autonomy) and display high adherence, accompanied by concrete impacts on their everyday life.

Affective Computing: Implementation of emotional computing algorithms can employ wearables to elicit stress-related physiological information based on EDA, HR and temperature sensors, augmented via the use of voice analysis seeking patterns hinting at elevated stress levels and depression. In this regard, efficient training and validation of machine learning (ML) models heavily depends on the successful simulation of stress. For instance, studies have utilized emotion-inducing pictures like the International Affective Picture System (Zangróniz et al., 2017) (IAPS). In addition, reproduction of stressful situations in research has incorporated the social aspect through evaluation of users' responses during a task. The Trier Social Stress Test (Mozos et al., 2017) (TSST), as well as its derivatives like the Montreal Imaging Stress Task (Tolba et al., 2018) (MIST) induce stress by gauging derived versus expected responses, representing social evaluation. Data analysis under these tasks should also be able to differentiate the effects of stress from those attributed to mere mental workload. Emotional/affective computing for automated stress detection can benefit from the exploitation of AI tools to identify and interpret the emotional state of the user in order to better modulate the lifestyle parameters and the interventions needed. This is an important tool not only for the emotional support of the older individuals, but also for providing the means to encourage adherence, acceptance and increase self-esteem and independence. Furthermore, information extracted by the emotional/affective computing module can be combined with the motivational strategies described above, incorporated as a supplementary software component, providing information exploited by AI-DSS towards an optimized regime design for adjusting goals, difficulty levels and social interaction plans. On this premise, incorporation of motivational/nudge strategies in parallel to emotional/affective computing identifies key motivational elements in initiating and maintaining a healthier lifestyle and maintaining prospects of autonomy, self-efficacy, participation, social acceptance and self-satisfaction.

User Interfacing: Application interfacing can be implemented via multilingual voice-based interaction and a simplistic user-friendly GUI, familiarizing older adults with modern technology without demanding particularly complex skills. Voice-based interaction comprises a viable and effective pathway for accessing intangible non-measurable information and managing inputs that under current application settings are mostly handled manually (such as activity and diet regimes). The optimal approach consists of a fully voice-based and gesture interface for easy adaptation towards the ageing population, as part of a general effort to minimize manual information insert, that has often proven to be challenging and by extension frustrating and discouraging, especially for older adults. On the flip side, presentation of output information should preferably be conducted via simplistic charts and voice exemplars with minimum need for typing or button pressing, with the users having immediate access to assessment results and providing the required actions and feedback during training via gestures and voice commands.

Figure 3. Detailed components and capabilities of the envisioned environment for successful ageing

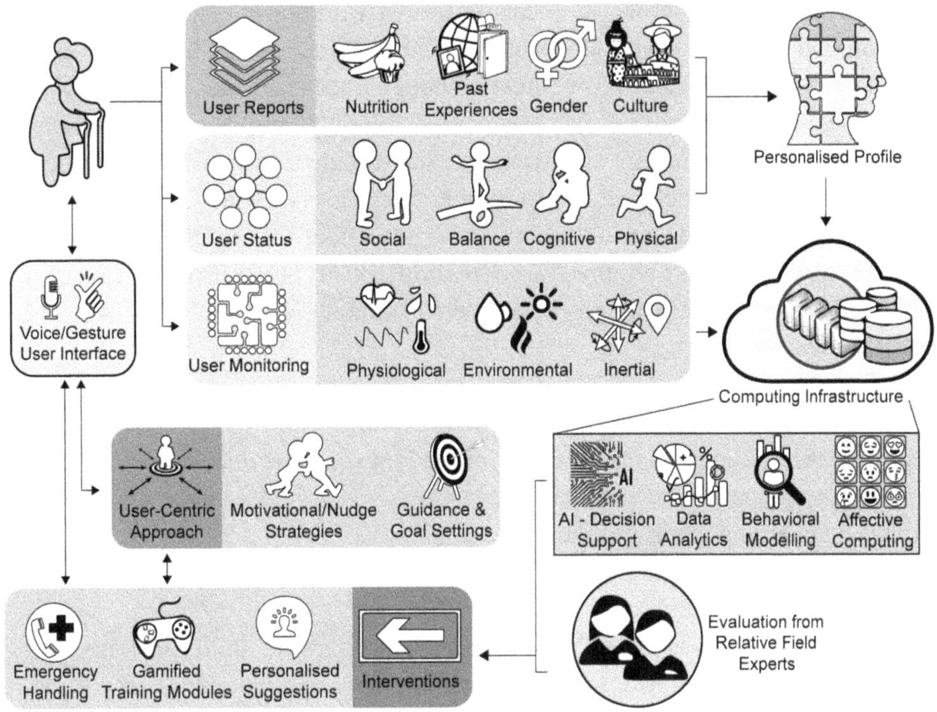

Real-Life Objectives and Impact

Summarizing the impacts of the proposed environment on older individuals and healthcare administrators, the following aspects are targeted:

- Independent living and quality of life of older population
- Personalized recommendations and follow-up in terms of preserving physical, cognitive, mental and social well-being
- Fostering social participation and reducing social exclusion's risks
- Validation of non-obtrusive technology personalized and cost-effective remote healthcare
- Contribution on sustainable healthcare service provision through enhanced resource allocation and improve monitoring towards proactive healthcare (i.e. prevent outcomes and hospitalizations before the imposed burden on infrastructures escalates)
- Promoting behavioral training and healthy lifestyle adaptations, with the benefits easily ascertained by users in their everyday life, fostering adoption of such behaviors even outside the scope of the proposed system and target group

Data Security and Privacy Aspects

On top of the methodological soundness regarding service provision, privacy issues bear utmost importance regarding both actual data security (mechanisms employed) and perceived security, i.e. how safe the end-users feel about their own data regardless of the measures applied. Specifically, the proposed environment entails wearable sensors, ambient sensors at workplace and at home, activity logging, voice recognition, inference on health prospects and capacity, as well as intelligent cloud-based data management and processing. Within this context, although unobtrusiveness is easily verifiable from a hardware perspective, these capabilities may be regarded as highly obtrusive from a privacy perspective.

Continuous monitoring, history logs, personal accounts, data transmission and processing outside their own devices, all constitute deterring reasons for potential users, even more for older people. Within a digital era where privacy rights are – rightfully so – a trending matter, concerns are raised. On that account, dedicated mechanisms are required for addressing security issues. Privacy-Enhancing Technologies (PETs) include data-oriented and process-oriented strategies (Hoepman, 2014), covering aspects such as the following:

- Amount of personal data processed (should be minimal)
- Compartmentalization of personal data (for minimizing leakage due to a security breach)
- Transparency on personal data processed
- Individuals maintaining control of their data
- Compatibility with legal requirements

The most important factors are encryption, storage sites and the ability to trace data back to the owner. Within the proposed platform, management and processing take place mainly on cloud-based infrastructures, in the context of providing high-end services to low-end devices. However, it is highly recommended that identifying data are compartmentalized, with most of the information kept within the user's device. End-to-end authentication approaches should be favored, establishing user-defined keys and encoding information in a way untraceable to the owner by a third party. Particularly, GDPR clearly states that pseudonymized data are still considered as personal data, so health-related data should be separated from identifying information for minimizing risks (Finck & Pallas, 2020).

From a client perspective, users should be fully informed on personal data collected and why they are necessary. Moreover, as opposed to the "maximum time window" rule applied by most platforms for data maintenance, users should have the opportunity to withdraw all personal data "here and now". Such measures aim at establishing "user-perceived data safety", which should be addressed along with the actual data safety, referring to technical aspect of implementation.

Challenges, Barriers and Recommendations

This section briefly outlines a number of critical barriers with regard to the implementation and effective penetration of the described smart-living environment, along with indicative pathways.

Behavioral Barriers: For older individuals, interaction with automated technology systems can be rather difficult, often finding themselves unable to trust or comprehend technology products. Although an enormous number of lifestyle-related and health-related interventions are available, older population displays low acceptance for behavior modifications (Vrdoljak et al., 2013) despite the range of functionalities provided. This can be attributed to a number of individual-level factors, ranging from interactional elements such as the lack of involvement in decision making, to psychosocial factors that may include negative attitudes and beliefs about mental and physical health issues (Ronteltap & van Trijp, 2007). Race, ethnicity and cultural factors also play a significant role in the engagement processes with fear of involuntary hospitalization, concerns about social stigma and reluctance to divulge personal information. Moreover, mistrust toward health care professionals may reflect powerful remnants from earlier health care abuses (Alexopoulos & Kiosses, 2011). All these factors add up and enhance the existing issue of technology utilization skills by older people and their often limited capacity for learning new information. The associated challenges can be overcome through simplistic interfaces that require minimum actions utilizing gesture-based and voice-based interactions, as well as through the combination of affective computing and motivational modules for individualized, feasible and meaningful goal-setting.

Data Management and Availability Barriers: The global and personalized data analysis required for the effective implementation and smooth operation of the presented design entails demanding data processing and information transfer capabilities, which in turn necessitate high-end resources, not always available in the low-end devices (smartphones) used in the everyday life. Moreover, it is not advised to transfer computational burden to distinct wearable sensors since this would increase the financial cost of the per-person modules. Furthermore, healthcare adoption of personalized AI-based solutions for successful ageing is hampered by reliability/trust issues attributed to a lack of a large and diverse data pool for model training and validation in the direction of accounting for variability on factors including gender, age, culture, socioeconomics, status, etc. Both of these issues may be resolved via the proposed utilization of a back-end cloud-based platform that can provide data transfer, management and processing, making high-end services available to low-end user devices whilst limiting the overall financial cost since the cloud platform is in position to conduct centralized analysis, letting the end-user devices handle only data input and output within user interaction.

Interoperability Barriers: One of the main challenges that need to be addressed en route to successful integration corresponds to the interoperability issue existing on device, network, syntax, semantics and platform levels, affecting all system layers. Cloud-based data management and processing is one step towards addressing this problem, minimizing the workflow conducted on individual sensors and devices. However, there still exists an undeniable need for dedicated application programming interfaces (APIs) and data standardization modules, bridging sensors with devices, processing modules and existing platforms (e.g. hospital ICT systems).The basic architecture for interoperable component integration varies based on model design, the fundamental scaffold consists of three layers (Ma et al., 2018): a) Data Acquisition Layer (incorporation of all input data and their conformation into unified structures through basic preprocessing, b) Data Management Layer (storage, distribution, retrieval and analytics services) and c) Application Service Layer (user interfaces and APIs). Interoperability issues affect successful integration in all framework layers existing in device, network, syntax, semantics and platform levels (Noura et al., 2019). Regarding the data acquisition layer, interconnection of input nodes with the main data collection device prerequisites the establishment of suitable data transfer protocols, applicable to multiple devices with different software/hardware specifications (Aloi et al., 2016). To that end, a highly interoperable framework should employ limited use of local hardware resources and prerequisite low-end hardware specifications, thus broadening the range of devices that can be utilized. Interoperability issues also exist in the data management and application service layers, where the proposed solutions of middleware and protocol proxies bear scalability concerns when performing updates/additions.

Privacy and Regulatory Barriers: Ethical and legal factors play an important role in digital health and care systems, with securing the privacy rights being an integral and important aspect. On top of the mistrust by end-users regarding their data recording and management, practical issues on anonymization & security of data management, processing and sharing (i.e. data transfer to clinicians) pose significant threats within a digital era. In this light, the challenge for both developers and policy makers is heavy.

The former are burdened with the employment of a privacy and security-by design approach regarding data recorded by sensors, smart applications and cloud platforms. From a regulatory perspective, policy makers have already progressed on this field (e.g. General Data Protection Regulation, GDPR, Directive on Security of Network and Information Systems, NIS directive) by introducing strict standards that allow the individuals to maintain control on their data and the ways in which they are utilized and shared. As opposed to the other barriers mentioned, privacy-related barriers should not be overcome, rather enhanced and efficiently enforced without compromising the functionality provided to end-users.

Knowledge Communication and Translation Barriers: These barriers do not refer to knowledge of the ageing population on new technologies, rather to the under-exploitation issue of current knowledge and advances, as mentioned in a previous section. In their majority, general population and healthcare professionals are not fully aware of existing capabilities on behavioral and medical data measurement, processing and inference. Even in cases where companies and research institutions reach out to healthcare service providers for collaboration and novel service/product deployment, the benefits fail to reach the public and thus translate into tangible impact on healthcare. Moreover, a number of operational cutting-edge systems are isolated from real healthcare requirements, poorly translating into clinical practice (Kaplan & Harris-Salamone, 2009). This missing link leads to adoption issues, raising the matter of providing context-aware systems. Presenting healthcare and general population users with a holistic system with high usability (rather than separate modules), as well as actively seeking enhanced communication regarding the capabilities and potential benefits could comprise a decisive step towards knowledge communication and therefore wide exploitation of the full potential offered by technological innovations.

Financial Barriers: The last challenge referred in this chapter concerns the fundamental financial issue. Obviously, the development, deployment and continuous application of the proposed architecture in real settings demands major strategic decisions regarding investment in time and money by governments and health systems. The high short-term costs may certainly seem demoralizing for governing bodies, however the current numbers and future projections clearly show that healthcare faces a huge unsustainability threat on the long term. Thereupon, existing studies and economic scenarios show that major reforms are needed towards sustainable healthcare, with the proposed system ticking all boxes with reference to dealing with age-related burden on individuals and infrastructures.

Existing Technologies and Future Directions

Feasibility of the presented design is ascertained by the fact that, from both technological and regulatory aspects, the current status on individual components is at a high maturity level, albeit progress in each domain is still needed. Some comments on state of the art and research directions are provided below.

Motivational Strategy Domain: Regarding behavioral modeling and motivational theories, the necessary background currently exists and has been validated in several different contexts and environments. The addition needed by future research is the full integration of multimodal data (physiological and behavioral) into the established theories, so that they can be implemented within a personalized AI-based framework.

Medical Knowledge Domain: With reference to the information needed within the proposed environment, there is not much room for improvement regarding existing knowledge, the aim for current research being the incorporation of medical knowledge into AI models in an explainable way for providing meaningful decision support. On the other hand, this domain should expect the generation of new knowledge regarding further insight into ageing and its impacts through system implementation in clinical practice.

Training Games Domain: A large number of computerized physical and cognitive training games have been already proposed and successfully implemented. These have shown the potential to enhance functions, demonstrating tangible transfer into everyday tasks (Anguera & Gazzaley, 2015). Research for their deployment within a successful ageing environment should focus on optimizing these games for the older people as well as on establishing interoperability with wearable sensors and supporting voice-based and gesture-based interaction through dedicated APIs. Moreover, from a significantly large pool of available interventions, real-life application will highlight those with the best impact and acceptance by the target population.

Interfacing Domain: Despite the advances in computer vision and speech recognition algorithms, the application of universal standards and the realization of truly interoperable algorithms for everyday large-scale applications, accessible by the general public is yet to be fully achieved. More specifically, since voice-based interaction is mainly based on the recognition of predefined word sets, multilingual and multi-vocal support and establishment of general speech standards are essential. Parameters such as language, dialect variants, accent, gender and voice pitch heavily affect the performance of an AI model (Mary Zarate et al., 2015). Similarly, age, ethnicity and gesture expressions present a significant impact on the results of computer pattern recognition algorithms (Benalcázar et al., 2017).

Smart-Living Domain: Smart homes (and their possible extensions in outdoor settings) for older adults' healthcare, either in the form of prototype or in the form of commercially available solutions, have been presented both in the literature and in the market (Kon et al., 2017). The reported environments focus on different aspects of daily living of older people including safety, health and nutrition, physical activity, personal hygiene and care, social engagement and leisure. However, although these solutions offer excellent services for safety, and control of home environment, they still lack comprehensive healthcare monitoring services and outdoor implementation.

ICT Domain: Issues on multimodal data analytics, software/hardware interoperability, data transmission and computational resources have been greatly assisted via cloud computing, providing remote access to resources and allowing data upload for computationally demanding processing pipelines. In that way, wearable sensors and mobile devices are able to provide users with high-quality services within low-cost set-ups. However, the constantly increasing amount of mobile devices that connect to the internet and the need for real-time processing may limit those benefits due to latency/bandwidth issues that compromise the establishment of reliable connectivity (Andriopoulou et al., 2016). To that end, approaches like edge and fog computing have been implemented in order to shorten the gap between the devices and cloud services (Dolui & Datta, 2017), i.e. bring the resources closer to the data. Such methods, combined with the emergence of 5G networks (Han et al., 2015) are expected to dramatically alter the IoT landscape. The primary enabling requirements for the adoption of ICT are security, confidentiality, interoperability, scalability and standards.

AI Domain: In the digital age, AI applications have been apparent in almost all aspects of everyday life. In the most advanced-level systems of interrelated computing devices, many techniques have been applied in various fields in recent years. Deep structured learning and Artificial Neural Networks (ANNs) (Fogel & Kvedar, 2018) bear the potential of identifying patterns and relations without supervision and explicit rules, albeit functioning within a "black-box" framework, while many health professionals prefer more explainable AI models. Within this context, building a personalized AI-DSS for effectively assessing and managing smart-living environments and providing reliable medical prognosis and decision support, still comprises a filed under research.

CONCLUSION

In a nutshell, the smart-living environment presented in this chapter takes advantage of modern technology for decisively contributing to healthy and successful ageing, reduction of frailty risk, as well as boosting self-confidence and independence. The described modules constitute part of a general lifestyle altering scheme, promoting healthy behaviors by the means of behavioral modelling and application of user-specific motivational strategies. As such, a "safe" environment for the ageing population on the basis of risk prevention is submitted, both regarding health implications through compliance to a healthy way of life and by reducing the incidence probability of events known for triggering severe outcomes, such as arrhythmias or falls. By implementing fast data transfer and real-time processing through cloud computing capabilities, provision of high-end AI and IoT services to low-end user devices is feasible, rendering the proposed platform and its projected benefits accessible to the target audience and applicable to their everyday lives. The expected impact on healthcare is also invaluable, contributing towards long-term sustainability. Therefore, from a humanitarian perspective, the corresponding outcomes tackle pressing issues on the quality of life for citizens as well as on infrastructure sustainability on regional and global scales.

ACKNOWLEDGMENT

This research received no specific grant from any funding agency in the public, commercial, or not-for-profit sectors.

REFERENCES

Aguilar, G., Rozgic, V., Wang, W., & Wang, C. (2019). Multimodal and Multi-view Models for Emotion Recognition. *Proceedings of the 57th Annual Meeting of the Association for Computational Linguistics*, 991–1002. doi:10.18653/v1/P19-1095

Alberro, A., Iribarren-Lopez, A., Sáenz-Cuesta, M., Matheu, A., Vergara, I., & Otaegui, D. (2021). Inflammaging markers characteristic of advanced age show similar levels with frailty and dependency. *Scientific Reports*, *11*(1), 4358. https://doi.org/10.1038/s41598-021-83991-7

Alexopoulos, G. S., & Kiosses, D. (2011). Geriatric Psychiatry, An Issue of Psychiatric Clinics (Illustrated edition). Saunders.

Aloi, G., Caliciuri, G., Fortino, G., Gravina, R., Pace, P., Russo, W., & Savaglio, C. (2016). A Mobile Multi-Technology Gateway to Enable IoT Interoperability. *2016 IEEE First International Conference on Internet-of-Things Design and Implementation (IoTDI)*, 259–264. doi:10.1109/IoTDI.2015.29

Andriopoulou, F., Dagiuklas, T., & Orphanoudakis, T. (2016). In G. Keramidas, N. Voros, & M. Hübner (Eds.), *Integrating IoT and Fog Computing for Healthcare Service Delivery* (pp. 213–232). Springer. https://link.springer.com/chapter/10.1007%2F978-3-319-42304-3_11

Anguera, J. A., & Gazzaley, A. (2015). Video games, cognitive exercises, and the enhancement of cognitive abilities. *Current Opinion in Behavioral Sciences*, *4*, 160–165. https://doi.org/10.1016/j.cobeha.2015.06.002

Annele, U., Satu, K. J., & Timo, E. S. (2019). Definitions of successful ageing: A brief review of a multidimensional concept. *Acta Bio Medica : Atenei Parmensis*, *90*(2), 359–363. https://doi.org/10.23750/abm.v90i2.8376

Baltes, P. B., & Baltes, M. M. (1990). Psychological perspectives on successful aging: The model of selective optimization with compensation. In Successful aging: Perspectives from the behavioral sciences (pp. 1–34). Cambridge University Press. https://doi.org/10.1017/CBO9780511665684.003.

Benalcázar, M. E., Jaramillo, A. G., Zea, A., Páez, A., & Andaluz, V. H. (2017). Hand gesture recognition using machine learning and the Myo armband. *2017 25th European Signal Processing Conference (EUSIPCO)*, 1040–1044. doi:10.23919/EUSIPCO.2017.8081366

Cheng, X., Yang, Y., Schwebel, D. C., Liu, Z., Li, L., Cheng, P., Ning, P., & Hu, G. (2020). Population ageing and mortality during 1990–2017: A global decomposition analysis. *PLoS Medicine*, *17*(6), e1003138. https://doi.org/10.1371/journal.pmed.1003138

Cotten, S. R., Ford, G., Ford, S., & Hale, T. M. (2014). Internet Use and Depression Among Retired Older Adults in the United States: A Longitudinal Analysis. *The Journals of Gerontology: Series B*, *69*(5), 763–771. doi:10.1093/geronb/gbu018

Cristea, M., Noja, G. G., Stefea, P., & Sala, A. L. (2020). The Impact of Population Aging and Public Health Support on EU Labor Markets. *International Journal of Environmental Research and Public Health*, *17*(4), E1439. https://doi.org/10.3390/ijerph17041439

Demiris, G., Thompson, H. J., Reeder, B., Wilamowska, K., & Zaslavsky, O. (2013). Using informatics to capture older adults' wellness. *International Journal of Medical Informatics*, *82*(11), e232–e241. https://doi.org/10.1016/j.ijmedinf.2011.03.004

Dolui, K., & Datta, S. K. (2017). *Comparison of edge computing implementations: Fog computing, cloudlet and mobile edge computing*. doi:10.1109/GIOTS.2017.8016213

Finck, M., & Pallas, F. (2020). They who must not be identified—Distinguishing personal from non-personal data under the GDPR. *International Data Privacy Law*, *10*(1), 11–36. doi:10.1093/idpl/ipz026

Fogel, A. L., & Kvedar, J. C. (2018). Artificial intelligence powers digital medicine. *NPJ Digital Medicine*, *1*(1), 1–4. https://doi.org/10.1038/s41746-017-0012-2

Ghosh, D., & Dinda, S. (2020). Determinants of the Quality of Life Among Elderly: Comparison Between China and India. *The International Journal of Community and Social Development*, *2*(1), 71–98. https://doi.org/10.1177/2516602620911835

Giuli, C., Papa, R., Lattanzio, F., & Postacchini, D. (2016). The Effects of Cognitive Training for Elderly: Results from My Mind Project. *Rejuvenation Research*, *19*(6), 485–494. https://doi.org/10.1089/rej.2015.1791

Gold, M. R., Stevenson, D., & Fryback, D. G. (2002). HALYS and QALYS and DALYS, Oh My: Similarities and differences in summary measures of population Health. *Annual Review of Public Health*, *23*, 115–134. https://doi.org/10.1146/annurev.publhealth.23.100901.140513

Guo, H. J., & Sapra, A. (2021). *Instrumental Activity of Daily Living*. StatPearls Publishing. https://www.ncbi.nlm.nih.gov/books/NBK553126/

Han, Q., Liang, S., & Zhang, H. (2015). Mobile Cloud Sensing, Big Data, and 5G Networks Make an Intelligent and Smart World. *IEEE Network*, *29*, 40–45. https://doi.org/10.1109/MNET.2015.7064901

Hoepman, J.-H. (2014). Privacy Design Strategies. In N. Cuppens-Boulahia, F. Cuppens, S. Jajodia, A. Abou El Kalam, & T. Sans (Eds.), ICT Systems Security and Privacy Protection (pp. 446–459). Springer. https://doi.org/10.1007/978-3-642-55415-5_38.

Hung, J.-Y., & Lu, K.-S. (2014). Research on the Healthy Lifestyle Model, Active Ageing, and Loneliness of Senior Learners. *Educational Gerontology*, *40*(5), 353–362. https://doi.org/10.1080/03601277.2013.822200

Ismail, H. (2015). Gait and Postural Sway Analysis, A Multi-Modal System. *Proceedings of the 2015 ACM on International Conference on Multimodal Interaction*, 629–633. doi:10.1145/2818346.2823310

Kaplan, B., & Harris-Salamone, K. D. (2009). Health IT Success and Failure: Recommendations from Literature and an AMIA Workshop. *Journal of the American Medical Informatics Association*, *16*(3), 291–299. https://doi.org/10.1197/jamia.M2997

Kastner, M., Hayden, L., Wong, G., Lai, Y., Makarski, J., Treister, V., Chan, J., Lee, J. H., Ivers, N. M., Holroyd-Leduc, J., & Straus, S. E. (2019). Underlying mechanisms of complex interventions addressing the care of older adults with multimorbidity: A realist review. *BMJ Open*, *9*(4), e025009. https://doi.org/10.1136/bmjopen-2018-025009

Khalili, P., Vazirinejad, R., Jamalizadeh, A., Nakhaee, P., Fatehi, F., & Fallah, A. (2014). The impact of healthy lifestyle education and integrated care on the quality of life of the elderly. *Journal of Occupational Health and Epidemiology*, *3*, 81–87. https://doi.org/10.18869/acadpub.johe.3.2.81

Kon, B., Lam, A., & Chan, J. (2017). *Evolution of Smart Homes for the Elderly*. doi:10.1145/3041021.3054928

Labbe, J. A. (2010). Health-Adjusted Life Expectancy: Concepts and Estimates. In V. R. Preedy & R. R. Watson (Eds.), Handbook of Disease Burdens and Quality of Life Measures (pp. 417–424). Springer. https://doi.org/10.1007/978-0-387-78665-0_23.

Lee, M., Kim, M. J., Suh, D., Kim, J., Jo, E., & Yoon, B. (2016). Feasibility of a Self-Determination Theory-Based Exercise Program in Community-Dwelling South Korean Older Adults: Experiences from a 13-Month Trial. *Journal of Aging and Physical Activity*, *24*(1), 8–21. https://doi.org/10.1123/japa.2014-0056

Li, R., Cheng, X., Schwebel, D. C., Yang, Y., Ning, P., Cheng, P., & Hu, G. (2021). Disability-adjusted life years associated with population ageing in China, 1990-2017. *BMC Geriatrics*, *21*(1), 369. https://doi.org/10.1186/s12877-021-02322-7

Longobucco, Y., Benedetti, C., Tagliaferri, S., Angileri, V., Adorni, E., Pessina, M., Zerbinati, L., Cicala, L., Pelà, G., Giacomini, V., Barbolini, M., Lauretani, F., & Maggio, M. (2019). Proactive interception and care of Frailty and Multimorbidity in older persons: The experience of the European Innovation Partnership on Active and Healthy Ageing and the response of Parma Local Health Trust and Lab through European Projects. *Acta Bio-Medica: Atenei Parmensis*, *90*, 364–374. https://doi.org/10.23750/abm.v90i2.8419

Lunenfeld, B., & Stratton, P. (2013). The clinical consequences of an ageing world and preventive strategies. *Best Practice & Research. Clinical Obstetrics & Gynaecology*, *27*(5), 643–659. https://doi.org/10.1016/j.bpobgyn.2013.02.005

Ma, X., Wang, Z., Zhou, S., Wen, H., & Zhang, Y. (2018). Intelligent Healthcare Systems Assisted by Data Analytics and Mobile Computing. *Wireless Communications and Mobile Computing*, *2018*, e3928080. https://doi.org/10.1155/2018/3928080

Martinez-Martin, P., Prieto-Flores, M.-E., Forjaz, M. J., Fernandez-Mayoralas, G., Rojo-Perez, F., Rojo, J.-M., & Ayala, A. (2012). Components and determinants of quality of life in community-dwelling older adults. *European Journal of Ageing*, *9*(3), 255–263. https://doi.org/10.1007/s10433-012-0232-x

Mary Zarate, J., Tian, X., Woods, K. J. P., & Poeppel, D. (2015). Multiple levels of linguistic and paralinguistic features contribute to voice recognition. *Scientific Reports*, *5*(1), 11475. https://doi.org/10.1038/srep11475

McMorrow, K., & Roeger, W. (n.d.). *The economic consequences of ageing populations.* European Commission. Retrieved September 23, 2021, from https://ec.europa.eu/economy_finance/publications/pages/publication_summary11156_en.htm

Meekes, W., & Stanmore, E. K. (2017). Motivational Determinants of Exergame Participation for Older People in Assisted Living Facilities: Mixed-Methods Study. *Journal of Medical Internet Research*, *19*(7), e238. https://doi.org/10.2196/jmir.6841

Moran, A., Gu, D., Zhao, D., Coxson, P., Wang, Y. C., Chen, C.-S., Liu, J., Cheng, J., Bibbins-Domingo, K., Shen, Y.-M., He, J., & Goldman, L. (2010). Future cardiovascular disease in china: Markov model and risk factor scenario projections from the coronary heart disease policy model-china. *Circulation: Cardiovascular Quality and Outcomes*, *3*(3), 243–252. https://doi.org/10.1161/CIRCOUTCOMES.109.910711

Mozos, O. M., Sandulescu, V., Andrews, S., Ellis, D., Bellotto, N., Dobrescu, R., & Ferrandez, J. M. (2017). Stress Detection Using Wearable Physiological and Sociometric Sensors. *International Journal of Neural Systems*, *27*(2), 1650041. https://doi.org/10.1142/S0129065716500416

Noura, M., Atiquzzaman, M., & Gaedke, M. (2019). Interoperability in Internet of Things: Taxonomies and Open Challenges. *Mob. Networks Appl.* doi:10.1007/S11036-018-1089-9

Prince, M. J., Wu, F., Guo, Y., Gutierrez Robledo, L. M., O'Donnell, M., Sullivan, R., & Yusuf, S. (2015). The burden of disease in older people and implications for health policy and practice. *Lancet*, *385*(9967), 549–562. https://doi.org/10.1016/S0140-6736(14)61347-7

Raggi, A., Corso, B., Minicuci, N., Quintas, R., Sattin, D., De Torres, L., Chatterji, S., Frisoni, G. B., Haro, J. M., Koskinen, S., Martinuzzi, A., Miret, M., Tobiasz-Adamczyk, B., & Leonardi, M. (2016). Determinants of Quality of Life in Ageing Populations: Results from a Cross-Sectional Study in Finland, Poland and Spain. *PLoS One*, *11*(7), e0159293. https://doi.org/10.1371/journal.pone.0159293

Ronteltap, A., & van Trijp, H. (2007). Consumer acceptance of personalised nutrition. *Genes & Nutrition*, *2*(1), 85–87. https://doi.org/10.1007/s12263-007-0003-z

Schoene, D., Heller, C., Aung, Y. N., Sieber, C. C., Kemmler, W., & Freiberger, E. (2019). <p>A systematic review on the influence of fear of falling on quality of life in older people: Is there a role for falls?</p>. *Clinical Interventions in Aging*, *14*, 701–719. https://doi.org/10.2147/CIA.S197857

Schwarzer, R., & Luszczynska, A. (2008). How to overcome health-compromising behaviors: The health action process approach. *European Psychologist*, *13*(2), 141–151. https://doi.org/10.1027/1016-9040.13.2.141

Sims, T., Reed, A. E., & Carr, D. C. (2017). Information and Communication Technology Use Is Related to Higher Well-Being Among the Oldest-Old. *The Journals of Gerontology: Series B, 72*(5), 761–770. doi:10.1093/geronb/gbw130

Talarska, D., Tobis, S., Kotkowiak, M., Strugała, M., Stanisławska, J., & Wieczorowska-Tobis, K. (2018). Determinants of Quality of Life and the Need for Support for the Elderly with Good Physical and Mental Functioning. *Medical Science Monitor*, *24*, 1604–1613. https://doi.org/10.12659/MSM.907032

Tolba, A., Al-Makhadmeh, Z., & Hussein, A. (2018). Montreal Imaging Stress Task Based Human Stress and Strain Analysis Using Image Processing Techniques. *Journal of Medical Imaging and Health Informatics*, *8*(4), 775–780. https://doi.org/10.1166/jmihi.2018.2349

Urtamo, A., Huohvanainen, E., Pitkälä, K. H., & Strandberg, T. E. (2019). Midlife predictors of active and healthy aging (AHA) among older businessmen. *Aging Clinical and Experimental Research*, *31*(2), 225–231. https://doi.org/10.1007/s40520-018-1100-0

Vo, N. X., Vo, T. Q., Watanapongvanich, S., & Witvorapong, N. (2019). Measurement and Determinants of Quality of Life of Older Adults in Ho Chi Minh City, Vietnam. *Social Indicators Research*, *142*(3), 1285–1303. https://doi.org/10.1007/s11205-018-1955-7

Vrdoljak, D., Marković, B., Puljak, L., Lalic, D., Kranjcevic, K., & Vucak, J. (2013). Lifestyle intervention in general practice for physical activity, smoking, alcohol consumption and diet in elderly: A randomized controlled trial. *Archives of Gerontology and Geriatrics, 58*. doi:10.1016/j.archger.2013.08.007

Wallace, E., Salisbury, C., Guthrie, B., Lewis, C., Fahey, T., & Smith, S. M. (2015). Managing patients with multimorbidity in primary care. *BMJ (Clinical Research Ed.)*, *350*, h176. https://doi.org/10.1136/bmj.h176

Walther, D., Curjuric, I., Dratva, J., Schaffner, E., Quinto, C., Schmidt-Trucksäss, A., Eze, I. C., Burdet, L., Pons, M., Gerbase, M. W., Imboden, M., Schindler, C., & Probst-Hensch, N. (2017). Hypertension, diabetes and lifestyle in the long-term—Results from a Swiss population-based cohort. *Preventive Medicine*, *97*, 56–61. https://doi.org/10.1016/j.ypmed.2016.12.016

Wang, C., Lu, W., Narayanan, M., Chang, D., Lord, S., & Lovell, N. (2016). Low-Power Fall Detector Using Triaxial Accelerometry and Barometric Pressure Sensing. *IEEE Transactions on Industrial Informatics*, *12*, 1–1. https://doi.org/10.1109/TII.2016.2587761

Wiesner, G., & Bittner, E. K. (2004). Life expectancy, potential years of life lost (PYLL), and avoidable mortality in an East/West comparison. *Bundesgesundheitsblatt, Gesundheitsforschung, Gesundheitsschutz*, *47*(3), 266–278. https://doi.org/10.1007/s00103-003-0793-0

Zangróniz, R., Martínez-Rodrigo, A., Pastor, J. M., López, M. T., & Fernández-Caballero, A. (2017). Electrodermal Activity Sensor for Classification of Calm/Distress Condition. *Sensors (Basel)*, *17*(10), 2324. https://doi.org/10.3390/s17102324

KEY TERMS AND DEFINITIONS

Ageing Population: Usually people over 60 years old, which is considered by healthcare as a critical age limit for the emergence of age-related impacts on everyday life.

Disability-Adjusted Life Years (DALYs): Life years either lost due to premature mortality or lived under poor health.

Emotional Computing: Development of software and hardware modules for recognizing human emotion and general psychological status.

Heath-Adjusted Life Expectancy (HALE): The lifespan of a person, mathematically adjusted for years lived under health burden.

Lifestyle Interventions: Nonpharmacological interventions revolving around behavioral changes towards the adoption of new habits, usually aiming for a positive impact on quality of life.

Potential Years of Life Lost (PYLLs): The number of years a person has not lived with reference to their average expected lifespan.

Quality-Adjusted Life Years (QALYs): Years lived while maintaining functional capacity and a high degree of independence despite potential health issues.

Successful Ageing: A holistic concept representing the maintenance of adequate physical, cognitive, social, and emotional functioning levels over the course of ageing, adopting a healthy and active lifestyle for mitigating the impacts of natural decline.

Chapter 11
Contact Tracing Apps in the COVID-19 Pandemic:
Exploring the Underlying Personal Data Processing

Natalia Baxevanou
University of Peloponnese, Greece

Sotiria Triantafyllia Sotirhou
https://orcid.org/0000-0003-3293-8397
Open University of Cyprus, Cyprus

Konstantinos Limniotis
University of West Attica, Greece & Open University of Cyprus, Cyprus

ABSTRACT

Contact tracing apps for the COVID-19 pandemic, from a personal data protection point of view, are discussed in this chapter. More precisely, having the relevant European legal framework as a basis, the authors investigate the underlying personal data processing that occurs from these apps in relation with the relevant privacy risks that are inherent in the smart mobile ecosystem. The analysis indicates that there exist several different approaches across the world, whereas even in countries with unified legal framework there exist some discrepancies in the design philosophy and the operation of these apps. Since user trust is essential for the effectiveness of such types of apps, they discuss the importance of respecting the fundamental right of privacy when developing and operating such applications. The ultimate goal is to illustrate the need to strike the proper balance between substantial public interests and the right to the protection of personal data.

INTRODUCTION

The year 2020 has been marked by an unexpected pandemic throughout the world. The new coronavirus,

DOI: 10.4018/978-1-7998-9190-1.ch011

known as COVID-19, is spreading out without facing any borders, constituting a global pandemic, as WHO (World Health Organisation) declared on March 11th, 2020. Under this new humanitarian crisis, governments put much effort on employing appropriate technologies to help striking the virus dissemination (Nabben et al 2020). In this framework, a technology that has been adopted by several countries throughout the world is the so-called contact tracing applications. The main motivation under such smart apps is to ensure an automatic detection of the contacts of an infected individual, thus allowing the competent health authorities to trace the chain of infection and proceed with appropriate actions.

There are several different approaches that can be followed in developing such smarts apps. A main classification rests with the fact of their centralized nature – i.e. some of them are centralized, in the sense that non-permanent identifiers of all smart devices in proximity to the user are stored and further processed on a centralized server operated by the public health authorities, whereas other apps have a de-centralized flavor, since the non-permanent identifiers of all devices in contact are stored and further processed on the user's device. However, the distinction between the two classes is becoming less and less obvious, since a backend server is actually always present. Another way to classify the contact tracing apps is based on the underlying technology that is being used to find devices in proximity, since there are two main approaches: a Bluetooth-based (i.e. Bluetooth Low Energy – BLE) and a location-based approach.

A major concern that arises in the use of such contact tracing apps is the users' privacy. This is actually an issue spanning the whole mobile ecosystem for any application, since there exist well-known privacy threats that are related with the lack of transparency (i.e. the users do not have adequate information on which of their personal data are being processed and for what purposes), as well as with the overall permission model of the smart applications (i.e. users are often "forced" to allow permissions to the smart apps even if the necessity of such permissions is questionable, whereas the provision of such a permission actually allows access not only to the smart app provider but also possibly to third-party trackers). All these well-known privacy issues are clearly further accentuated in the context of contact tracing, since they process sensitive health data. Without careful consideration, tracking contacts may become highly intrusive in terms of privacy, thus resulting in high risks for the rights and freedoms of individuals.

In this chapter, an extensive study on the currently most known Android contact tracing apps throughout the world is presented, in terms of their privacy features. The study will be based mainly on examining these apps into a real-time operation, through appropriate software tools that allow for their dynamic analysis, with the aim to determine what these applications do in real time. The outcome of the analysis indicates several discrepancies amongst the solutions adopted by several countries, which actually reflect somehow the corresponding legal framework that is applicable each time. However, even in countries with a unified legal framework, different applications with different characteristics are being employed, with different levels of privacy protection. A main threat that is present in some of them is the use of third party trackers without being transparent to the users, whereas some apps require specific permissions to operate smoothly whose necessity is not justified or clear. Bearing in mind that citizens' trust is essential to ensure the effectiveness of such apps, whereas personal data protection is indispensable to build trust, it becomes evident that respecting the individuals' privacy is of utmost importance; to this end, a discussion on how the possible privacy issues could be alleviated is also given.

It should be pointed out that this chapter does not aim to evaluate the effectiveness of apps in terms of addressing the virus dissemination; in the same framework, practical issues such as their usability are out of our scope. Hence, this chapter should not be considered as a comparative study of the contact tracing apps. Our ultimate goal is to examine their privacy features, regardless their type and their underlying technologies, taking into account known privacy issues that span the entire mobile ecosystem.

BACKGROUND

COVID-19 is a highly contagious virus, whereas a COVID-19 patient may be infectious even if she/he has not any symptoms. Hence, once an individual finds out that she/he is positive to this virus (e.g. through a COVID-19 test), she/he may have already pose at risk several other persons who came in contact with her/him recently. That's why the notion of *contact tracing* has been introduced, to describe the process to identify individuals who had close contact with the positive carrier, as these individuals themselves may now be infected (Ahmed et al 2020). Contact tracing can be accomplished through a manual interview of the infected individuals, under the supervision of the competent health authorities. However, this manual approach has its own limitations (e.g. the carrier may have forgotten some of her/his contacts or may not actually know their exact identities) and, thus, deployment of automated solutions for this task has been initiated. To this goal, smart applications being able to "capture" and "store" such contacts have been developed, being called *contact tracing apps.*

Due to their importance, contact tracing apps have attracted much attention since 2020. A comprehensive survey of such apps in terms of a wide range of their features, such as system architecture, data management, privacy, security, proximity estimation, and attack vulnerability has been performed (Ahmed et al 2020). Another survey that provides a classification of such apps regarding security and privacy criteria has been also presented (Reichert et al 2020). The dilemma of whether such apps should have centralized or decentralized nature has been also studied (Vaudenay 2020); interestingly enough, it is stated therein that none offers a sufficient level of privacy protection. Generic security and privacy risks of contact tracing apps have been also discussed (Baumgärtner et al 2020), whereas for the specific technology that is jointly developed by Apple Inc. and Google Inc., which is supposed to put emphasis on users security and privacy, a security analysis has been conducted (Gvili 2020).

This chapter studies privacy issues of contact tracing apps, in terms of identifying, through appropriate software tools, the underlying personal data processing that takes place. As a basis, we consider the corresponding legal framework in Europe, in which the main legal instrument is the General Data Protection Regulation (GDPR), establishing a framework that harmonises data protection across the European Union. The GDPR also sets the so-called European Data Protection Board (hereinafter EDPB), which is an independent European body whose purpose is to ensure consistent application of the Regulation, as well as to issue guidelines for important matters towards promoting cooperation among the EU's data protection authorities. With regard to contact tracing apps, the EDPB has issued guidelines on the use of location data and contact tracing tools in the context of the COVID-19 outbreak (European Data Protection Board 2020); therefore, in this work we use these guidelines as a baseline. Up to our knowledge, such an extensive technical analysis of these apps, aiming to analyze the underlying personal data processing, has not been conducted yet – an exemption being a work which describes, from a technical perspective, the data transmitted to backend servers by specific contact tracing apps utilizing the so-called Google/Apple Exposure Notification (GAEN) service on Android devices (Leith & Farrell 2021) – identifying therein that this component is "*extremely troubling from a privacy viewpoint*".

PRIVACY ISSUES IN CONTACT TRACING APPS

The Notion of Personal Data Protection

The right to privacy has been recognized as a fundamental human right by the United Nations Declaration of Human Rights, the International Covenant on Civil and Political Rights, the Charter of Fundamental Rights in European Union and other international treaties. Privacy is strongly related with the personal data protection; as stated in the Charter of Fundamental Rights, personal data (i.e. data relating to an identified or identifiable natural person) must be processed fairly for specified purposes and on the basis of the consent of the person concerned or some other legitimate basis laid down by law under Article 8(2) of the Charter.

In Europe, as stated above, the main legal instrument for the personal data protection is the General Data Protection Regulation (GDPR), being applicable since 25th of May, 2018. The GDPR also applies to the processing of personal data by organizations not established in the Union, once they process data of individuals being in the European Union and the processing activities are related to either the offering of goods or services or the monitoring of their behavior (as far as their behavior takes place within the European Union).

According to the GDPR, the term *personal data* refers to any information relating to an identified or identifiable natural person, that is a person who can be identified; as it is explicitly stated in the GDPR, an identifiable natural person is one who can be identified, directly or indirectly, in particular by reference to an identifier such as a name, an identification number, location data, an online identifier or to one or more factors specific to the physical, physiological, genetic, mental, economic, cultural or social identity of that natural person. *Personal data processing* means any operation that is performed on personal data, including the collection, recording, structuring, storage, adaptation or alteration, retrieval, use, disclosure by transmission, dissemination, combination and erasure. The entity that, alone or jointly with others, determines the purposes and means of the processing of personal data, is the so-called *data controller*, whereas the entity which processes personal data on behalf of the controller is the *data processor*.

The notion of the personal data is wide. Indeed, device identifiers are generally being considered as personal data since they may allow – in combination with other information - the identification of a user. For example, for an Android device, the Android ID which is a permanent identifier, as well as the Google Advertising ID (GAID), which is a 32-digit alphanumeric identifier that can be reset at any time, constitute personal data, despite the fact that such single sequences of alphanumeric characters by themselves do not (normally) allow to identify the device's owner. Similarly, other device or network identifiers are also considered as personal (and not anonymous) data.

The GDPR codifies the basic principles that need to be in place when personal data are being processed, setting specific obligations to data controllers/data processors. According to the article 5 of the GDPR, the personal data shall be processed fairly and in a transparent manner, being adequate, relevant and limited to what is necessary in relation to the purposes for which they are processed (purpose limitation principle), whereas the purpose shall be specified, explicit and legitimate. Moreover, any processing of personal data requires a lawful basis, based on those described in the article 6 of the GDPR. In case that such a lawful basis is the individual's consent, then consent must meet certain requirements in order to be considered as valid – i.e. consent needs to be a freely given, specific, informed and unambiguous indication of the individual's agreement to the processing of his or her personal data, whereas it must be given by a statement or a clear affirmative action (art. 4 of the GDPR).

Some types of personal data are being mentioned as *special categories of personal data* (also known as *sensitive data*); for such data, there exist stricter requirements for their legitimate processing (see article 9 of the GDPR). It should be pointed out that health data – such as COVID-19 related data – are sensitive.

An important requirement for a legal personal data processing according to the GDPR is the so-called data protection by design principle (also known as privacy by design). According to this principle, the data controller shall, both at the time of the determination of the means for processing and at the time of the processing itself, implement appropriate technical and organisational measures, such as pseudonymisation, which are designed to implement data protection principles, such as data minimisation, in an effective manner and to integrate the necessary safeguards into the processing in order to meet the legal requirements and protect the rights of individuals. In simple work, privacy should be taken into account from the very first beginning of the design of a data processing – and this actually covers the design of smart apps.

It should be stressed that, depending on the techniques used, tracking of a mobile user may fall into the scope of the legal framework of the so-called ePrivacy Directive (Directive 2002/58/EC); this applies only to the European Union. Again, the notion of informed consent is also introduced therein as a possible legal basis for such a processing.

Privacy Issues in Android Apps

The mobile ecosystem, despite its well-known advantages in terms of the provided services, also raises several privacy concerns (European Union Agency for Cybersecurity 2017). We next focus on the Android environment, which constitutes the most popular platform (Statista 2021). A smart application explicitly requests the permissions needed either at install-time or at run-time. Permissions are divided into three protection levels, as follows:

- **Normal permissions** cover areas where the app needs to access data or resources outside the app's sandbox, but where there's very little risk to the user's privacy or the operation of other apps. For example, permission to set the time zone is a normal permission. If an app declares in its manifest that it needs a normal permission, the system automatically grants the app that permission at install time. The system doesn't prompt the user to grant normal permissions, and users cannot revoke these permissions
- **Signature permissions** where the system grants these app permissions at install time, but only when the app that asks for the permission is signed by the same certificate as the app that defines the permission. Some signature permissions aren't for use by third-party apps.
- **Dangerous permissions** cover areas where the app requests data or resources that involve the user's private information, or could potentially affect the user's stored data or the operation of other apps. For example, the ability to read the user's contacts is a dangerous permission. If an app declares that it needs a dangerous permission, the user has to explicitly grant the permission to the app. Until the user approves the permission, the app cannot provide the functionality that depends on that permission.

Only dangerous permissions require user agreement. The way Android asks the user to grant dangerous permissions depends on the version of Android running on the user's device, and the system version targeted by the app.

Granting a (dangerous or not) permission to an app is somehow related with provision of the user's consent since, if the user does not grant the relevant permissions, the app cannot access the corresponding resources, including device/personal data. However, in practice, the user is actually "forced" to grant permissions, since otherwise the app cannot operate properly; therefore, the consent cannot constitute a legal basis for such a processing, since it is not freely given. In addition, it is not always clear whether the permissions granted are indeed strictly necessary for the functionality that the user asks, since the apps may require more permissions than actually needed to functioning properly. For example, it has been shown that there exists a discrepancy between the permissions requested by different apps lying in the same category of GPS (Global Positioning System) navigation (Monogios et al 2019).

Apart from the above issues, it should be also taken into account that the majority of the smart apps utilize third-party libraries for several purposes - e.g. to facilitate the programming procedure. In an Android environment, these libraries obtain the same access rights with the host app. Unfortunately, this may not be transparent to the user and, thus, users have limited understanding of the associated risks of enabling permissions in apps (European Union Agency for Cybersecurity 2017). Clearly, this permissions model does not facilitate the provision of a legally valid consent for a third-party. Moreover, the use of several libraries by several different smart apps may result in the so-called *intra-library collusion* issue (Taylor et al 2017), that is the case that if a single library is embedded in several different apps on a device, then its third-party provider actually leverages the acquired privileges and is able to gather (a possibly large amount of) personal information without the user being aware of this. More specifically, the current Android security model, which does not support the separation of privileges between apps and the embedded libraries, facilitates the following relative privacy threats without the user's consent (Taylor et al 2017):

1. Libraries may abuse the privileges granted to the host applications.
2. Libraries may track the users.
3. Libraries may aggregate multiple signals for detailed user profiling

Concluding, since many smart apps will need to rely on users' consent for the processing of certain personal data, the requirement of consent deserves special attention, in particular as it relates to the issue of permissions (European Union Agency for Cybersecurity 2017).

Analyzing Contact Tracing Apps

As it is stated previously, the main legal instrument for the personal data protection in Europe is the General Data Protection Regulation (GDPR). The GDPR sets the so-called European Data Protection Board, which is an independent European body whose purpose is to ensure consistent application of the GDPR, issuing also guidelines for important matters towards promoting cooperation among the EU's data protection authorities.

With regard to contact tracing apps, the EDPB has issued guidelines on the use of location data and contact tracing tools in the context of the COVID-19 outbreak (European Data Protection Board 2020). As it is stated in these guidelines, the EDPB firmly believes that, when processing of personal data is necessary for managing the COVID-19 pandemic, data protection is indispensable to build trust, create the conditions for social acceptability of any solution, and thereby guarantee the effectiveness of these measures. Moreover, data and technology used to help fight COVID-19 should be used to empower, rather than to control, stigmatise, or repress individuals. As it is also stated in these guidelines, the EDPB takes position on the fact that the use of contact tracing applications should be voluntary and should not rely on tracing individual movements but rather on proximity information regarding users. In addition, the EDPB emphasizes that when it comes to using location data, preference should always be given to the processing of anonymised data rather than personal data, whereas special emphasis is being put on how cautious one should be in order to characterize data as anonymous; more precisely, at is it explicitly stated therein,

Indeed, a large body of research has shown that location data thought to be anonymised may in fact not be. Mobility traces of individuals are inherently highly correlated and unique. Therefore, they can be vulnerable to re-identification attempts under certain circumstances (European Data Protection Board 2020)

The EDPB also gives specific recommendations for developing and using contact tracing apps, towards respecting the protection of personal data. An indicative sub-list of these recommendations follows (the reader may see (European Data Protection Board 2020) for the whole list):

1. The application must be able to broadcast and receive data via proximity communication technologies like Bluetooth Low Energy so that contact tracing can be carried out.
2. This broadcast data must include cryptographically strong pseudorandom identifiers, generated by and specific to the application.
3. Pseudorandom identifiers must be renewed regularly, at a frequency sufficient to limit the risk of re-identification, physical tracking or linkage of individuals, by anyone including central server operators, other application users or malicious third parties. These identifiers must be generated by the user's application, possibly based on a seed provided by the central server.
4. According to the data minimization principle, the application must not collect data other than what is strictly necessary for the purpose of contact tracing.
5. The application must not collect location data for the purpose of contact tracing. Location data can be processed for the sole purpose of allowing the application to interact with similar applications in other countries and should be limited in precision to what is strictly necessary for this sole purpose.
6. Users must be informed of all personal data that will be collected. This data should be collected only with the user authorization.

It should be pointed out that many contact tracing apps claim that they process anonymous data. However, this is not accurate, according to the GDPR. More precisely, the Recital 26 of the GDPR states:

To determine whether a natural person is identifiable, account should be taken of all the means reasonably likely to be used, such as singling out, either by the controller or by another person to identify the natural person directly or indirectly. To ascertain whether means are reasonably likely to be used to identify the natural person, account should be taken of all objective factors, such as the costs of and the amount of time required for identification, taking into consideration the available technology at the time of the processing and technological developments.

Therefore, one should be very careful when characterizing data as anonymous - i.e. non-personal (Chatzistefanou & Limniotis 2017). That's why the EDPB, as it is described above, refers to "pseudorandom identifiers"; these actually play the role of pseudonyms and pseudonymous data, as the GDPR also explicitly states, should be considered as personal data (despite the fact that pseudonymisation is a nice privacy technique that should be considered as a possible option towards fulfilling the data protection by design principle). In this chapter, we do not discuss the strength of the pseudonymisation performed by the contact tracing apps; however we point out that if the app states that it processes anonymous information, then this information may be somehow misleading.

Testing Environment

For our analysis, we utilized two tools. The first one is the so-called Exodus Privacy (Exodus Privacy 2021), supported by a French non-profit organization, which analyses Android applications by looking for embedded trackers and lists them as an output report. Each report contains which trackers are embedded into the application. For every tracker, the user is able to view more details on this tracker. A report also provides the list of permissions the application requests to operate on the smartphone. Some of these permissions are tagged as dangerous in terms of privacy. Exodus does not decompile applications.

The second tool has been created by the Privacy International (PI), being an interception environment released to the public. PI a registered charity based in London that works at the intersection of modern technologies and rights. This interception environment has been used by PI during its research on apps interactions with Facebook on Android, where interesting findings were published (Privacy International 2021). This environment will be our basic tool during our research for perfoming dynamic analysis of the applications. This environment actually constitutes a virtual machine. Two Android handsets, running Android 7.1.2 and Android 6.0.1 were also used. The Android mobile operating system (OS), developed by Google, is the most popular smartphone operating system currently in use (Statista 2021). Note that the permissions that each app requires, as well as its embedded trackers, are not dependent on the version of the Android host system.

This toolkit has been built taking into account the trust paradigm used extensively on the Internet, which is based on encrypted - mainly through the Transport Layer Security (TLS) protocol - connections. When secure connections are established, the client checks against an internal store of "trust anchors" in its "trust store" - known as certificate authorities (CAs for short). CAs are trusted in most operating systems through a number of methods, predominantly via commercial agreements. This toolkit introduces a CA that will be added to the "trust store", allowing us to intercept – for our research purposes - secure traffic in transit as the client now trusts this CA in addition to the preconfigured ones. To this end, an open-source proxy (intermediary) server written in Python, being called mitmproxy, is being used. In essence, it takes connections on one side, opens (decrypts) them up for analysis, and then forwards them (encrypted) to the other side. Its key feature is that it generates any required certificates in real-time. This ensures that the incoming connection (from the client), believes that the proxy is the real destination. In this pre-configured environment, mitmproxy is configured in "transparent" mode. This means that no additional settings need to be changed on the client, other than the installation of mitmproxy's certificate to the Android devices. This is where the term man-in-the-middle (MITM) comes from, as we are somehow putting ourselves between the client and the server, intercepting and analyzing any communications between them. There is also a number of other components inside this toolkit which work together to allow for interception of data in transit.

Main Findings

We first used a sample of fifty three (53) Android apps from Google Play Store that are COVID-19 related, in order to perform an analysis for each app keeping count of the embedded trackers and permissions access requests. These apps are being presented in Appendix. While reviewing the information available in Google Play, we identified three main categories to which each app can be associated – although there exist some overlaps amongst them:

1. **Information broadcast:** Apps in this category aim to provide users with various types of information, like general guidelines, infection statistics or general Covid-19 news. Although such apps are not always officially released by government bodies, they often relay official information from authorities to users. This category does not actually include contact tracing apps.
2. **Upstream collection:** Apps in this category collect information from users and make it available to the developer and/or an official body, such as a country's health authorities.
3. **Tooling:** Apps in this category serve as tools with functionalities that directly deal with daily aspects of the Covid-19.

After our first analysis, which took place in September 2020, it was observed that 20 different trackers were found that are being used by some of these 53 apps. The most popular trackers were Google Firebase Analytics (being met in 35 apps), Google AdMob (being met in 23 apps) and Google CrashLytics (being met in 16 apps). The maximum number of trackers per app is 7 and this was the case of two apps. It should be also pointed out that 17 apps (about 32% of the apps studied) were found with no embedded trackers; these are described in Table 1, where the number of permissions required from each of them is also presented; a special reference to the so-called dangerous permissions is also given.

Interestingly enough, as stated above, 23 apps (i.e. the 43% of the apps examined) included – in September of 2020 - the Google AdMod, a free mobile advertising platform, along with other trackers; all these 23 apps also simultaneously incorporated the Google Firebase Analytics. These are shown in Table 2. Apart from the total number of trackers, the dangerous permissions required by each of them are also presented.

However, in subsequent versions of the aforementioned applications (i.e. those in Table 2) that were examined in March 2021, the Google AdMod tracker has been removed in almost all of them, thus illustrating that the app providers seem to put some effort on reducing privacy risks (although it is unclear why such a tracker was initially in place and which was the underlying data processing that took place due to this). Only two apps still include the Google AdMod tracker, namely the "COVID-19" from Vietnam and the "Track Virus" from Israel. On the other side, the tracker Google Firebase Analytics (which is an analytics service, providing insight on app usage and user engagement) is still present in all apps presented in Table 2 – an exemption being the Spanish app "Radar Covid", which does not have trackers any more. The Google Firebase Analytics tracker is also present in several other apps not shown in Table 2 (see the subsequent discussion).

It should be pointed out that if an app asks for a dangerous permission, this by itself does not mean that the app does not respect user's privacy. However, according at least to the European legal framework, the necessity of such permissions with respect to the desired purpose should be fully justified under the privacy by design principle, as well as to be clearly notified to the users; this also includes the case that third parties obtain – due to the incorporation of third party libraries – the same permissions. Therefore,

Table 1. Apps with no third-party trackers

App	Authority	# of Permissions	Dangerous permissions
Corona-Warn-App	Robert Koch-Institut, Germany	8	Camera
COVID-19!	Nemocnice Milosrdnych bratri p.o., Czech	3	Access Coarse Location Access Fine Location
COVIDWISE	VDH, USA	6	-
COVID Tracker Ireland	Health Service Executive (HSE), Ireland	5	-
COVID Alert Malta	Government of Malta	7	-
StopCovid France	Government of France	11	Camera Access Coarse Location Access Fine Location
CoronaCheck	Aga Khan University, Pakistan	1	-
Coronavirus Support App (UK)	Expert Self Care Ltd, UK	3	Call Phone
COVID Alert South Africa	Department of Health - South Africa	6	-
STAYAWAY COVID	FCT FCCN, Portugal	7	-
NHS COVID-19	Department of Health and Social Care, England & Wales	7	Camera
COVID-19 Sounds	cam-cl-mobile-systems, UK	11	Access Coarse Location Access Fine Location Get Accounts
TrackTogether	TrackTogether	9	Access Coarse Location Access Fine Location Write External Storage Read External Storage
COVID Alert	Health Canada I Santé Canada	8	-
SwissCovid	Federal Office of Public Health FOPH, Swiss	7	-
Contact Tracing	Interactive Health Solutions, Global	4	-
Immuni	Italian Extraordinary Commissioner for the COVID-19 Emergency, in collaboration with the Ministry of Health and the Ministry for Innovation Technology and Digitalization	6	-

Table 2 should not be understood as a list of apps that do not respect user's privacy; however, these apps clearly raise some concerns and, thus, each of them needs careful assessment (with the exemption, as already stated, of the Spanish app "Radar Covid", which does not have any tracker now neither requires dangerous permissions). For example, the necessity of accessing the camera that is required by many of them seems to be questionable. Such privacy concerns become more obvious from the subsequent analysis.

Table 2. Apps initially including both the Google AdMob and the Google Firebase Analytics trackers

App	Authority/Organization	# of trackers (including the two Google trackers)	Dangerous permissions
WHO Info	World Health, Global Organization	2	Read External Storage Write External Storage Read Calendar Write Calendar
OpenWHO	HPI Knowledge Engineering Team, Global	3	Read External Storage Write External Storage
COVID-19	Electronic Health Administration, Vietnam	7	Write Settings Read External Storage Write External Storage System Alert Window Record Audio Read Phone State Camera Access Coarse Location Access Fine Location
Stop COVID-19	Ministry of Health Croatia	3	-
Track Virus	Panda O.S. Ltd., Israel	3	Write External Storage Camera Call Phone Access Coarse Location Access Fine Location
Service NSW	Service NSW, Australia	3	Write Settings Read External Storage Write External Storage Camera
COVID19 - DXB Smart	Dubai Health Authority - DHA	4	Write External Storage System Alert Window Record Audio Camera Access Coarse Location Access Fine Location Write Calendar Read Calendar
CDC	Centers for Disease Control and Prevention, USA	4	Write External Storage Write Settings
Radar COVID	Ministry of Economic Affairs and Digital Transformation, Spain	2	-
Covid SafePass	Medek Health Systems LLC, USA	3	Read External Storage Write External Storage Access Coarse Location Access Fine Location Call Phone
Canada COVID-19	Health Canada | Santé Canada	4	Read External Storage Access Coarse Location Access Fine Location Camera
Āwhina - informing health workers	Ministry of Health New Zealand	3	Write Settings Read External Storage Write External Storage
Covid19-MY	Qualitas Medical Group, Malaysia	2	Read External Storage Access Coarse Location Access Fine Location Camera
TraceTogether	Government Technology Agency, Singapore	3	Read External Storage Access Fine Location Camera
CovTracer	RISE Research Centre of Excellence, Cyprus	2	Write Settings Write External Storage Access Coarse Location Access Fine Location
Covid Locator	Intugine Technologies	3	Read Phone State Access Fine Location
myGovID	Australian Taxation Office	2	Camera
ZeroCOVID	Athleten Foundation	7	Access Coarse Location Access Fine Location Camera Read External Storage Write External Storage
Stop COVID	Ministry of IDPs, Labour, Health, Social Affairs, Georgia	4	Access Coarse Location Access Fine Location
Rakning C-19	Icelandic Government	5	Access Coarse Location Access Fine Location Read External Storage Write External Storage

continued on following page

Table 2. Continued

App	Authority/Organization	# of trackers (including the two Google trackers)	Dangerous permissions
BC COVID-19 Support	Province of British Columbia, Canada	4	Access Coarse Location Access Fine Location Camera Read External Storage
COVID-19 Andhra Pradesh	Entro Labs, India	3	Access Coarse Location Access Fine Location Read External Storage Write External Storage Camera
JamCOVID19	opmjm mobile	3	Call Phone Access Coarse Location Access Fine Location Read External Storage Write External Storage Camera

Further Issues from the Dynamic Analysis of Apps

We subsequently proceeded in analyzing some important contact tracing apps by monitoring their behavior in real time. This second round of analysis took place within March 2021. The choice of this particular set of apps has been done so as to ensure that all the main categories of contact tracing apps are being represented (i.e. either centralized or not, European or not, with several contact tracing protocols etc.).

Our results are next given, for each app examined.

1. **STOP COVID – ProteGO Safe:** This is a decentralized app from Poland. There is no registration to a server. A risk score, in terms of probability of infection, is being computed locally. The app sends cryptography-based pseudonyms to other devices through the BLE. In case that the user is identified – through an official COVID-19 test – as infected, she/he may enter this information to the app. The data is being stored for 14 days, whereas the app is applicable to European Union.
   The app does not ask for any dangerous permission. The app though utilizes the Google Firebase tracker. Our analysis through the PI environment illustrates that info such as our country and our device, including also the Android ID, are being transmitted to Google.
2. **Covid19-DXB Smart App:** This is a centralized app from Dubai, in which the user is being registered by entering her/his personal data (first name, last name, passport number and phone number). This is a simple smart application that helps user get information about COVID-19 symptoms and what treatments can be adopted to reduce the risk.
   As already described in Table 2, the app asks for several dangerous permissions, whereas it still incorporates the Google Firebase tracker. Our analysis through the PI environment illustrates that info such our country and our device, including also the Android ID, are being transmitted to Google. By also clicking the "get help" option, the GPS service is starting.
3. **NOVID:** This is a decentralized app from the United States and it does not allow any user to declare that she/he is COVID-19 positive unless a certified diagnostic Center provides such information. NOVID is mainly based on BLE for informing user whether a nearby user is/was infected and it does not collect user identifying information, as it claims. The app asks the "Record Audio" permission which is considered as dangerous, as well as the typical permissions for location (i.e. "Access Coarse Location" and "Access Fine Location"), despite the fact that GPS is not being used for contact tracing.
   The app also incorporates the Google Firebase tracker. Our analysis through the PI environment illustrates that the model of our device, the version of the Android operating system as well

as the Google Advertising ID, are being transmitted to Google. We also noticed that the app sends information of our Wi-FI (SSID), our MAC address, as well as - surprisingly - of the nearby SSIDs. The app also gives the option to the user to enter information of one or more of her contacts (name and e-mail address); we entered such dummy information and we noticed that this has been forwarded to the app's provider (we don't know if such a contact becomes aware of this processing of her data).

4. **SwissCovid:** This is a decentralized app from Switzerland, based on the open protocol DP-3T (Decentralized Privacy-Preserving Proximity Tracing) which uses BLE to track and log encounters with other users. The app does not allow any user to declare that she/he is COVID-19 positive unless a certified diagnostic Center provides such information. Pseudonyms of other users are being stored locally for 14 days.
The app does not ask for any dangerous permission, whilst it does not use any third party tracker. This was confirmed from our analysis through the PI tool.
5. **Immuni:** This is a decentralized app from Italy, based on BLE and making use of the Apple/Google Exposure Notification system. There is no registration to a server; all data is stored locally. The app does not use GPS; the device is connected once a day to a server to download the pseudonyms (cryptographically generated) of infected users.
The app does not ask for any dangerous permission, whilst it does not use any third party tracker. This was confirmed from our analysis through the PI tool.
6. **Corona-Warn-App:** This is a decentralized app from Germany. Similarly to the Immuni app, it is based on BLE through the Apple/Google Exposure Notification system and there is no registration to a server.
The app does not use any third party tracker; however, it asks for accessing the camera of the user, which raises some concerns. In any case though, our analysis through the PI tool did not detect any other privacy issue.
7. **TraceTogether:** This is a centralized app from Singapore, using the so-called privacy-preserving protocol BlueTrace for community-driven contact tracing using Bluetooth devices. In order to use the app, the user needs to provide either an ID card number or a passport number, in order to be checked whether the user resides in the country. This put some limitations in our research with respect to this app and, thus, we performed a dynamic analysis of this application up to the extent that it was possible. We noticed that all our personal data that we enter (name, id card number, birth date, nationality) are being transmitted to the application provider, in conjunction with device information such as the telephone number, the version of the Android and the device model. We also noticed that our gmail address is also being sent to Google, whilst connections with Facebook were also identified through the Facebook's Graph API. Moreover, connections with the analytics company MixPanel were also seen through our analysis with the PI tool.
8. **PathCheck SafePlaces:** This is a decentralized app from the United States, using location information for contact tracing in a privacy-friendly manner. The app asks for the "Access Coarse location" and "Access Fine Location" permissions which are necessary for location information, but it also requires the "Write External Storage" and "Write Settings" permissions. It has only one embedded tracker, the well-known Google Firebase tracker (which is present, as we have already discussed, in many contact tracing apps).

Our analysis through the PI environment illustrates that device info such as its language and the Android ID, are being transmitted to Google. Moreover, our gmail address is also transmitted, through this app, to Google.

SOLUTIONS AND RECOMMENDATIONS

The above analysis confirms that contact tracing apps, as any other smart application, may suffer from privacy issues if they are not properly designed. In our study we didn't explore other possible security or data protection issues, like, e.g, whether pseudonymous data that are being collected may allow re-identification of an individual (i.e. whether a strong pseudonymisation is being developed, in cases that a pseudonymisation takes place), but we pointed out on the number and nature of the trackers used, taken into account the intra-library collusion issue (in relation with the relevant permissions that each tracker inherits from the host application), as well as the actual device/personal information that is being "leaked" to third parties. Note that large companies, such as, e.g., Google, may be also a "third party" and, thus, since such companies already have much personal information of users due to their own main applications, they can easily enrich the personal information that they possess, without the users being aware of this. Although the results are not surprising – i.e. we confirm our intuition that some apps raise more data protection concerns than others, as actually expected – they have their own importance since such apps are being used in a pandemic period, in which some temporary restrictions on human rights (e,g. restrictions on free movement), for the sake of public health, became already a reality.

From the above it becomes evident that the data protection by design principle should be present in humanitarian technologies – including the challenging case of smart applications used in this context. All the relevant stakeholders – namely app developers, library providers, operating system providers, app stores, but also regulators and authorities that define the purposes of these apps – have a crucial role in this. Full transparency of the underlying personal data processing, as well as fulfillment of the data minimization principle, are essential for building users trust on such applications. To this end, we next give some recommendations on the matter:

- The necessity of each data processing should be fully justified; the users should understand why a permission required by such an app is absolutely necessary for the app's operation. Denying permissions that are not necessary should not lead to stopping the typical app's operation.
- The users should be fully aware of what personal data processing takes place, including possible data leakages to third parties. Such information should be exact – for example, erroneous use of the term "anonymous data" should be avoided.
- Data leakages to third parties should respect the data minimization principle, as well as the purpose limitation principle. Hence, technology should be further promoted in order to pave the way for privacy friendly solutions. For example, privilege separation strategies for apps and their embedded libraries is an essential issue that needs to be carefully considered. Moreover, if third party libraries are being incorporated for statistical analysis, such third parties should not have access to information that could possibly lead to identification of a user.

It seems that some steps towards enhancing privacy have indeed taken place. For example, in the framework of contact tracing apps, our analysis indicated that the vast majority of the apps that initially incorporated the Google AdMob tracker removed this tracker between the period September 2020 – March 2021. In addition, as a more general mitigation measure, it is known that, starting from the version 8.0 of Android, no app has access to the Android ID (therefore, even for the apps that described above that were found to process Android ID, this will not hold for newer Android versions). However, there is clearly still room for improvement; for example, the Google Advertising ID is still accessible from apps and it is questionable whether the users reset it periodically, as well as how easy is to identify an individual from its device GAID, taking into account all the additional "traces" that the user's device leaves.

CONCLUSION

The scale and severity of the COVID-19 pandemic imposed – as a public health threat - restrictions on certain rights, such as the free movement. At the same time, technology has been used to help striking this pandemic, which in turn needs attention in terms of respecting other human rights such as non-discrimination and human dignity; contact tracing apps, which is the main focus of this chapter, constitute such a technology example.

The effectiveness of these apps, in terms of impeding the virus dissemination, has not been evaluated yet; in addition, there are several scientists from varying fields stating that such technology tools are of limited value for striking the COVID-19. For example, the famous cryptographer Bruce Schneier expressed the opinion that the errors due to false positive alarms are quite high (Schneier 2020); as he states, false positives could e.g. arise when someone was within two feet of someone who has Covid-19 for eight hours, but there was a wall between them because they were in adjacent hotel rooms. Similar concerns have been also expressed in an essay from the Security Group of the University of Cambridge (Anderson 2020). Concerning the centralized approach, the famous security engineer Ross Anderson writes in this essay:

I recognise the overwhelming force of the public-health arguments for a centralized system, but I also have 25 years' experience of the National Health Service being incompetent at developing systems and repeatedly breaking their privacy promises when they do manage to collect some data of value to somebody else. (Anderson, 2020)

On the other side, a large number of governments have chosen to utilize such apps, taking into account the severity of this new virus. The EDPB, in the relevant guidelines, states that

Such applications need to be part of a comprehensive public health strategy to fight the pandemic, including, inter alia, testing and subsequent manual contact tracing for the purpose of doubt removal. (European Data Protection Board, 2020)

In any case, it is essential that personal data protection should be fully respected by these apps (as in any humanitarian technology); this in turn necessitates a very cautious design and implementation, as well as full transparency on their underlying personal data processing. By these means, individuals will trust these apps and such a trust is prerequisite for their effectiveness. However, since these apps lie in the mobile ecosystem, there exist inherent privacy risks that are not present in other technologies, which should be additionally taken into account when designing such an app. This chapter presented an overview of these risks by studying in practice a significant number of such apps, pointing out their current status and their potential drawbacks, whereas some recommendations are formulated, which span all the relevant stakeholders. Even if this pandemic becomes – as we all hope – fully treatable in the near future, the conclusions of this research are of high importance since they will also be applicable in other cases in the future, taking into account the wide range of applicability of smart applications.

ACKNOWLEDGMENT

This research received no specific grant from any funding agency in the public, commercial, or not-for-profit sectors.

REFERENCES

Ahmed, N., Michelin, R. A., Xue, W., Ruj, S., Malaney, R., Kanhere, S. S., Seneviratne, A., Hu, W., Janicke, H., & Jha, S. K. (2020). A Survey of COVID-19 Contact Tracing Apps. *IEEE Access: Practical Innovations, Open Solutions*, *8*, 134577–134601. doi:10.1109/ACCESS.2020.3010226

Baumgärtner, L., Dmitrienko, A., Freisleben, B., Gruler, A., Höchst, J., & Kühlberg, J. Mezini. M., Mitev, R., Miettinen. M., Muhamedagic. A., Nguyen, T. D., Penning, A., Pustelnik, D. F., Roos, F., Sadeghi, A., Schwarz. M., & Uhl, C. (2020). *Mind the GAP: Security & Privacy Risks of Contact Tracing Apps.* https://arxiv.org/abs/2006.05914

Chatzistefanou, V., & Limniotis, K. (2017). On the (non-)anonymity of anonymous social networks. *e-Democracy – Privacy-Preserving, Secure, Intelligent E-Government Services Communications in Computer and Information Science*, *792*, 153–168. doi:10.1007/978-3-319-71117-1_11

European Data Protection Board. (2020). *Guidelines 04/2020 on the use of location data and contact tracing tools in the context of the COVID-19 outbreak.* https://edpb.europa.eu/sites/default/files/files/file1/edpb_guidelines_20200420_contact_tracing_covid_with_annex_en.pdf

European Union Agency for Cybersecurity. (2017). *Privacy and data protection in mobile applications - A study on the app development ecosystem and the technical implementation of GDPR.* https://www.enisa.europa.eu/publications/privacy-and-data-protection-in-mobile-applications

Exodus Privacy. (2021). *What Exodus Privacy Does.* https://exodus-privacy.eu.org/en/page/what/

Gvili, Y. (2020). *Security Analysis of the COVID-19 Contact Tracing Specifications by Apple Inc. and Google Inc.* Cryptology ePrint Archive, Report 2020/428. https://eprint.iacr.org/2020/428

Leith, D. J., & Farrell, S. (2021). Contact Tracing App Privacy: What Data Is Shared By Europe's GAEN Contact Tracing Apps. In *IEEE INFOCOM 2021 - IEEE Conference on Computer Communications* (pp. 1-10). IEEE.

Monogios, S., Limniotis, K., Kolokotronis, N., & Shiaeles, S. (2019). A case study of intra-library privacy issues on Android GPS navigation apps. *e-Democracy 2019: Safeguarding Democracy and Human Rights in the Digital Age Communications in Computer and Information Science, 1111*, 34–48. doi:10.1007/978-3-030-37545-4_3

Nabben, K., Gardner-Stephen, & Poblet, M. (2020). Identifying and Mitigating Humanitarian Challenges to COVID-19 Contact Tracing. In *2020 IEEE Global Humanitarian Technology Conference (GHTC)* (pp. 1-8), IEEE. 10.1109/GHTC46280.2020.9342852

Privacy International. (2018). *Investigating Apps interactions with Facebook on Android.* https://privacyinternational.org/taxonomy/term/552

Reichert, L., Brack, S., & Scheuermann, B. (2020). *A Survey of Automatic Contact Tracing Approaches Using Bluetooth Low Energy.* Cryptology ePrint Archive, Report 2020/672. https://eprint.iacr.org/2020/672

Statista. (2021). *Android - Statistics & Facts.* https://www.statista.com/topics/876/android/

Taylor, V. F., Beresford, A. R., & Martinovic, I. (2017). *Intra-Library Collusion: A Potential Privacy Nightmare on Smartphones*. https://arxiv.org/pdf/1708.03520.pdf

Vaudenay, S. (2020). *Centralized or Decentralized? The Contact Tracing Dilemma.* Cryptology ePrint Archive, Report 2020/531. https://eprint.iacr.org/2020/531

ADDITIONAL READING

Chitkara, S., Gothoskar, N., Harish, S., Hong, J. I., & Agarwal, Y. (2017). Does this App Really Need My Location? Context-Aware Privacy Management for Smartphones. *Proceedings of the ACM on Interactive, Mobile, Wearable and Ubiquitous Technologies, 1*(3), 1–42. doi:10.1145/3132029

Claesson, A., & Bjørstad, T. E. (2020). *Out of Control – A review of data sharing by popular mobile apps*. Technical Report, Norwegian Consumer Council. https://fil.forbrukerradet.no/wp-content/uploads/2020/01/mnemonic-security-test-report-v1.0.pdf

Kaminski, M. (2020). A recent renaissance in privacy law. *Communications of the ACM, 63*(9), 24–27. doi:10.1145/3411049

Mustafa, N. (2021). Research and Statistics: Coronavirus Disease (COVID-19). *International Journal of System Dynamics Applications, 10*(3), 67–86. doi:10.4018/IJSDA.20210701.oa1

Scassa, T. (2021). COVID-19 Contact Tracing: From Local to Global and Back Again. *International Journal of E-Planning Research, 10*(2), 45–58. doi:10.4018/IJEPR.20210401.oa4

KEY TERMS AND DEFINITIONS

Bluetooth: A wireless technology networking protocol used to exchange data over short distances.

Contact Tracing: A disease control methodology that lists all people who have been in close proximity to a carrier of the virus so as to check whether they are at risk of infection and take the appropriate sanitary measures towards them.

Device Identifier: A unique identifier (e.g., number, sequence of characters etc.) used to distinguish a device.

Location Data: Data processed in an electronic communications network or by an electronic communications service indicating the geographical position of the terminal equipment of a user of a publicly available electronic communications service, as well as data from potential other sources, relating to 1) the latitude, longitude, or altitude of the terminal equipment; 2) the direction of travel of the user; 3) the time the location information was recorded.

Personal Data: Any information relating to an identified or identifiable natural person.

Personal Data Processing: Any operation or set of operations which is performed on personal data or on sets of personal data, whether or not by automated means, such as collection, recording, organisation, structuring, storage, adaptation or alteration, retrieval, consultation, use, disclosure by transmission, dissemination or otherwise making available, alignment or combination, restriction, erasure or destruction.

Third Party Tracking: The practice by which an entity (the tracker), other than the website directly visited by the user or the provider of the mobile application utilized by the user, tracks or assists in tracking the user's behavior across multiple digital services.

APPENDIX 1

Table 3. List of smart apps examined during our research

App	Authority/Organization
WHO Info	World Health Organization
OpenWHO	HPI Knowledge Engineering Team
COVIDSafe	Australian Department of Health
Corona-Warn-App	Robert Koch-Institut
COVID-19!	Nemocnice Milosrdnych bratri, p.o.
COVID-19	Electronic Health Administration
COVIDWISE	VDH
NZ COVID Tracer	Ministry of Health NZ
COVID Tracker Ireland	Health Service Executive (HSE)
Stop COVID-19	Ministry of Health Croatia
Track Virus	Panda O.S. Ltd.
COVID Radar	LUMC Leiden
CoronaSAFE	EnayaTech
Service NSW	Service NSW
COVID19 - DXB Smart	Dubai Health Authority - DHA
PathCheck SafePlaces	Path Check, Inc
COVIDTrace	Sarawak Government
CDC	Centers for Disease Control and Prevention
Radar COVID	Ministerio de Asuntos Económicos y Transf. Digital
Covid SafePass	Medek Health Systems LLC
Canada COVID-19	Health Canada I Santé Canada
Āwhina - informing health workers	Ministry of Health NZ
Covid19-MY	Qualitas Medical Group
TraceTogether	Government Technology Agency, Singapore
COVID Alert Malta	Government of Malta
StopCovid France	Gouvernement
CoronaCheck	Aga Khan University
CovTracer	RISE Research Centre of Excellence
Coronavirus Support App (UK)	Expert Self Care Ltd
COVID Alert South Africa	Department of Health - South Africa
Covid Locator	Intugine Technologies
STAYAWAY COVID	FCT FCCN
NHS COVID-19	Department of Health and Social Care
myGovID	Australian Taxation Office
COVID-19 Sounds	cam-cl-mobile-systems
TrackTogether	TrackTogether
COVID Alert	Health Canada I Santé Canada
South African Government	GCIS
World Health Organization (WHO) Academy	World Health Organization
ZeroCOVID	Athleten Foundation
CoronaReport - COVID-19 reports for Social Science	SPOTTERON
SwissCovid	Federal Office of Public Health FOPH
Contact Tracing	Interactive Health Solutions
Stop COVID	Ministry of IDPs, Labour, Health, Social Affairs
Rakning C-19	Icelandic Government
BC COVID-19 Support	Province of British Columbia, Canada
Coronavírus - SUS	Governo do Brasil
COVID-19 Andhra Pradesh	Entro Labs
JamCOVID19	opmjm mobile
Corona Tracking and Response App	Punjab IT Board
Stop COVID – proteGO Safe	Chancellery of the Prime Minister of Poland
NOVID	Expii, Inc.
Immuni	Italian Extraordinary Commissioner for the COVID-19 Emergency, in collaboration with the Ministry of Health and the Ministry for Innovation Technology and Digitalization

Chapter 12
Catching Learners Early in Humanitarian and Sustainable Principles Through Chemistry Education

Ruby Hanson
University of Education, Winneba, Ghana

Charles Hanson
Kwame Nkrumah University of Science and Technology, Ghana

ABSTRACT

The purpose of this chapter is to describe a possible best practice to teaching chemistry from a humanitarian engineering perspective. The interest in teaching chemistry by focusing on humanitarian engineering arises from the economic and environmental concerns that the country of this study faces, some of which are poverty, climatic changes, food crisis, inadequate healthcare, water crisis, and pollution. As an educator, there is an interest in educating future generations to be able to cope with environmental changes that face their countries and the world at large. This exposition of a possible new approach with appropriate pedagogies that is presented here may be an answer that underdeveloped, developed, and emergent economies may adopt to close the gap between themselves and other industrialised nations.

INTRODUCTION

Science and technology have greatly enhanced some life activities, careers and longevity, due to improved healthcare and useful products that make life pleasurable. However, these feats have also resulted in the impairment and depletion of many natural resources and paradoxically man's health and general welfare. Other occurrences such as ethnic conflicts, over population, poor mining practices, destruction of lands for agriculture, pollution of water bodies from mining activities and effluents from industries cause human crisis, directly or indirectly. Natural disasters such as floods, volcanic and forest fires, tornadoes,

DOI: 10.4018/978-1-7998-9190-1.ch012

hurricanes and landslides have occurred without man's intervention in the past and have been with us since creation. These environmental disasters, though life threatening, restored and healed themselves. However, in recent times, the intervention of adverse human activities, have resulted in man-made disasters that would require, once again, the intervention of humans to correct. Nearly two-thirds of the natural systems that provide services for humans are in distress and on the verge of collapse, or past retrieval (United Nations Statistics Division, 2005). Quite often, and unfortunately, it is the marginalised in society, and those with limited resources to fight the menace of natural disasters and industrialised wastes, that suffer the brunt or negative impacts of humanitarian challenges, even though they do not gain much from the benefits of science and technological applications. On a global scale, these negative impacts result in food crisis, financial crisis, water scarcity, poor health, migration issues, struggle for power and urbanisation in limited places that exceed carrying capacities with its attendant problems. Attempts at mitigating these challenges could be through science (chemistry) education and humanitarian engineering, especially as the identified challenges were also caused by humans in the first place. Collaboration between disciplines and stakeholders have attempted to provide positive solutions to human challenges.

Inequities in the distribution of resources such as food, shelter and healthcare (Nolet, 2009) have also led to poverty and diseases. The unique feature of this chapter is the concern for how to practice humanitarian engineering through a discipline such as chemical education, rather than its evaluation, so that countries with marginal economies will be inspired to adopt the approach to prepare their communities to cope squarely with environmental exploitation and humanitarian challenges. In this document, I will provide an outcome of an earlier study where microchemisty was integrated as a laboratory-based pedagogy in teaching chemistry. Adaptations that were made to the regular chemistry curriculum to produce an integrated microchemistry curriculum would be narrated briefly as it is not the object of this current exposition.

Humanitarian engineering is the use of engineering to improve the well-being of marginalised people who are often found in disadvantaged, under-developed, emergent, and developing communities. Such people and their governments often have limited resources for their use and cannot package the few resources that they have in such a manner to get value for them in trade- even if they should want to export them for value and such. They cannot adapt to disasters and tend to be more dependent on resources affected by environmental change such as local water and food supplies. They solve problems in superficial and temporary measures as they are capable of at any one moment, as they do not have the capacity to address long-term and near permanent solutions. Shelter, education, clean water, road infrastructure, and cleanliness of the environment remain perennial problems. This is where engineering and particularly humanitarian engineers could come in handy. Engineers must first build for humanity and then use the engineering profession to affect positive change for marginalised communities globally; something called 'capital age humanitarianism'. However, this chapter seeks to explore how chemistry education could be integrated with some concepts of humanitarianism and engineering to do same- solve humanitarian challenges. This quest is because humanitarian engineering explores the role and application of engineering to disadvantaged, marginalised and vulnerable communities to improve quality of life and support empowerment. It seeks to placing human well-being at the centre of life, provide collaborative skills, knowledge, and experiences that could be applied not only in development and humanitarian contexts, but all aspects of life. Thus, learners could be equipped to work in cross-disciplinary teams, undertake stakeholder engagement, think creatively, assess and evaluate the impacts from science, engineering and technology. Mathematics and technology are applied to solve problems practically, often in real contexts and concerns itself with the design, building, and use of engines, machines, and structures (Campbell &

Wilson, 2011) for humanitarian aid purposes such as natural disaster recovery or improving the prospects of people in impoverished areas.

There is a paradigm shift in many disciplines towards the application of engineering theories to improve the well-being of marginalised societies. In order to emphasise the paradigm shift in education and to perpetuate it, there is a need for teaching of interdisciplinary development skills in all disciplines, chemistry inclusive, to help us to break free from wars, sicknesses and abject poverty. A powerful developmental tool for gain of knowledge, understanding of change in one's environment, and frees one from want is education. This could be employed to create awareness of humanitarian challenges and how the application of science, technology, engineering and mathematics (STEM) could be functional in an integrative and multidisciplinary manner to instil in learners the principles of sustainability and humanitarian engineering. Directed or purposeful chemistry education in schools could be an initial turning point of call to conscientize and impact young learners with the principles of humanitarianism and sustainability. Thus, if we wish to change the way we think, then we need to change our educational system, and, the practices that teachers employ to contextualise learning and make it meaningful. Infusing engineering education into chemistry education would provide the learning required for successful, sustainable and humanitarian chemists and engineers with technical expertise, social awareness, and a bias toward innovation (Crawley, Malmqvist, Ostlund, Brodeur, & Edstrom, 2014), for the acquisition of cognitive, aesthetic, lifelong and transferrable skills to make life activities pleasantly sustainable for our global village. This must be an important goal for all educators, especially science, technology, engineering and mathematics (STEM) educators as science and its applications drive the world now.

Chemistry education (or chemical education) is the study of teaching and learning chemistry. It is one subset of STEM education or discipline-based education research. Chemistry education is important because the field of chemistry is fundamental to our world. Described as the central science, chemistry connects physical sciences with the life sciences and applied sciences. It plays a central role in creating innovative technologies and products for comfortable life, though our current practices of production and consumption are not sustainable. We need future chemists to be able to think holistically and systematically about chemistry in order to maximize resource efficiency while minimizing hazards and pollution with evidence-based decisions about science-related policy (Orgill, et al., 2019) and engineering issues for the benefits of humankind. This holistic underlying concept for systems thinking implicitly seeks to unify the natural and social sciences in dealing with real life issues.

According to Orgill, York and MacKellar (2019), in chemistry education, a 'systems thinking' perspective could be achieved by investigating environmental, social and economic factors in addition to the chemical content of a specific case. For example, a 'systems thinking' approach would ask students to consider NO_2 and photochemical smog from yet another perspective: one that focuses not just on chemical context, but on dynamic behaviour of chemicals; the causes of dynamic behaviour; feedback systems; and social, environmental, and economic contexts. The role of teachers in this regard is of vital importance because they are expected to teach chemistry education for the development of concepts and skills that bear on issues such as the sustainability and humanitarianism in integrated ways for lifelong living, care for others, care for the world, and peace. Unfortunately, many training institutions for teachers do not pay attention to the training of teachers for the acquisition of skills for understanding humanitarianism and the need to sustain the environment for posterity and well-being and mankind (Okanlawon, 2018).

Chemistry laboratories and industries generate a large amount of toxic liquids and solid wastes from chemical reactions like nitration, chlorination, sulphonation with evolution of noxious gases that cause pollution and impact the health of students, teachers, other school workers and the ecosystem. This chapter seeks to contribute knowledge to the pool of data on humanitarian challenges and mitigation of existing and other possible challenges that could arise through chemistry education. It further seeks to give an exposition of how microchemistry and green chemistry (with emphasis on microchemistry) could be pathways to project how humanitarian engineering could begin in a small way through young learners and cascade for positive effects in underprivileged communities such as Ghana and other sub-Saharan African countries.

One way of promoting engineering in teacher training and general education would be through the integration of STEM early in the lives of learners as is commonly known (DeJarnette, 2018; Hargroves, 2011; Karpudewan, Ismail, & Roth, 2012a; Resources for thinking, 2021). There is a growing need in the world, with Ghana being no exception, to produce more skilled persons in science, technology, engineering and mathematics (STEM) as such knowledge acquisition go without saying that the nature of matter and how the world systems run would be better understood.

The exposition will look at possible topics that could be taught in interdisciplinary ways at various levels of education and the approaches that could be adopted to impart skills and knowledge for the creation of technologies that could lead to further research and ingraining of human skills. Chemistry courses such as physical, analytical, organic and materials chemistry would be applicable for all these applications for safer and healthier environments with gratified inhabitants and where adequate resources could be equitably spread for good life. This kind of education, which is project-, problem- and context-based would help students in impoverished communities to further analyse and design appropriate technology to support the well-being and welfare of their communities.

In summary, the purpose of this paper is to present a simple microchemistry approach that was used in chemistry lessons in a tertiary institution as an approach to introduce and engage learners in the principles of humanitarianism and humanitarian engineering to propel the concepts of sustainability.

The Role of Chemistry Education in Sustainability and Humanitarian Issues

In Ghana, in particular, chemistry teacher preparation is geared more towards production of knowledgeable, competent, professional chemistry teachers as they are more exposed to content knowledge in organic, physical, inorganic, materials, and environmental chemistry. Communication skills, computer studies, and liberal studies are also offered during the first two years of their 4-year teacher education. In addition, students study pedagogical courses a few which are the principles and practice of teaching, trends in education and methods of teaching. With the acquisition of the pedagogical theories, they engage in micro teaching on campus and follow that up with a semester of rigourous internship off campus and in real classrooms at various levels of education (excluding tertiary). A critical examination of the chemistry teacher education programme revealed that there are no standalone or integrative courses that take care of contemporary issues such as sustainability, humanitarianism and humanitarian engineering.

The implication is that chemistry teachers would graduate with little or no exposure to such contemporary issues and so would be less equipped to teach such knowledge to their future students competently or be able to manage challenges in such areas adequately, regardless the education that they have acquired. Most higher education curriculum is dominated by the mechanistic paradigm and would require a shift towards multidisciplinary, contextual and ecological paradigms (Stirling & Kerr, 2015; Szalacsi, 2015). With this inadequacy of the current teaching programme in mind, contemporary topical issues would have to be integrated into chemistry teacher training programmes. There are several pedagogies that could be employed to help with the integration of engineering, and humanitarian engineering in particular into existing chemistry curriculum as has been intimated. Pedagogies such as problem-based learning, project-based learning, inquiry-oriented classroom discussions could be adopted for such teaching and learning processes.

In this document, chemistry education at the tertiary level for preservice teachers would be considered as an educational channel towards the integration of integrating humanitarian engineering which borders on freedom from want, human security and economic peace into curricula. Human security, also defined as "freedom from want and fear and living in dignity", provides an overarching concept to address threats to human security dimensions such as health, food, economics, the environment and sustainable development, while placing the individual at the centre of attention in chemistry education (Matlin, Krief, Hopf, & Mehta, 2021). Matlin et al. (2021) go on to say and contest that, chemistry is central to addressing these challenges, but surprisingly its role and contributions to human security have hitherto not been explicitly set out. Higher education institutions have to respond now, in a systematic and coherent way, in terms of embedding humanitarian and engineering concepts, values, and skills into chemistry learning experiences (Szitar, 2013). Higher education has always had a special academic freedom, the critical mass, and diversity of skills to develop new ideas as well as comment on society and its challenges (Cortese, 2003). Changes sought in humanitarian issues involve deep cultural shifts that are difficult to achieve, so the most important point to leverage on for transformations could be through chemistry education.

CONCERN

The school chemical laboratory has already been presented as one pathway for endangering the ecosystem. Outside the school community, pesticide and fertiliser residues, flame retardants from electronics and oil spillage affect especially, the biosystem. Thus, scaling down chemistry activities or integrating microchemistry activities in school laboratories would not only create safer, more engaging practical lessons and work spaces, but prepare students for a humanitarian and sustainable future and allow them to think creatively through real life practical problems to make their environments safer, more productive, and healthier. As learners get to know about the nature of chemicals and devise good laboratory management plans, they will invariably employ cross-disciplinary concepts that align with societal needs and the United Nation's (UN) Sustainable Development Goals (SDGs) (United Nations, 2015). The SDGs are a series of measures to end poverty, protect the planet and ensure the global well-being of human beings (UNDP, 2020). The 2030 Development Agenda titled *Transforming our world: the 2030 Agenda for Sustainable Development* contains 17 goals for sustainable development that highlights how urbanization, economic development, climate change, and the need to produce more food for a growing population are limiting water availability for the already marginalized countries (International Water Management Institute, 2020). As reiterated earlier, scaling down activities to micro levels in schools and industries or going 'green' could save the world in small measures for a state, but culminate into well-deserved saves in the near future.

Microchemistry, is a branch of chemistry responsible for making school science (chemistry) less dangerous (green) to users and the environment, as it employs the use of miniature science ware called micro science or microchemistry equipment or kit, that require the use of minimal chemicals but give meaningful observable results. Microchemistry equipment and the practice of microscale science or chemistry have been found to possess the potential to change the way chemistry practical is organized,

Figure 1. Components of the RADMASTE microchemistry kit.

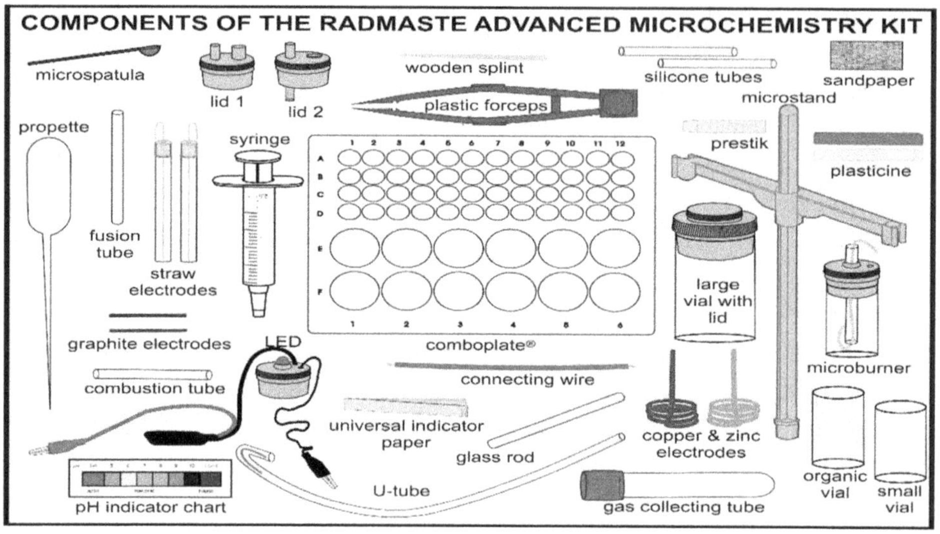

thus alienating the use conventional macro standard equipment with its attendant negative impacts and re-establish sustainable development (Hanson, 2017). The components of a microchemistry kit are shown as Figure 1.

It also attested to have the potential in enhancing teaching and learning of both secondary and tertiary level chemistry (Hanson, 2017; Mafumiko, 2012).

Green chemistry on the other hand refers to the use of chemistry that prevents pollution. It employs materials, processes and practices that reduce or eliminate the creation of pollutants or wastes such as hazardous and non-hazardous materials, energy, water, or other resources as well as protect natural resources through efficient use (Bhandari & Raj, 2017). The practice of green chemistry is guided by 12 principles that emphasise waste prevention *rather than treatment* and the design of safer chemicals, energy, degradation, accident prevention real-time analysis for population prevention (Karpudewan, Ismail, & Roth, 2012a). It is associated with the concept of sustainability in chemical education (Karpudewan, Ismail, & Roth, 2012b). Green chemistry has been applied in higher education (van Bergen, 2020) with success and so with its ecologically safe attributes it could be integrated into existing chemistry curricula to embrace the tenets of humanitarian engineering.

ISSUES, CONTROVERSIES, PROBLEMS

Teaching about Humanitarian Engineering

In order to have a more focused paradigm shift towards humanitarian engineering in chemistry education, and an ultimate sustainable development and peaceful living, there should be an increasing need for teaching of interdisciplinary and context-based development skills, especially in education. Besides the interdisciplinary context-based teaching, teachers must effectively repackage existing curricula into forms that could easily be assimilated by learners because when contents are not properly sequenced and distorted, that could lead to the formation of misconceptions. Besides, an understanding of contemporary issues and resequencing of content, based on teacher knowledge and understanding of social issues, there must be pedagogical transformation. In this wise, one could say that, humanitarian engineering pedagogical content knowledge must be devised in order for chemistry teachers to be able to integrate the content knowledge of humanitarian engineering and pedagogical knowledge for teaching humanitarian engineering

Ideas and underlying concepts of humanitarian engineering must permeate chemistry courses (Hargroves, 2011; Szitar, 2013; Talanquer, 2019), because it is a unifying discipline that employs all other disciplines as encompassed in both the humanities and the natural sciences to help learners make sense of the world and their place in it. This new paradigm will require rigorous review of curricula and the provision of credible appropriate teaching resources that could be contextualised to suit the needs of different communities, especially, marginalised ones. The idea and paradigm look laudable, but the question to how best to execute it pedagogically comes to bear. Are teachers going to use the same old teacher-focused approach that they are used to? How do learners best construct and retain constructed knowledge?

Some Teaching Approaches to Attain Sustainability and Humanitarianism Concepts

Problem-based learning (PBL) is a teaching method in which students learn principles and concepts through complex and open-ended real-world problems (Ali, 2019) could be employed to equip learners with critical thinking skill, problem solving abilities, communication skills and lifelong learning as was found in a study by Mc Donnell, O'Connor and Seery through problem-based mini projects in chemistry (2007). Project-based learning also helps students in constructing knowledge (Tsaparlis & Gorezi, 2007) as observed in teaching of pharmacy students by Strohfeldt and Khutoryanskaya (2015). By adopting such an approach as problem-based learning in chemistry education, it is hoped that learners will not only come to appreciate humanitarian challenges and how to engineer their solutions from a multidisciplinary view point, but fully engage in their learning processes, acquire transferable, concept and lifelong learning skills, independence in problem solving, understand best practices for healthy work and improve communal environments. They will come to further understand the concept of sustainability and care for humanity through a more systems-thinking approach (Talanquer, 2019). Bell (2010) sees project-based learning as an innovative approach to learning that teaches a multitude of strategies critical for success in the twenty-first century. Here, learners learn inquiry, as well as work collaboratively to research and create projects that reflect acquired knowledge. Context-based learning (CBL) refers to the use of real-life and fictitious examples in teaching environments in order to learn through the actual, practical experience with a subject rather than theoretical parts. The hypothesis that underpins CBL is that if a learner can understand why they are learning what they are learning, it will be of significance to them and so retain and transfer it in other situations (Arroio, 2010).

Introducing Humanitarian Engineering Issues in Chemistry Education- A Baseline Case Study

In the past couple of years, the regular traditional method of teaching environmental chemistry theory and practice in the third year of study in the Department of Chemistry Education to chemistry and physics education major students in teaching institutions in Ghana have changed in order to introduce concepts of sustainability, humanitarianism, and humanitarian engineering as approaches towards development and improvement in communities and eventually the nation (Burmeister, Rauch, & Eilks, 2012; Campbell & Wilson, 2011). Some courses that have employed this integration are environmental chemistry, analytical chemistry and periodic chemistry, to name a few.

The Case Study with Environmental Chemistry

Introduction to the Case

Scaling down chemistry activities or integrating microchemistry activities does not only create safer, more engaging lessons, but prepares students for the future and allows them to think creatively through real life practical problems (Hanson & Acquah, 2014). It also allows teachers to teach cross-disciplinary concepts that align with societal needs. The main theme in environmental chemistry, which was the curricula integrated with humanitarian concepts and humanitarian engineering, about is 'safer and healthier living conditions'. It involves the study of chemicals and chemical processes within the air, water, soil, humans and their effects. It also involves studying how these chemicals get there, what they do, and how humans are intertwined in all of this. Thus, how these chemicals, which are matter, can affect people, especially in poverty-stricken areas and how solutions could be engineered became an object of study.

The theory that underpinned this case study was the social constructivist theory proposed by pioneers such as Vygotsky and Leontiev as well as Brown, Collin, and Duguid (Alanazi, 2016), who argue that learning is an interactive, social process. The theory states that language and culture are the frameworks through which humans experience, communicate, and understand reality. According to Vygotsky, language and culture play essential roles both in human intellectual development and in how humans perceive the world. Thus, knowledge and meaning are socially constructed (Akpan, Igwe, Ikechukwu, Mpamah, & Okoro, 2020). It also deemphasises teacher-monotony in the classroom, but encourages active interaction among learners, the teacher and other components of the teaching learning process. It also concretises learning and knowledge by making students retain the facts that they discover and construct by themselves than those they are told by the teacher. Social constructivism emphasises the collaborative nature of learning under the guidance of a facilitator or in collaboration with other students. It is also called collaborative learning because it is based on interaction, discussion and sharing among students (Akpan, Igwe, Ikechukwu, Mpamah, & Okoro, 2020). In this study, most of their designs for solving problems had to be developed and tried on small scales in the laboratory using micro science equipment (Hanson & Acquah, 2014; Hanson, 2014).

Objective for the Case Study

The objectives set for this case study was to evaluate the integration of engaging pedagogical constructivist conceptual chemistry procedures into real-world solutions of challenges that required the demonstration of humanitarian engineering.

Methodology

A case study design with interpretive undertones was adopted for this study. Undergraduate classes in two universities were purposively sampled to participate in this study because they were enrolled in a course titled 'Environmental Chemistry' that was taught by the researchers in experiential manner. The major sections in the curriculum were environmental resources, urbanisation, industrialisation and the environment for Unit 1. Further on, Units 2, 3 and 4 were on atmospheric chemistry and air pollution, aquatic chemistry and water pollution and soil chemistry and pollution respectively. The laboratory activities/projects were obtained from various authentic sources as well as some newly developed ones that had been piloted to assess their feasibility for integration into chemistry curricula. Permission was obtained from the respective Institutional Review Boards, after which consent forms were processed by participating students. Instruments used to gather data were questionnaires to assess participants' views, observation ratings, semi-structured focus group guided conversations to collaborate pertinent issues in questionnaire or otherwise, and microchemistry activity laboratory reports on students' activities. Data were analysed descriptively, using SPSS Version 20 for the questionnaire and the observations, while the focus group conversations were analysed qualitatively after data saturation had been reached. Reports and oral presentations from projects /lab works were also assessed.

The first thirty minutes of each lab session consisted of a pre-lab where the chemical theories (chemical knowledge) in an environmental problem that had been presented a week ahead was discussed and possible procedures to solve the problems identified. Literature was reviewed and discussed within peer groups. This set the tone for the day and enabled students to gather the materials that could facilitate their assignments in real-world contexts. That was followed by two and half hours of practical laboratory work time which extended to the community for the enactment of real-time solutions. The objectives and rationale for each lab session were clearly spelled out to enable students to assume a project-like attitude towards their activities. Questions went beyond traditional ones to cover the impacts of students' observations and findings on individuals, economy and the environment. Before students engaged in practical activities or projects, a local community was identified where out-of lab activities could be carried out, using their mobile micro equipment where necessary. Specialists were sometimes invited to give talks on best practices in dealing with identified common challenges in deprived or poor communities, if this was required for a project. Students could work in big groups of 10 persons with a chemistry lecturer who was also a humanitarian engineering expert. After this, smaller groups of four persons were formed and time spent with specialists who had earlier given seminars on humanitarian engineering and sustainability in the communities where challenges such as poverty, hunger, diseases, poor housing and pollution had been identified.

Influenced by humanitarian engineering topics proposed by Micham and Munoz (2010) and concepts of sustainability (Hargroves, 2011), adjustments were made to the existing curriculum. Some integrated new topics centred on:

- Water, sanitation and hygiene in challenging communities
- Urbanisation and industrialisation through chemistry
- Atmospheric, aquatic and soil chemistries
- Economic impacts in pollution clean ups
- Health and disease
- Exploring application of emerging technologies in humanitarian and development contexts
- Adapting participatory design approaches to the context of humanitarian design and engineering projects
- Co-designing institutional arrangements for water and sanitation service provision in impoverished communities

Table 1. Examples of integrative topics for chemistry education

Area of study	Lesson topic	Interdisciplinary subtopics for consideration
Environmental sociology	Evaluating construction waste and its impact on the environment	Nature of waste (material) Chemical composition (Think of periodic table and positions of constituents of waste) Designing alternative/best compositions Food production and safety Effect on ecosystem Economic implications
Environmental Health and safety	Chemical detoxification in drinking water Chemical (elemental) levels in water and impact	Assay of water Comparison to WHO permissible levels Effect on organisms Diseases and implications on population Water safety Purification of safe water from contaminated water Model/design of simple water purification system for deprived communities
Environmental structures	Assessment of structures such as buildings, bridges and roads	Nature of structure Composition of building materials Effects of contaminants on metals (corrosion) Using local products and the dext kit to model sustainable bridges, roads and houses Economic implications
Economic implications of over population and pollution	Issues of over population and design of stable and durable amenities to sustain economies	Health Diseases Planning communities Design sustainable communities with sustainable materials

Source: Researchers' Integrated Design (2019)

- Strengthening the chemistry education and engineering professions through humanitarianism

For example, a humanitarian engineering integrated chemistry topic/project for discussion could be to address a community's identified basic needs. There had to be justification for the said identified need before solutions could be designed, taking into consideration such things as nature of materials for use, cost, effect on the ecosystem and aesthetics where necessary. Other examples were projects that create access to clean water, ensure adequate shelter and sanitation; and develop reliable pathways to markets and healthcare centres. In all of these activities, the chemical knowledge (theory) required its application and the effect of the service or product had to be paramount.

Four examples of topics for integrated lessons are shown in Table 1.

In all of the topics, the chemistry content in them were highlighted as the chemical knowledge was necessary for most of the projects that were attached to each lesson. Socio-cultural, ecological and economic components all came into play. Yet, students were expected to know about the nature of acids, water, air, soil, contaminants in drinking water, the structure of water, chemical equations and many more. Each lesson began with a story that was peculiar to the community in which the institution was situated or communities that were common to most of them. A story could begin like:

Today, let us extend our topic on water to a contaminated river between North and South Campus. Let us look at the community that lives along that river and access it. Access to good drinking water is a hu-

Figure 2. Outlook of a dext science kit

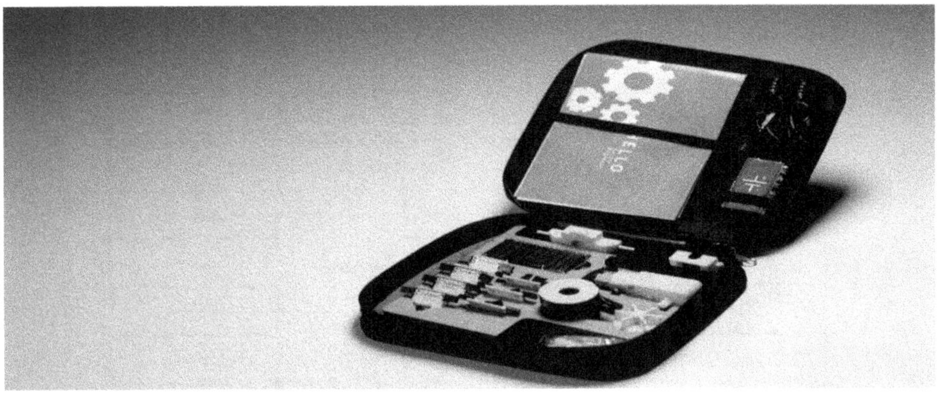

man right. Let us consider some of the risks that could be associated with the use of such contaminated water for gardening and other purposes. After a discussion, a project was then assigned.

Students were expected to use the microchemistry equipment for their lab works, because of the unique environmentally-friendly characteristics of the kit and their safety. This micro equipment was obtained from the University of Witwatersrand in South Africa (SOMERSET Educational Products and Systems, 2017). It was chosen for use because of its attributes of being fun and easy to use, being robust, helps to reach attainment of results in relatively faster times, uses very little chemicals and so is safe for use and kind on the environment, since little waste is generated and disposed of. The dext equipment was also employed for their modelling and engineering activities on small scale before they were translated on a bigger scale in parts of their work communities. Most of the chemicals that were required and the processes for their productions were mostly by green methods.

The micro science equipment was supported with a dext science equipment, that was used by participants who needed to build models to try them out first for their feasibility before real site work began. The science kit, designed and produced by Dext Technologies in Ghana, is a syllabus compliant, portable, highly scalable science laboratory designed to fit in the bags and on the desks of students. This is shown as Figure 2:

The Dext kit is a construction pack for curious people of all ages with 30 different exercises and instructions. It consists of more than 50 components and is used for learning and experimenting in various fields, such as electronics, energy, mechanics, or agriculture. Some activities that it could be used for without external support are making electromagnets, water pumps, automatic night lamp, visualising magnetic lines of force, demonstration of rectilinear propagation of light, building security alarms, function of resistors, capacitors, diodes, switches and many more. In this case study, students used the components in the dext kit to model how they would build water pumps and lay out electrical wiring in efficient ways for low-income households to mitigate their electrification cost. The components are shown below in Figure 3.

An example of a designed and fixed motor through the use of the dext kit is shown as Figure 4.

In the laboratory sessions, students were expected to outline their designs with respect to their identified and defined challenges and then explain how their proposed solution could tackle facets of the identified problems and the expected outcome. They were expected to apply their knowledge on the

Figure 3. Components within a science dext kit

Components within a science dext kit

Breadboard Kits			
QTY	Part	Symbol	Picture
5	330 Ohm Resistor		
5	1K Ohm Resistor		
5	10K Ohm Resistor		
5	22K Ohm Resistor		
1	10K Rotary Potentiometer		
1	Light Dependent Resistor (LDR)		
2	Green LED 5mm		
2	Red LED 5mm		
2	Yellow LED 5mm		
2	Push Button Switch		
2	10uF Electrolytic Capacitor		
2	100uF Electrolytic Capacitor		
2	0.1uF Ceramic Capacitor		
2	Transistor BC546 NPN		
1	Diode (1N4007)		
10	Breadboard Prototype Wires		
2	Alligator clip wires		
1	5V Regulator LM7805		
1	Battery Snap		
1	3 AA Battery supply 4.5V		
1	Breadboard 400 Hole		

55 Parts in total

chemical composition and nature of materials such as metals (cost, ductility, malleability, conduction, durability, reactivity with air) and propose best materials for construction purposes, where necessary in each unique situation. In this case, students were bound to think critically, analyse situations, produce workable strategies, select best solutions for maximum impact to solve problems, and deduce expected outcomes, all based on application of chemical concepts that had been learned in theory. Such a five-

Figure 4. A photo of a motor built from components in the dext kit

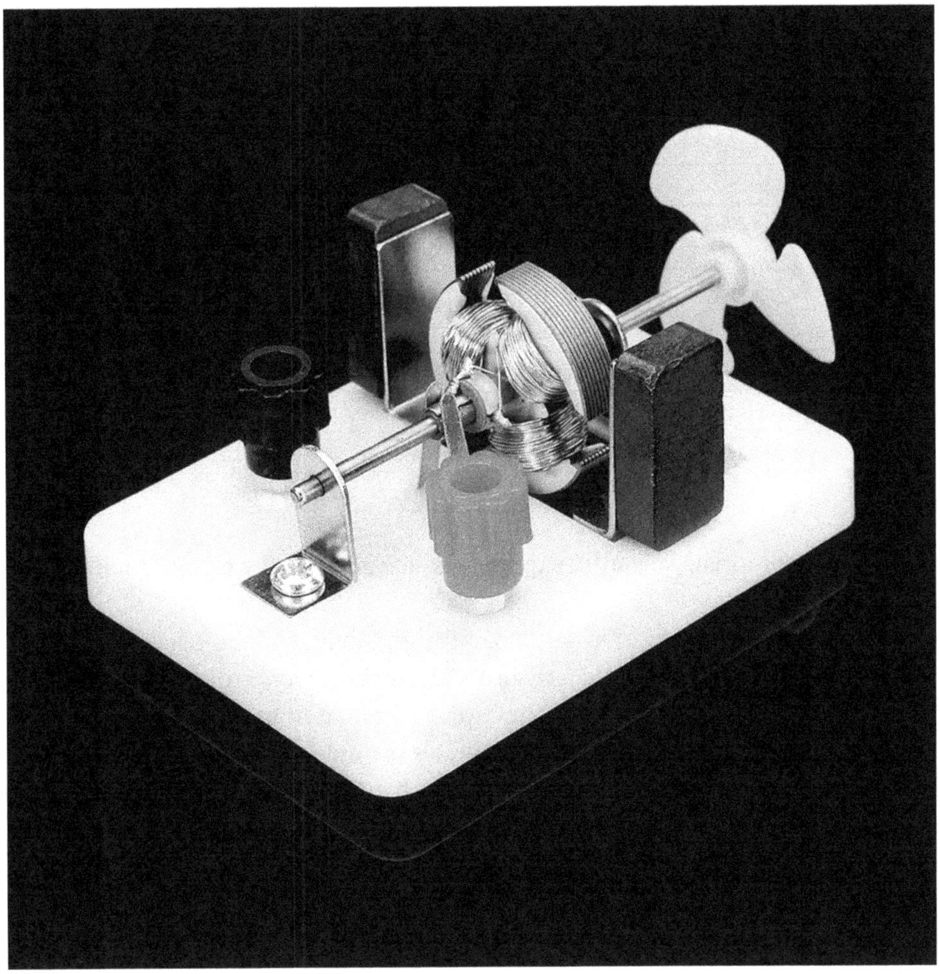

stage interactive technique to apply in chemistry education, engineering and humanitarian situations to solve challenges have been used unobtrusively by many teachers and other authors (Cerghit, 2006).

Students' Feedback in the Case Study

Table 2 presents a summary of the analysed 10-item questionnaire on a scale of one (1) to four (4) ranging from strongly disagree through to strongly agree.

From Table 2, an interpretive analysis would be that students enjoyed the use of the micro kits in performing chemistry activities and perceived that it could help to solve environmental humanitarian problems that cause challenges. The data gathered from observation showed that students were able to complete activities within relatively shorter periods than the regular in-door laboratory activities when micro equipment was employed, although they worked on longer projects outdoors. They were also observed to be more communicative as they collaborated with each other to suggest and solve identified

Table 2. Students' views of MCE and DSK as tools for enhancing sustainability and humanitarian education

Item	Statement	Mean	Interpretation
1	Activities motivate further chemistry and humanitarian studies	3.37	Strongly agree
2	Reduced chemical use leads to reduction in toxins in the ecosystem	3.49	Strongly agree
3	I did not enjoy the out-of-lab activities	2.08	Disagree
4	The micro equipment was fun and safe to use	3.45	Strongly agree
5	The integration of everyday life challenges into school chemistry brought chemistry alive for me	2.48	Agree
6	Preference for macro chemistry to microchemistry	1.44	Strongly disagree
7	Increase in understanding of chemical concepts	3.35	Strongly agree
8	Increase in understanding of sustainability concepts	2.85	Agree
9	Micro equipment is cost effective, efficient and useful for activities	2.78	Agree
10	Activities enhance research skills into mitigating human crisis	2.77	Agree

challenges, and used appropriate scientific expressions, meaning they were improving in communicating more scientifically.

Farther findings from the observation, sustainability and humanitarian concepts were gained while students were observed to be engaging ardently in and exploring diverse roles of chemistry in mitigating effects of chemical processes and products. They were observed to be engaged in discussions on how the reduction of chemical wastes from chemical activities in homes, laboratories and farms (agriculture) could reduce environmental pollution as well as lead to better climate, health and food supply for all. They demonstrated increased desire to contribute to the mitigation of environmental challenges. Generally, best work practices were adopted in all projects as observed by other researchers engaged in such studies. There was better engagement with the chemistry course.

Some of their comments that were captured in interviews (conversations) were analysed thematically (ensuring trustworthiness) supported the observed. The findings appeared to support what other researchers had found out about social constructivism (Akpan, Igwe, Ikechukwu, Mpamah, & Okoro, 2020), collaborative work (Bell, 2010) and the benefits that societies could gain from humanitarian engineering (Campbell & Wilson, 2011), micro and green chemistry (Bhandari & Raj, 2017) as well as acquisition of sustainability (Burmeister, Rauch, & Eilks, 2012) concepts. Most of the findings from the questionnaire and observation were corroborated in the semi-structured conversations with students. Some few captured comments were:

- *Working in groups makes learning easy*
- *The micro equipment is fun and easy to work with and give faster results than conventional equipment*
- *Humanitarian engineering enables me to see the environment I live in differently and more as my responsibility*
- *I was able to build a water pump that I engineered for a community using safe materials*
- *I am confident I can contribute to solve most of my community's social and physical problems now.*

- *I can understand now why government hammers on the illegal mining activities in my village because the pollution affects many other systems and harms us, as well as trees and other animals that we can feed on. This could lead to climatic effects and hunger.*
- *I can construct a durable ridge over rivers in hinterlands for easy transport of food and services if I can get the needed support*
- *Teaching chemistry is safe and easier now as natural safe materials from the environment could be used to teach concepts with effect just as we could do with inorganic chemicals that don't degrade quickly in the environment*
- *We must attempt to prevent or reduce the generation of waste and challenges to the barest minimum, instead of treating or cleaning them up. Thus, micro, green or 'regenerative' chemistry should be adopted*
- *Soil and water samples from farms, houses and water bodies were quickly analysed before they could deteriorate*

Analysis of students' laboratory reports, demonstrations of chemical language used, and the sequencing (narration) of their oral presentations also supported their gains in conceptions about chemical, sustainability and humanitarian concepts. For example, the choice of use of a particular soil type, metal or chemical were explained with respect to chemical principles and reasoning and how they could best be applied in a project to solve problems more efficiently. There was evidence of the application of chemical principles and humanitarian principles in trying to solve humanitarian challenges in real life situations as alluded by other researchers (Matlin, Krief, Hopf, & Mehta, 2021; Mc Donnell, O'Connor, & Seery, 2007; Mitcham & Munoz, 2010).

Conclusion to the Baseline Case Study

It was concluded in the presented case study that exposing students to real life experiences in a safe, fun-filled environment, with simple, cost-effective, but appropriate resources such as the MCK and DSK not only made them feel like real scientist, chemists or engineers with a human touch, but enabled them to see the link between chemistry, sustainability of the environment and humanitarianism as an integral part of everyday life that could be used to mitigate humanitarian challenges. Awareness to produce as minimal waste as possible and to dispose of them chemically, and humanitarianly appropriately was heightened. The SDGs became real and workable to the students. It was also observed that class participation, self-directed learning, and engagement in field/outdoor practice of chemistry increased as their class morale improved as confirmed by feedback in post questionnaire and guided conversations. Embedding or integrating sustainability principles, tinged with humanitarianism and humanitarian engineering into chemistry education was possible and has the potential to make the study of chemistry real and its practitioners also responsible for themselves, their neighbours and the world as a whole. It was deduced interpretively that practice of chemistry education beyond the walls of school laboratories could enable learners to imbibe the principles of sustainability and humanitarianism early enough to combat ecological degradation, ensure sustainability and mitigate humanitarian challenges sustainably.

FURTHER DISCUSSION OF CASE PRESENTED

The environmental chemistry lessons that were integrated with humanitarian and sustainability concepts in the microchemistry equipment, explored the role of chemistry in reducing the toxicity of products and services so as to provide better ways to empower life styles through new designs. The activities carried out mirrored real-life situations as they were mostly performed on the field (communities). It was found that humanitarian and sustainability concepts could be integrated into chemistry lessons where by students apply their chemical knowledge on the nature and interaction of matter to solve problems in low-income communities (Campbell & Wilson, 2011). A similar observation was made by Karpudewan et al. (2012b) in their studies on green chemistry where the concepts of sustainability and the need to sustain the environment was learned through chemistry laboratory practice. This was important as it exposed students to real-life challenges and exploration of possible solutions. In an implicit way, provision was made for motivation and full engagement in the learning process to solve real life problems by applying chemical knowledge of matter in this current study. Many advantages were identified with the group interactive techniques and especially with the use of interdisciplinary and multidisciplinary approach to solve humanitarian challenges in identified communities.

Findings from the rated observation and questionnaire data showed that participants acquired manipulative, process, transferrable and concept skills for lifelong learning and sustenance of the environment, especially through the use of the microchemistry equipment. Hanson and Acquah (2014) had made that observation in research studies that involved the use of microchemistry equipment. The activities stimulated and motivated the participating students to solve real time and complex tasks. They discovered their capabilities and limits. More importantly they learned that when they work in groups, they are able to analyse and synthesise ideas in varied ways for deep learning. What was more important and exciting was the incidental involvement of some community members in the students' pre-project and actual on-site project activities.

According to Szitar (2013), teamwork helps in the development of multiple intelligence, stimulates and develops divergent thinking, critical thinking, tolerance, cooperative skills and minimises the emotional blockage of creativity. Each of the work groups in the current research engaged as specialists in their area of challenge. It was observed that better cohesion was created in the small groups and proved to be beneficial as active learning, reflective practice and personal accountability of actions for the welfare of others was paramount (Alanazi, 2016; Ali, 2019; Arroio, 2010). UNESCO (2012) notes and recommends that pedagogies that could be considered effective in transforming teaching and learning as well as entire curricula progressively should be capable of stimulating learners to ask questions, analyse situations, think critically and make worthwhile decisions. Such learning situations must spur collaboration. It is imperative that in the introduction and application of humanitarian engineering as well as sustainability concepts, people work together in concert to arrive at solutions which are often of environmental, political, social and economic origins. In this feat, pedagogical techniques to employ for major transformations as with the integration or introduction of humanitarian engineering must prepare learners to enter the workforce on collaborative terms.

Some alien (non-traditional) disciplines are economics, humanities, sociology, cultural studies, health sciences, water sustainability, climatology, peace and security, and agricultural science all became part of the new integrated chemistry course and exposed students to many new facets of life engagements. From this exposition, it becomes obvious that successful integration of chemical education with sustainability and humanitarianism would require developing partnerships with these disciplines to form a contemporary, united educational platform in order to move towards eco-friendly sustainability and humanitarianism as observed by Matlin et al. (2021).

High chemistry content knowledge and principles of engineering were important factors that aided the feasibility of the study. According to Orgill, et al. (2019) and Okanlawon (2018), content knowledge was what teachers would require to be able to sequence letters appropriate and be able to weave in other social issues with proper coordination and this was observed to be an important factor. As reiterated earlier, the researchers would want to deduce interpretively that practice of chemistry education beyond the walls of school laboratories could enable learners to imbibe the principles of sustainability and humanitarianism early enough to combat ecological degradation, ensure sustainability and mitigate humanitarian challenges sustainably.

A weakness that was observed in this study was the difficulty of cooperation in the early phases of the project where participants had to be in larger groups, possibly from unanimity of opinions, lack of skills to collaborate in a group or far of having one's contribution side-lined. This weakness and factors that initiate it so that it could be mitigated, could be investigated in another research.

SOLUTIONS AND RECOMMENDATIONS

Humanitarian challenges are mostly human-caused and could be solved if we could change our attitudes with increase in content knowledge on initiators of the challenges. As reiterated in earlier texts stand-alone courses that relate to socio-scientific and environmental issues within national and international contexts could also be developed if not integrated. Besides that, pedagogies that involve student engagement in the integration of engineering, and humanitarian engineering would promote concepts of humanitarianism and sustainability for a better, sustainable, resilient, and safer world. In this regard, pedagogies such as problem-based learning, project-based learning, inquiry-oriented classroom discussions are recommended for contemporary chemistry teaching and learning processes. It is further recommended that institutions acquire mobile laboratories for use during-out-door or on-field activities so that samples are not necessarily kept for long periods before they are transported to school laboratories for analysis, as some samples deteriorate over time.

FUTURE RESEARCH DIRECTIONS

The future of integration of the tenets of humanitarian engineering and issues of sustainability have bright prospects on 21st Century learners and the positive repercussions on the future of the world. The theme of this book and this chapter in particular that provides a pathway on how chemistry education could be managed from humanitarian and sustainability point views would provide the both the natural and social sciences with integrative and multidisciplinary approaches to teaching and learning in more contextual ways. In the area of science education, all the distinctive disciplines avail themselves for this new integrative way of teaching. A similar situation is obvious with the social sciences and humanities where the integration of the natural sciences, and technology in particular, could be applied to solve problems evident in local communities and national fronts through integrative studies in multidisciplinary ways. It would be useful to conduct similar research among secondary and primary school learners to really 'catch them earlier' in the sensitisation of sustainability and humanitarian concept acquisition.

CONCLUSION

The exposition, supported by the case study showed that the integration of humanitarian engineering in concert with sustainability concepts had tremendous potential in changing the face of engineering and especially the study of chemistry and marginally, physics. In addition, it had the potential to change attitudes and motivate learners to act pro-humanitarian and exhibit pro-humanitarian values. Practice of chemistry theories studied even in earlier years became useful as students were observed discussion the nature of metals with respect to periodic trends and chemical justifications as to why metal should be chosen as against another due to their reactivities- based on positions on the periodic table. This implies that the integration of project-based laboratory work could build connection between regular chemistry theories and labs, and real world challenges.

I fundamentally hope- though this remains to be studied further- that the endeavour to integrate humanitarian engineering either as an entire concept or on the wings of sustainability principles, would lead to positive humanitarian change and development of chemistry students of University of Education, cascade to the entire university (especially the humanities faculties) and the nation as a whole. Besides, it is anticipated that the design for effecting humanitarian engineering would assist people to contribute to sustainable development, care for the marginalised, strive for peace in immediate communities where they find themselves and endeavour for a planet that is more habitable because each one of us behave responsibly as an individual and contribute to collective decision making for a better and safer world. This chapter, additionally, puts out a call for redesigning practical activities to reorient engineering, physics, chemistry, geography, political science and all teachers of other disciplines to address humanitarian engineering issues in a more multidisciplinary manner through conferences and workshops as outlets or platforms for training teachers.

ACKNOWLEDGMENT

Chemistry students from the University of Education, Winneba as well as engineering students from the College of Engineering in the Kwame Nkrumah University of Science and Technology, are acknowledged for their immense support through participation.

REFERENCES

Akpan, V. I., Igwe, U. A., Ikechukwu, B., Mpamah, I., & Okoro, C. O. (2020). Social constructivism: Implications on teaching and learning. *Brock Journal of Education*, *8*(8), 49–56.

Alanazi, A. (2016). A critical review of constructivist theory and the emergence of constructionism. *American Research Journal of Humanities and Social Sciences*, *2*, 1–8.

Ali, S. S. (2019). Problem based learning: A student-cemtred approach. *English Language Teaching*, *12*(5), 73–78. doi:10.5539/elt.v12n5p73

Arroio, A. (2010). Context based learning: A role for cinema in science education. *Science Education International*, *21*(3), 131–143.

Bell, S. (2010). Project-based learning for the 21st century: Skills for the future. *The Clearing House: A Journal of Educational Strategies, Issues and Ideas*, *83*(2), 39–43. doi:10.1080/00098650903505415

Bhandari, M., & Raj, S. (2017). Practical approach to green chemistry. *International Journal of Pharmacy and Pharmaceutical Sciences*, *9*(4), 10–26. doi:10.22159/ijpps.2017v9i4.15640

Burmeister, M., Rauch, F., & Eilks, I. (2012). Education for sustainable development and chemistry education. *Chemistry Education Research and Practice*, *13*(2), 59–68. doi:10.1039/C1RP90060A

Campbell, R. C., & Wilson, D. (2011). The unique value of huanitarian enginering. In *2011 ASEE Annual Conference & Exposition* (p. 12). ASEE.

Cerghit, I. (2006). *Metode de invatamant*. Polirom.

Cortese, A. D. (2003). The critical role of higher education in creating a sustainable future. *Planning for Higher Education*, 15–22.

Crawley, E. E., Malmqvist, J., Ostlund, S., Brodeur, D. R., & Edstrom, K. (2014). Introduction and motivation. In E. F. Crawley (Ed.), *Rethinking engineering education* (pp. 1–9). Springer. doi:10.1007/978-3-319-05561-9_1

DeJarnette, N. K. (2018). Impleenting STEM in the early childhood classroom. *European Journal of STEM Education*, *3*(3), 1–9. doi:10.20897/ejsteme/3878

Hanson, R. (2014). The impact and challenges of integrating micro chemistry experiments into e-learning. *International Journal for Cross-Disciplinary Subjects in Education*, *4*(1), 1884–1892.

Hanson, R. (2017). Enhancing students' performance in organic chemistry through context-based learning and microactivities: A casestudy. *International Journal of Research and Reflection in Educational Sciences*, *5*(6), 7–20.

Hanson, R., & Acquah, S. (2014). Enhancing concept understanding through the use of micro chemistry equipment and collaborative activities. *Journal of Education and Practice*, *5*(12), 120–130.

Hargroves, K. (2011). *Teaching sustainability in high schools.* Retrieved February 14, 2021, from Resources for rethinking: https://resources4rethinking.ca/en/resource/chemistry-innovations-in-sustainable-development

International Water Management Institute. (2020, October 20). *Annual Report 2020.* Retrieved 2021, from Summary of sustainable development goals: https://www.iwmi.cgiar.org

Karpudewan, M., Ismail, Z., & Roth, W.-M. (2012a). Ensuring sustainability of tomorrow through green chemistry integrated with sustainable development concepts (SDCs). *Chemistry Education Research and Practice*, *13*(2), 120–127. doi:10.1039/C1RP90066H

Karpudewan, M., Ismail, Z., & Roth, W.-M. (2012b). Fostering pre-service teachers' self-determined environmental motivation through green chemistry experiments. *Journal of Science Teacher Education*, *23*(6), 673–693. doi:10.100710972-012-9298-8

Linn, M. C., Bell, P., & Davis, E. A. (2004). Specific design principles: Elaborating the scaffolded knowledge integration framework. In M. C. Linn, E. A. Davis, & P. Bell (Eds.), *Internet environments for science education* (pp. 315–339). Lawrence Erlbaum Associates Publishers.

Mafumiko, F. M. (2012). The potential of micro-scale chemistry experimentation in enhancing teaching and learning of secondary chemistry: Experiences from Tanzanian classrooms. *NUE Journal of International Cooperation*, *3*, 63–79.

Matlin, S. A., Krief, A., Hopf, H., & Mehta, G. (2021). *Re-imagining priorities for chemistry: A central science for 'freedom from fear and want* (International Edition). Angewandte Chemie. doi:10.1002/anie.2021

Mc Donnell, C., O'Connor, C., & Seery, M. K. (2007). Developing practical chemistry skills by means of student-driven problem-based learning mini-projects. *Chemistry Education Research and Practice*, *8*(2), 130–139. doi:10.1039/B6RP90026G

McCullough, J. E., & Hayles, C. S. (2009). Education for sustainable development- Interim paper for COST C25. Sustainability of Construction-Integrated Approach to Lifetime Structural Engineering, 290-297.

Mitcham, C., & Munoz, D. R. (2010). *Humanitarian engineering*. Morgan & Claypool Publishers. doi:10.2200/S00248ED1V01Y201006ETS012

Nolet, V. (2009). Preparing sustainability-literat teachers. *Teachers College Record*, *111*(2), 409–442. doi:10.1177/016146810911100207

Okanlawon, A. E. (2018). Reorienting chemistry teacher education towards integration of sustainability into classroom instructions. In G. Bello, M. T. Mustapha, & M. M. Osokoya (Eds.), *Contemporary isues in science,technology, engineering, arts and mathematics education in Nigeria* (pp. 226–236). Department of Science Education.

Phillips, D. (2007). The good, the bad, and the ugly: The many faces of constructivism. *Educational Researcher*, *24*(7), 5–12. doi:10.3102/0013189X024007005

Resources for Thinking. (2021). *Sustainability Education for high schools*. Retrieved February 2, 2021, from https://resources 4 rethinking.ca/en/sustainable development

SOMERSET Educational Products and Systems. (2017). RADMASTE Microchemistry experiments. University of Witwatersrand.

Stirling, A. E., & Kerr, G. A. (2015). Creating meaningful co-curricular experiences in higher education. *Journal of Education & Social Policy*, *2*(6), 1–7.

Strohfeldt, K., & Khutoryanskaya, O. (2015). Using problem-based learning in chemistry practical class for pharmacy students and engaging them with feedback. *American Journal of Pharmaceutical Education*, *79*(9), 1–11. doi:10.5688/ajpe799141 PMID:26839430

Szalacsi, A. (2015). Paradigm shift in 21st century: A call for action. In *Culture and Education, ISCHE 37* (pp. 1-2). Istanbul, Turkey: *Annual Conference of the Association of Caribbean Higher Education Administrators*.

Szitar, M.-A. (2013). Learning about sustainability community development. *Procedia-Social and Behavioural Sciences, 116*(2014), 3462-3466.

Talanquer, V. (2019). Some insights into assessing chemical systems thinking. *Journal of Chemical Education*, *96*(12), 2918–2925. doi:10.1021/acs.jchemed.9b00218

Tsaparlis, G., & Gorezi, M. (2007). Addition of a project-based component to a conventional expository physical chemistry laboratory. *Journal of Chemical Education*, *84*(4), 668–670. doi:10.1021/ed084p668

UNDP. (2020). *Sustainable development goals*. Retrieved May 14, 2020, from https://www.undp.org/content/en/home/sustainable-development-goals

UNESCO. (2012). *Education for sustainable development in action. Learning and training*. UNESCO Education Sector.

United Nations. (2015). *Sustainable development goals*. Retrieved from Take action for the sustainable development goals: https://www.un.org/sustainabledevelopment/sustainable-development-goals/

United Natons Statistics Division. (2005). *Living beyond our means: Natural assets and human well-being*. UNESCO.

van Bergen, S. (2020). *Green chemistry in higher education*. Retrieved February 02, 2021, from https://ecology.wa.gov/waste-toxics/reducing-toxic-chemicals/greenchemistry

Chapter 13
Humanitarian Needs and Challenges Related to the Disabled Population:
A Socio-Epidemiological Study

Anastasia Kadda
National and Kapodistrian University of Athens, Greece

Yiannis Koumpouros
https://orcid.org/0000-0001-6912-5475
University of West Attica, Greece

Theodora Psaltopoulou
National and Kapodistrian University of Athens, Greece

Thomas Toulias
University of Thessaly, Greece

ABSTRACT

People with disabilities raise important and special humanitarian approaches. There is a strong necessity for an epidemiological and sociological study of disabled population with regard to disease distribution and the patients' social profile based on the bio-psycho-social model of health and illness. Adverse economic data, low social cohesion, and consequently, social exclusion are often observed within the specific population. According to the quantitative research conducted, the population's nosological profile has serious effects on patients' everyday lives and is directly related to social parameters. Fundamental social factors and access to health services affect people with a degree of disability of over 67% health in terms of disease prevention and treatment. At the same time, social determinants seem to negatively affect their health and treatment. There is a clear need for an information technology system that will monitor and produce appropriate key performance indicators to help design appropriate policies for the specific population.

DOI: 10.4018/978-1-7998-9190-1.ch013

INTRODUCTION

Lately, there has been an increasing interest towards health and ill regarding their social dimension. Moreover, the World Health Organization (WHO 1958) definition of health given by the World Health Organization (WHO 1958) has made clear that health is intertwined with social well-being. The changes of the nosological model during the last few decades -the eradication of several infectious diseases together with the appearance of new chronic diseases requiring long-lasting health care- demand holistic medical treatment. In fact, according to the new bio-psycho-social model used for the interpretation of health and illness and in contradiction to the previous biomedical health model (Brandby, 2010), there are not only biological factors that are being involved (genetic predisposition, viral or germ infections) but also psychological (cognitive processes, emotions, behaviors) and social factors (social norms, social values concerning health, social status, nationality etc.) (Nettleton, 2020; Phellas, 2009). Without any doubt, there is a decisive necessity of an epidemiological study of the disease distribution and the morbidity within a population. More specifically, a socio-epidemiological study of diseases is considered to be more than necessary, within the target population of West Athens, based on results derived from researches and indicate that West Athens concentrates on socially disadvantaged groups, such as refugees, economic migrants, Roma, etc., who mainly face poverty, unemployment, lack of access to products and services, low social cohesion and consequently, social exclusion. Such a study would bring to light significant humanitarian needs and challenges closely related to the disabled population.

Based on the observations mentioned above, and since there have not been any similar studies, it was considered essential to conduct a survey that aims at analyzing the social-epidemiological dimension of the diseases among the population of West Athens in Attica with a degree of disability of over 67%.

The objectives of this study were: a) the analysis of demographic characteristics of the respondents (year of birth, area of residence, nationality, religion), b) the study of the crucial social factors of their health/disease (education level, working status, economic status, living conditions, eating habits, alcohol-coffee-tea consumption, smoking habits, sleep habits, television consumption, physical activity, anxiety or social stress, delinquency, official or unofficial social support, accessibility to health services c) the epidemiological study of the diseases bared by the respondents and due to which might have been hospitalized.

The survey was based on the quantitative research method. The target population of the survey consisted of approximately 1000 individuals. Among them, only a sample of 203 individuals was selected, which were suffering from a disease at that time and were coming to the Primary Health Committee of West Athens to get evaluated and rated according to their disease.

The method we used to collect data was through an appropriate questionnaire in web-based and paper formats. The questionnaire was anonymous and consisted of 87 closed-ended and open-ended questions respective to the study's purpose and objectives.

The survey was conducted in the Primary Health Committee of West Athens premises. The respondents were evaluated to obtain a disability degree of over 67%, based on their disease. The survey process lasted for about three months, and the results were collected and edited through the Statistical Package for the Social Sciences (IBM SPSS version 19).

Two hundred three answers were collected in total, and target-population consisted of all categories of people, irrespective of their sex, age, education level etc.

The difficulties and limitations experienced during this survey concern the following: concise time invested in answering the questionnaire on behalf of the respondents, some of the open-ended questions had been left incomplete and the fact that individuals with a severe motor disability did not participate in the survey due to lack of accessibility (they were evaluated by the Health Committee in another place separately).

A Sociological Study of Health and Illness: Key Social Factors

Health protection and illness depend on personal-behavioral factors (*Anatomical, physiological and metabolic* characteristics, genetic predispositions, personality structure and behavior) as well as environmental factors (Sarris, 2001) (social, economic, cultural, political and ecological factors), which interact directly with one another throughout individual's lifetime and affect his or hers health condition (Ioannidi & Manti, 1999; Irwin& Solar, 2010; Samuels-Kalow et al., 2020). According to World Health Organization (WHO), social determinants of health consist of "the conditions under which people are born, grow, work, live and the total number of external parameters and systems which affect 'individual's everyday life". These crucial social factors are causally related to health and treatment and prevent disease and maintain health. Social institutions such as family, education level, social status, social structure and adverse social conditions affect health status (Cockerham et al., 2017).

More specifically, family plays a primary role in health status, as it has been observed that apart from safety, guidance and help that may provide to its members, it can also positively affect their health and lay the foundation for the protection and promotion of mental and physical health, especially of children. On the contrary, certain studies have shown that individuals who have never been married, living individually or being divorced, become more vulnerable, with reduced readiness in their immune system and increased mortality rates. For example, it has been found that children from divorced families may experience physical and mental diseases (gastric ulcer, colitis, cognitive and behavioral disorders) (Tountas, 2000; Petridou, 1988). On the other hand, should a severe disease in a family become more widely known, it may severely affect social interactions like marriage negotiations (Davou & Christakis, 1994).

Additionally, an individual's *educational and cultural level* affects his or her health. At the same time, they shape corresponding attitudes and behaviors, equipping the individual with the appropriate knowledge and skills in issues such as proper nutrition, sex education, and rational use of health services. It is well known that men and women with low educational status, working in manual labor jobs are expected to participate less in health education programs, smoke more often, have a worse nutrition profile, and increase body weight. In England, for example, research results have shown that the life expectancy of a child born from University educated parents with a respective social status is five years longer than the life expectancy of a child born from unskilled manual workers.

Moreover, *income* and income distribution are significant determinants of health and illness. This finding lies in substantial differences in life expectancy; infant mortality and morbidity have been observed between the lower and upper social classes. Low income may affect health through lack of necessities -inability to acquire adequate nourishment, housing and heating-. It may also have social effects: incapability to participate in community life, social isolation, absence of support networks and lack of self-esteem, which all have a mostly negative impact on an 'individual's health (Frank et al., 2020). At the same time, social status also seems to determine an 's health (Avlund et al., 2003), since individuals with larger resources dispose of better social, psychological, and economical means to handle less favorable situations. The process between health and social stratification is defined through several stages, such as power (power and control at home and in the workplace), social advantages (social networking) and human resources (skills, education). Results from a relevant study indicate a 93% chance of survival for men with a higher socioeconomic status within the next five years (based on their income and education level) and an 84% chance of survival for men with a lower socioeconomic status.

Social and geographical mobility may also affect health, as it has been confirmed that healthy individuals are more likely to display upward mobility. On the other hand, individuals with poor health are more likely to fall into lower social classes. Additionally, several studies have been conducted among populations with a special internal or external migration rate concerning coronary insufficiency, lung cancer, sarcoidosis and other physical or mental illnesses. Those studies have demonstrated that frequent changes of occupation, residence, and living conditions were very early associated with an increase in morbidity among the respective population groups.

Social stress is another determinant (Aneshensel, 1992) that negatively affects health, as it *decreases the defenses* and makes the individual vulnerable to one or more severe diseases (back pain, palpitations, muscle spasms, cardiovascular disease, hypertension), and prone to unhealthy habits (cigarette smoking, drug use, alcohol consumption) as well as to psychosomatic problems and symptoms. Other studies have proved the association between stressful life events (change of residence, life psychological traumas, frequent professional changes, death of a relative, childbirth) and disease appearance.

Furthermore, *social and support systems* can play a positive or a negative role in a country's health level (Uckino, 2006). This depends on social cohesion, social equality, solidarity, and consequently on social capital (social security and support, solidarity networks, support groups for people with chronic illness, etc.). Social group inclusion and support (family, community, professional frame, formal or informal social organizations) provide access to knowledge, reinforce the incentives for selecting an adaptive behavior, encourage behaviors that promote health and well-being and increase an individual's ability to survive and adjust to the social environment. It has derived from a multitude of studies that there is a positive correlation between individuals' psychosocial health (regardless of gender and social status) and social incorporation – participation and conversely, a positive correlation between social isolation and disease (Pontisidis & Mpilali, 2015; Fakiolas, 2012; Peirce et al., 2000).

The relationship between health and employment (Wang et al., 2008) is equally essential: Employment may have indirect health effects (it adjusts income level which respectively affe*cts health level, also has a* severe *impact o*n an individual's psychosocial development, and his or hers basic life orientation and socialization). But it may also have direct health effects, mild or not, concerning a clinical entity. This applies to work accidents, which can cause damage to some parts of the body or even death, and that of occupational diseases such as occupational diseases of the skin, the liver, blood, the nervous system, sense organs, and the urinary and the cardiovascular system. Work-related stress also has a severe impact on health. A significant correlation has been found between workload and certain diseases such as cardiovascular disease (Price, 2004).

On the other hand, *unemployment* may cause poor health, premature death, and mental health deterioration. At the same time, it may raise the risk of suicide, especially within vulnerable populations (single mothers, low-income families, unskilled workers, less educated) and long-term unemployed individuals (Bambra & Eikemo, 2009). The social and psychological importance of unemployment has now clearly extended beyond its economic dimension since unemployment deprives individuals of their social role. Additionally, as the unemployment level rises and 'peoples' living conditions deteriorate, morbidity and mortality indicators (Tselepi, 2000). Most epidemiological studies conducted in this area tend to point out a negative role of unemployment in the overall function of an individual, based on morbidity and mortality 'rates' difference between employed and unemployed.

Moreover, living conditions, especially housing, are critical to health (Raphael, 2016), as the connection between poor housing conditions and poor health has been established long ago. Such conditions constitute a problem for non-privileged groups which are unable to have a decent house or a flat and additionally for thousands of homeless people that, even in our days, are met in the streets of many European countries. Several studies have indicated that socially disadvantaged areas are characterized by high rates of drug use, crime, social problems, unemployment, a weak system of social support and also insufficient social services and social activities, which are combined with depression, hypertension, cardiovascular diseases, multiple accidents, complications in pregnancy, use of harmful to health substances and even suicides.

At the same time, research has proved that a country's socioeconomic level affects morbidity and mortality rate -a country's economic and social development level is associated with morbidity and deaths caused by infectious diseases. A low socioeconomic status relates to high rates of mortality caused by malignant neoplasms and ischemic heart disease; and the level of industrialization positively correlates with the rise of deaths due to diabetes. Consequently, a country's social and economic development contributes to improving individuals' health levels. In contrast, industrialization, which creates a natural and social "micro-environment" is considered as a harmful factor for health (Kyriopoulos et al., 1983).

Healthcare access (Kanuganti et al., 2015) and *Quality of medical and hospital care* are also essential to health, given the fact that people in blighted areas who live under challenging conditions are very likely to be in a poor health condition. Also, they are not expected to receive a high level of healthcare services. For example, analysis of geographical distribution of services, often indicates a poor provision of services in rural or economically disadvantaged areas. Economic analyses they suggest that there is a poor provision of services to low-income groups. Furthermore, there are several financial obstacles to the establishment of new services and cultural barriers to the access of services.

Moreover, 'individuals' physical health may be affected by the *culture of a society* (customs, manners, lifestyle, values and beliefs) in several ways (e.g., it has been observed that the Mediterranean diet contributes to longevity), since culture shapes ideas and practices which may affect health. Consequently, we can observe different behaviors and attitudes towards health, illness and treatment among individuals from different ethnic backgrounds and cultures.

The *political system* -the political, legal and administrative services and the institutions- affects health similarly, since its parameters control the urban and the work environment. For example, alcohol prohibition for minors may prevent and decrease alcoholism, and a public information hotline on sexual practices may contribute to the prevention of sexually transmitted diseases.

Environmental and *geophysical* factors (natural resources, raw materials, climate conditions, protection level of ecosystems) also significantly impact health. At the same time, *the ecological imbalance* caused by environmental pollution (Lu et al., 2017) harms individuals' health through ozone depletion, for example, air pollution, nuclear accidents, chemical waste etc.

Furthermore, *lifestyle* and health-related behaviors constitute critical factors in protecting and promoting health. More specifically, eating habits, a key parameter of lifestyle, is causally related to cardiovascular diseases, malignant neoplasms, hypertension, diabetes, obesity etc. (Bleich et al., 2015); while malnutrition and undernutrition may lead to diseases such as Kwashiorkor, due to vitamin deficiencies, endemic goiter which is an iodine-deficiency disease, iron deficiency anemia, *rickets* -a skeletal disorder which is caused by a lack of vitamin D- and scurvy, a disease resulting from vitamin C deficiency.

Additionally, cigarette smoking has been associated with a significant number of premature deaths and at least 25 diseases such as cardiovascular diseases, many malignant neoplasms (*cancers of the lung, larynx, oral cavity,* esophagus, bladder, kidneys etc.), chronic respiratory diseases (chronic bronchitis, pulmonary emphysema, asthma) and many more diseases and disorders (Huttunen et al., 2011). According to WHO, cigarette smoking is also associated with poor life quality since 12.3% of disability-adjusted life years (DALYs) are attributed to smoking every year (Tountas, 2012). Moreover, consumption of alcoholic beverages is associated with cardiovascular diseases and malignant neoplasms (liver, mouth, pharynx and larynx cancers), mental illness, and accidents. At the same time, drug use causes mild or even more severe health damages, particularly in the brain structures and neurons.

On the other hand, physical activity reduces morbidity and mortality (Fletcher et al., 2018), while lack of physical activity is associated with cardiovascular diseases, diabetes, osteoporosis, and obesity. Healthy sex life has a positive effect on mental and physical health, while unsafe sex endangers health (risk of Sexually Transmitted Diseases, some of which may cause death). Also, proper sleep is essential for maintaining good mental and physical health, while unsafe driving behavior has been associated with injuries, disabilities and even deaths.

Additionally, *medicine consumption,* which is characterized by the polypharmacy phenomenon and the medical drug abuse and *overconsumption, often coming as a result of poor medical attention,* constitutes a health danger that may cause side-effects to the patient and frequently contributes to the appearance of “iatrogenic diseases”.

Finally, the psychological impact of life events on morbidity has been the subject of many retrospective studies. Such events are the change of residence, life psychological traumas, frequent professional changes, death of a relative and childbirth, which seem to be related to various nosological entities like eczema, coronary insufficiency, childhood diseases, pregnancy complications, etc.

Even though many epidemiological types of research have lately focused on individual health risk factors (diet, cholesterol level, exercise, etc.), scientists accept that social factors and especially the socioeconomic conditions and social support system are fundamental to maintaining health. Conversely, social factors may trigger diseases, as far as they involve accessibility to a vital resource. They may also affect the outcome of diseases through multiple processes, and consequently, they are still associated with the disease, despite the mechanisms of change.

Epidemiological Study of Diseases

Through the epidemiological study of the diseases among the human population, the unique features of a population and the factors that may affect it, e.g., age, sex, race, geographical region, profession, etc., we can acquire the essential knowledge to make an integrated evaluation of health within the human community. This knowledge forms the framework for practicing preventive and therapeutic medicine (Roumelioti, 2010; Tzoufras & Piperoglou, 2009).

More specifically, this study attempts to analyze the diseases among the population of West Athens in the Administrative Region of Attica, which has been certified with a degree of disability of over 67% by the competent Health Committee; therefore, an epidemiological study is conducted through all the critical phases.

The selection of West Athens’s population among the whole region of Attica has been made according to the unique characteristics observed in this particular area. More concretely, West Athens constitutes a regional unit of the urban complex of Athens (West basin of Attica), including seven municipalities: Agia Varvara, Haidari, Egaleo, Peristeri, Petroupoli, Ilion and Agii Anarghiri-Kamatero.

West Athens is the result of the residential development and the industrialization of the area beside the Kifissos River. Low-cost residences for workers, along with industrial and manufacturing buildings, sprawl over large amounts of land, resulting in degradation in many sectors (urban sprawl, high residential densities, lack of open spaces in residential areas, lack of social equipment and urban infrastructures, increased circulatory dysfunction, air and environmental pollution, etc.).

West Athens constitutes a relatively new society that has instead derived from the influx of "foreign" populations (Asia Minor Catastrophe refugees, economic migrants, repatriated populations such as expatriates from Albania, Albanians) from the evolution of local factors.

In social terms, the population of West Athens *includes socially disadvantaged groups* such as refugees, economic migrants, Roma, etc., sharing common characteristics which lead to social exclusion. Additionally, taking account of the rate of typically disadvantaged groups (people with disabilities, long-term unemployed individuals, drug users, homeless), which is slightly above the average, we can easily assume that social exclusion and low social cohesion and integration level are some of the significant problems for West Athens's population.

The population's education level is also related to the problems mentioned above. More specifically, it is very well known that the education level, which is also associated with other factors (poverty, lack of access to products and services, etc.), contributes to the social exclusion of certain groups and individuals.

Concerning the official unemployment rates, high unemployment rates are observed in the regional unit of West Athens, at the national level, especially among young people and women. The rising unemployment rates among economic migrants living in the area are not estimated in these total rates.

As far as enterprises are concerned, in recent years of socioeconomic crisis, the collapse of companies was often observed, particularly those of the secondary production sector and a fall in trade, which included the region's main economic activity.

Specialists who study the development path of the seven municipalities of West Athens estimate that unemployment is expected to sharpen social problems, especially the spread of drug use among youth, organized violence and crime. The rise in population ageing, which is observed mainly in Egaleo and Peristeri, is also a deteriorating factor since it is ascertained that it increases poverty so for the older adults as for their wider family environment (relatives that they may depend on).

In the health sector, observations concern *lack of material and technical equipment, lack of structure, nursing staff deficit, insufficient health clinics and services (IKA)* and *inadequate* Urban Health Centers.

RESULTS

Family Status

Investigating the marital status of the respondents, it was found that the most significant percentage of respondents (62.1%) are married while the rest of them belong to various marital status categories (20.7% unmarried, 10.8% divorced, 3.9% widowed, and 2.5% cohabit); see a pie chart of Figure 1.

The bar chart in Figure 1 presents the mean of the years in which they remain in the above statuses, with the married ones remaining the most (37.5 years) and the cohabit the less (8.8 years), where the dashed line corresponds to the total mean (30.5 years); see also the related boxplot of Figure 1 and the corresponding Descriptive tables of Table 1.

Figure 1. Family status

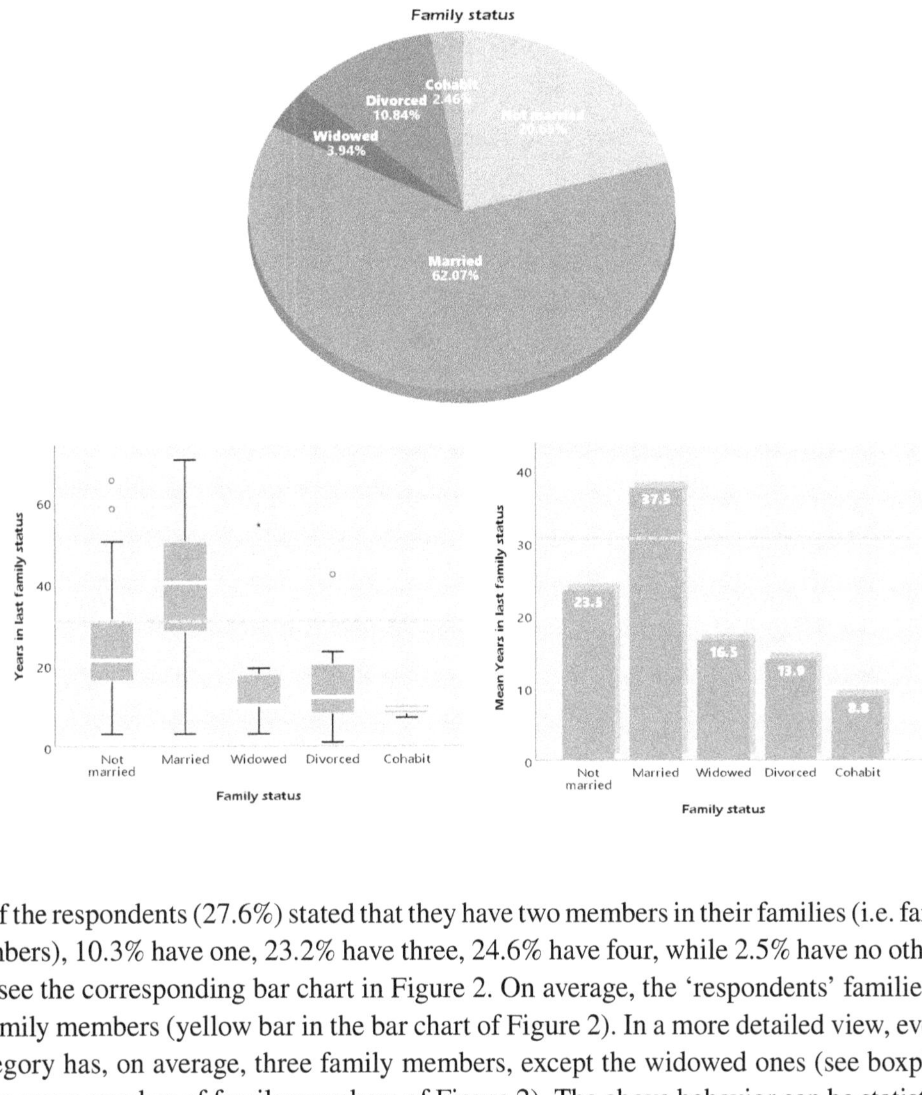

Most of the respondents (27.6%) stated that they have two members in their families (i.e. families with three members), 10.3% have one, 23.2% have three, 24.6% have four, while 2.5% have no other member (singles); see the corresponding bar chart in Figure 2. On average, the 'respondents' families consisted of three family members (yellow bar in the bar chart of Figure 2). In a more detailed view, every marital status category has, on average, three family members, except the widowed ones (see boxplot and bar chart of the mean number of family members of Figure 2). The above behavior can be statistically confirmed for the whole population of the disabled persons in West Athens (the Kruskal-Wallis test for the equality of the distribution of the number of family members and the distribution of the marital statuses, except the widowed, gives $p = 0.804 > 0.05$, i.e. we accept the equality of the distributions across the categories of family members except the widowed. In contrast, the corresponding Kruskal-Wallis test concerning all the categories of family statuses, including the widowed, gives $p = 0.023 < 0.0$, i.e. we reject the equality of the distributions across the categories of family members); see the Hypothesis Test Summary tables of Table 2.

Educational Level and Employment Status

As far as the educational level of respondents is concerned, the majority, 26.1%, have finished elementary education, 22.7% gymnasium, 14.3% lyceum, 11.3% failed at elementary but can read and write, 8.4%

Table 1. Years spend in last family status

Descriptives

			Statistic	Std. Error
Years in last family status	Mean		30.50	1.143
	95% Confidence Interval for Mean	Lower Bound	28.24	
		Upper Bound	32.75	
	5% Trimmed Mean		30.37	
	Median		30.00	
	Variance		265.439	
	Std. Deviation		16.292	
	Minimum		1	
	Maximum		70	

Descriptives

	Family status			Statistic	Std. Error
Years in last family status	Non-married	Mean		23.48	2.087
		95% Confidence Interval for Mean	Lower Bound	19.26	
			Upper Bound	27.69	
		5% Trimmed Mean		22.50	
		Median		21.00	
		Variance		182.938	
		Std. Deviation		13.525	
	Married	Mean		37.49	1.244
		95% Confidence Interval for Mean	Lower Bound	35.03	
			Upper Bound	39.95	
		5% Trimmed Mean		37.92	
		Median		40.00	
		Variance		194.876	
		Std. Deviation		13.960	
	Widowed	Mean		16.50	5.612
		95% Confidence Interval for Mean	Lower Bound	3.23	
			Upper Bound	29.77	
		5% Trimmed Mean		15.17	
		Median		10.00	
		Variance		252.000	
		Std. Deviation		15.875	
	Divorced	Mean		13.86	1.943
		95% Confidence Interval for Mean	Lower Bound	9.82	
			Upper Bound	17.90	
		5% Trimmed Mean		13.10	
		Median		12.00	
		Variance		83.076	
		Std. Deviation		9.115	
	Cohabit	Mean		8.80	.583
		95% Confidence Interval for Mean	Lower Bound	7.18	
			Upper Bound	10.42	
		5% Trimmed Mean		8.83	
		Median		9.00	
		Variance		1.700	
		Std. Deviation		1.304	

Figure 2. Family status and number of family members

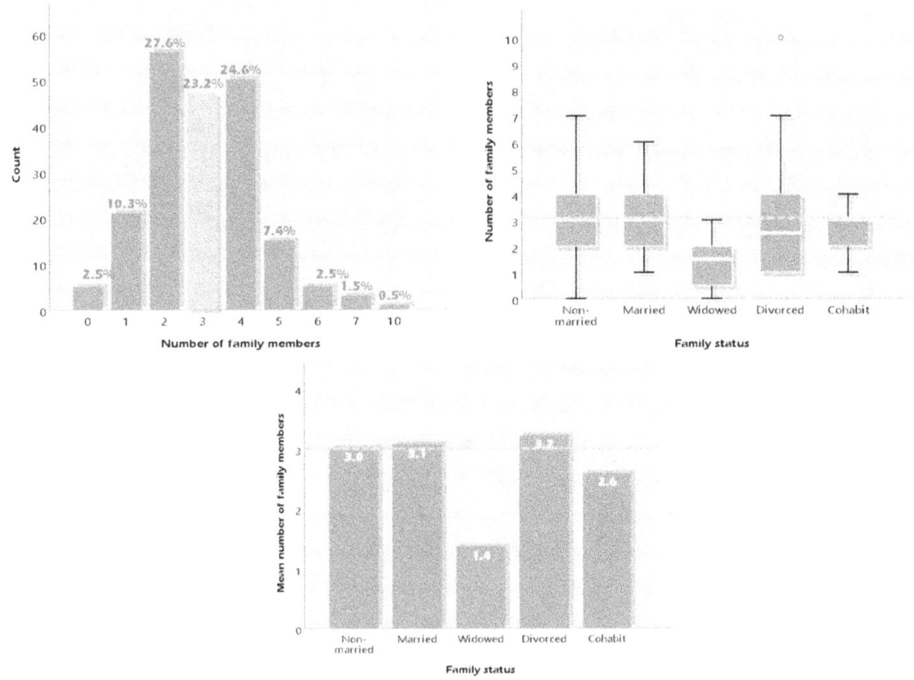

are illiterate. Also, 5.9% completed post-secondary studies, 6.4% have higher technological education (Technological Institutions), 3.4% have university tittles, while only 1.5% have postgraduate diplomas. For the years spent in educational institutes, the majority, 45.8%, of the respondents have 0-6 years of study, following those who have 7-9 years at a rate of 22.7%, 17.2% have more than 12 years of study and even fewer (14.3%) 9-12 years.

Concerning their professional status, only 13.7% are employed, with 68.2% employed in the past. It can be statistically inferred that one-tenth of the population (of the disabled persons in West Athens) is

Table 2. Years spend on last family status and number of family members

Descriptives

Number of family members

	N	Mean	Std. Deviation	Std. Error	95% Confidence Interval for Mean Lower Bound	95% Confidence Interval for Mean Upper Bound	Minimum	Maximum
Not married	42	3.02	1.660	.256	2.51	3.54	0	7
Married	126	3.08	1.100	.098	2.89	3.27	1	6
Widowed	8	1.38	1.061	.375	.49	2.26	0	3
Divorced	22	3.23	2.544	.542	2.10	4.36	1	10
Cohabit	5	2.60	1.140	.510	1.18	4.02	1	4
Total	203	3.00	1.471	.103	2.80	3.21	0	10

Statistics

Number of family members

N	Valid	203
	Missing	0
Mean		3.00
Median		3.00
Std. Deviation		1.471
Variance		2.163
Minimum		0
Maximum		10

Hypothesis Test Summary

	Null Hypothesis	Test	Sig.	Decision
1	The distribution of Number of family members is the same across categories of Family status.	Independent-Samples Kruskal-Wallis Test	.804	Retain the null hypothesis.

Asymptotic significances are displayed. The significance level is .05.

Hypothesis Test Summary

	Null Hypothesis	Test	Sig.	Decision
1	The distribution of Number of family members is the same across categories of Family status.	Independent-Samples Kruskal-Wallis Test	.023	Reject the null hypothesis.

Asymptotic significances are displayed. The significance level is .05.

Figure 3. Years spent in current and past professions and in education regarding the work categories

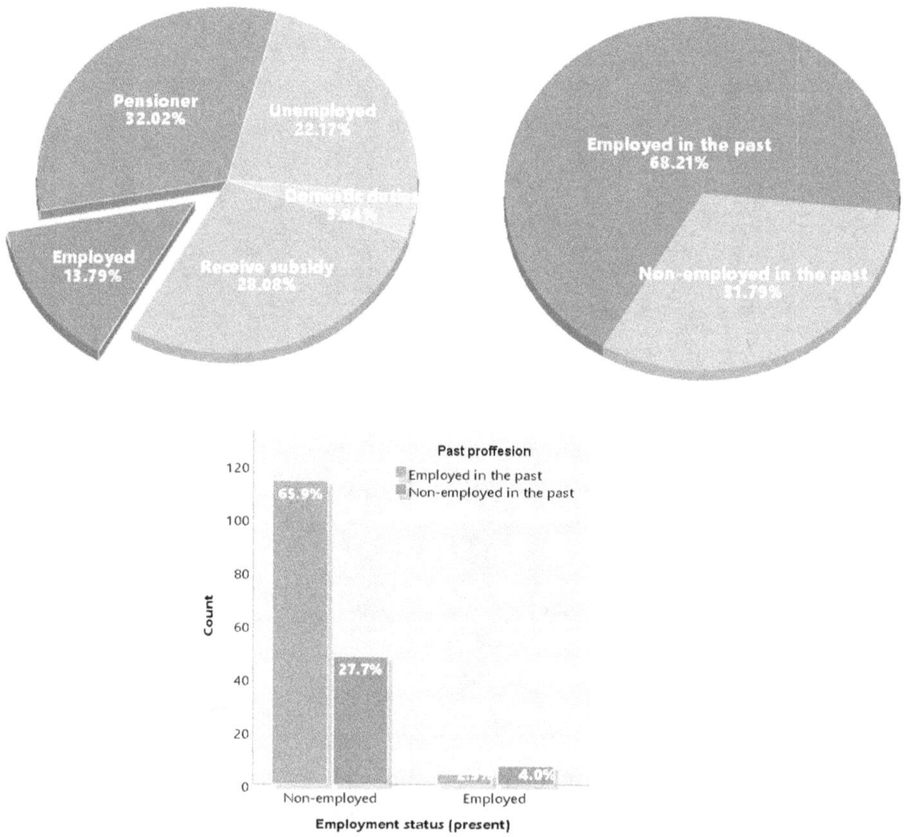

employed. In contrast, one third were employed in the past (the binomial tests in Table 3 confirms the above with $p = 0.051 > 0.05$ for the current employment and $p = 0.33 > 0.05$ for the past employment status). See also Figure 3 for a visual description of the above (pie charts).

The population of the current employees that are disabled persons have been working on average for 18.8 years (as their working years are typically distributed, with a mean of 18.81, since the K-S normality test shows a p-value of $0.431 > 0.05$ in the top Test of Normality table of Table 3); see upper Descriptives table of Table 3. The ones that worked in the past (and now being non-employment) had worked on average for around 21 (see the corresponding bottom tables of Table 3). Moreover, the employment years of the current and the past employed respondents do not seem to correlate with each other (the corresponding chi-squared test reveals $p = 0.19 > 0.05$); see the right sub-table of Table 3. However, it is interesting to mention that, in terms of risk analysis, **if a disabled person (from the whole population) worked in the past, it is 4.15 times more likely not working now** (odds ratio value 4.156 in Table 3).

Regarding the type and conditions of the employed disabled persons, the upper three pie charts present their responses, where most of them perform muscle work (48%), and their work conditions are mixed.

Table 3. Descriptives and tests regarding current and past profession

Descriptives

			Statistic	Std. Error
Years in last professional status - Employed	Mean		18.81	1.977
	95% Confidence Interval for Mean	Lower Bound	14.75	
		Upper Bound	22.88	
	5% Trimmed Mean		18.73	
	Median		20.00	
	Variance		105.541	
	Std. Deviation		10.273	
	Minimum		2	
	Maximum		38	
	Range		36	
	Interquartile Range		18	
	Skewness		.003	.448
	Kurtosis		-.945	.872

Descriptives

			Statistic	Std. Error
Years of past proffesion	Mean		21.442	1.1027
	95% Confidence Interval for Mean	Lower Bound	19.258	
		Upper Bound	23.626	
	5% Trimmed Mean		21.301	
	Median		20.000	
	Variance		142.256	
	Std. Deviation		11.9271	
	Minimum		2	
	Maximum		50.0	
	Range		49.8	
	Interquartile Range		19.5	
	Skewness		-.002	.224
	Kurtosis		-1.044	.444

Binomial Test

		Category	N	Observed Prop.	Test Prop.	Exact Sig. (1-tailed)
Employment status	Group 1	Not-employed	175	.9	.9	.051[a]
	Group 2	Employed	28	.1		
	Total		203	1.0		

a. Alternative hypothesis states that the proportion of cases in the first group < .9.

Chi-Square Tests

	Value	df	Asymptotic Significance (2-sided)	Exact Sig. (2-sided)	Exact Sig. (1-sided)
Pearson Chi-Square	5.493[a]	1	.019		
Continuity Correction[b]	4.037	1	.045		
Likelihood Ratio	5.037	1	.025		
Fisher's Exact Test				.038	.026
Linear-by-Linear Association	5.462	1	.019		
N of Valid Cases	173				

a. 1 cells (25.0%) have expected count less than 5. The minimum expected count is 3.50.

b. Computed only for a 2x2 table

Risk Estimate

	Value	95% Confidence Interval Lower	95% Confidence Interval Upper
Odds Ratio for Employment status (Non-employed / Employed)	4.156	1.163	14.858
For cohort Past profession = Employed in the past	1.935	.880	4.256
For cohort Past proffesion = Non-employed in the past	.466	.281	.772
N of Valid Cases	173		

Health and Safety Conditions at Work

As far as their 'work's health and safety conditions are concerned, the three middle pie charts of Figure 4 show a transition from worse to better conditions as we move from "manual or muscle" to "mental" work type. However, this sample observation cannot be inferred for the entire population of the disabled persons of West Athens since we have statistical evidence that the health and safety conditions are not correlated significantly with these two physical work types ($p = 0.08 > 0.05$ for the chi-squared test between the health and safety conditions and the muscle/mental work types; see Table 4). That is, the health and safety conditions are not affected by the physical style of work.

The physical work conditions (manual/mental labour) and the health and safety conditions do not reflect the years that the disabled persons spend in their current profession ($p = 0.865 > 0.05$ and $p = 0.932 > 0.05$ of the Kruskal-Wallis tests between the physical and the health/safety conditions respectively, shown in right sub-tables of Table 5). They tend to work the same number of years (on average 18.8), whether the work, manual or cognitive or the conditions. In Figure 5, the boxplots and bar charts depict the behaviour of the respondents regarding their current working years, categorized according to the physical conditions of the profession (top sub-figures) or the health and safety conditions (bottom sub-figures); the horizontal dashed lines correspond to the (total) average of working years of the population (18.8) and the samples median (20) of the working years.

Figure 4. Type of work and health and safety conditions about the physical kind of work

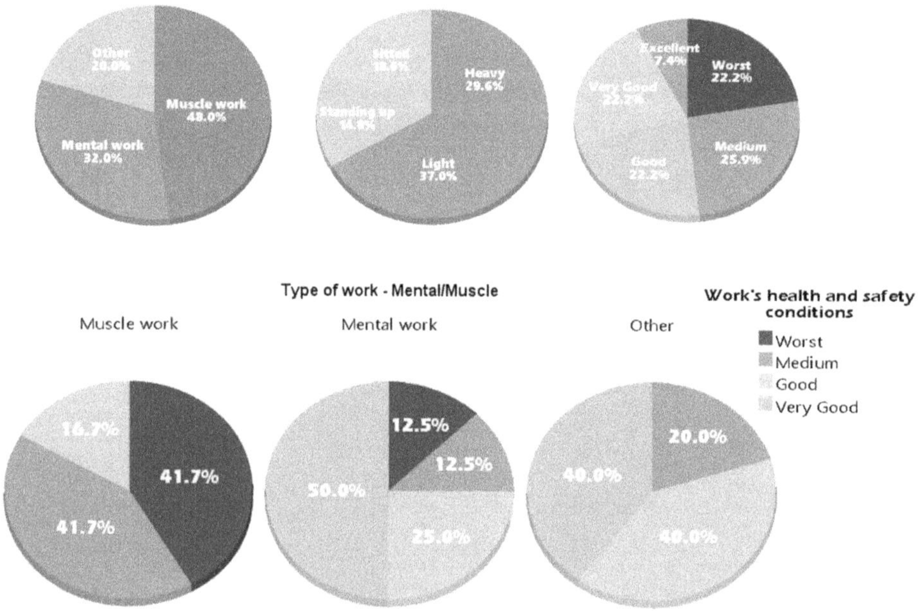

Educations vs. Work Status

The average level of education of the respondents is 3.95 (i.e. from secondary to lyceum; top descriptive sub-table of Table 6) on a scale from 0 (illiterate) to 10 (postgraduate studies); see left sub-figure of Figure 6. The average years spent by the respondents in educational institutes are 8.3. Moreover, the number of years spent in education is reflected in the physical type (mental or muscle) of work ($p = 0.014 < 0.05$

Table 4. Chi-squared test between the health and safety conditions and the physical type of work

Chi-Square Tests

	Value	df	Asymptotic Significance (2-sided)
Pearson Chi-Square	11.275[a]	6	.080
Likelihood Ratio	14.564	6	.024
Linear-by-Linear Association	8.062	1	.005
N of Valid Cases	25		

a. 12 cells (100.0%) have expected count less than 5. The minimum expected count is 1.20.

Figure 5. Years spent in current professions in relation to the physical nature and the health and safety conditions of the profession

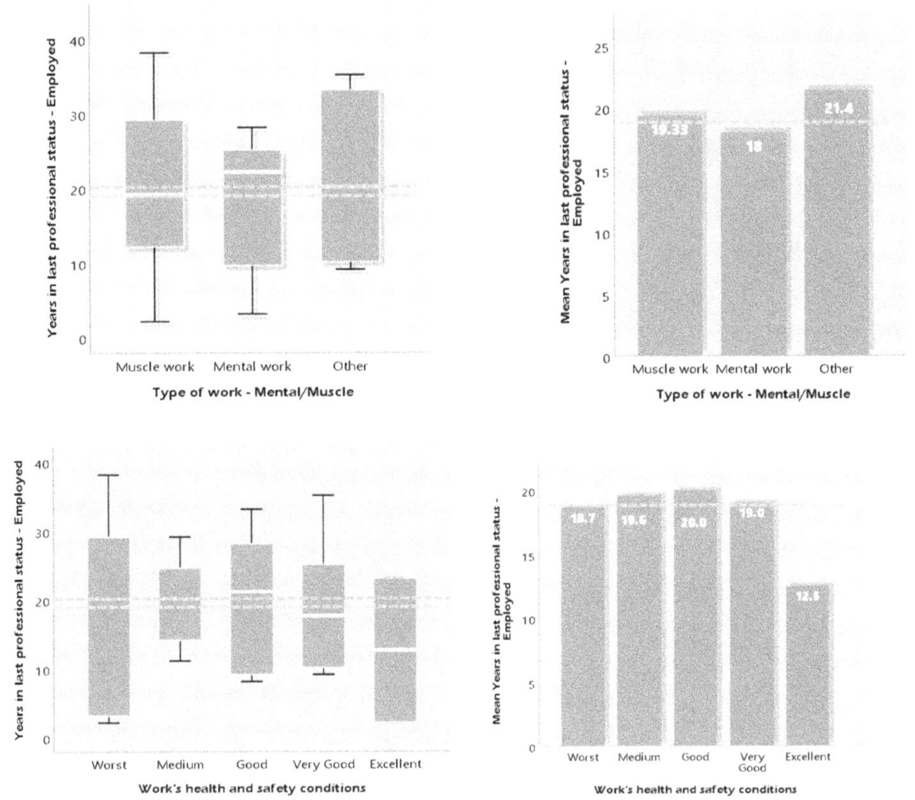

for the Kruskal-Wallis test, and $p = 0.048 < 0.05$ for the independent sample t-test of education years in relation to mental/muscle type of work; the independent samples t-test is valid since the muscle/mental categories are typically distributed, see normality test sub-table in Table 6). In the boxplots of Figure 6, note that the "mental" professionals have a university (and above) level of education.

Moreover, the employed spent almost three more years on their education (10.7) than the non-employed (7.9); see the corresponding mean bar chart of Figure 6. This differentiation is inferred to the whole population of the disabled persons ($p = 0.001 < 0.05$ for the Mann-Whitney U test, Table 6). For the mean of years spent in education across the non-employed categories, see the corresponding bar chart

Table 5. Distribution tests of working years in relation to the physical nature and the health and safety conditions of the profession

Hypothesis Test Summary

	Null Hypothesis	Test	Sig.	Decision
1	The distribution of Years in last professional status - Employed is the same across categories of Type of work - Mental/Muscle.	Independent-Samples Kruskal-Wallis Test	.865	Retain the null hypothesis.

Asymptotic significances are displayed. The significance level is .05.

Hypothesis Test Summary

	Null Hypothesis	Test	Sig.	Decision
1	The distribution of Years in last professional status - Employed is the same across categories of Work's health and safety conditions.	Independent-Samples Kruskal-Wallis Test	.932	Retain the null hypothesis.

Asymptotic significances are displayed. The significance level is .05.

Figure 6. Level and years spent in education in relation to the type of work

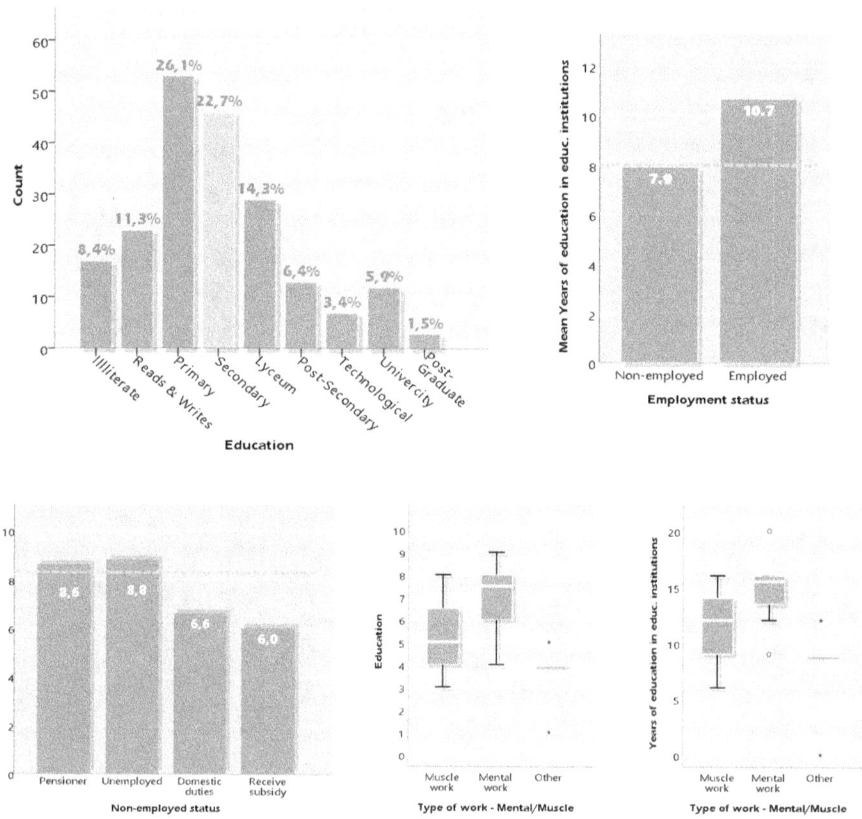

in Figure 6, deducing that the disabled persons in the four non-employed categories spent their years, on average, in educational institutes ($p = 0.001 < 0.05$ for the corresponding Kruskal-Wallis test in Table 6). In a more detailed view, pensioners and unemployed had the same average education (about two years more education) than the "domestic duties" and the "receiving subsidies" categories.

Financial Status

Concerning the financial data of respondents, the following results were obtained. Most of the respondents (70%) live in private property while the rest do not. Of the ones that do not have private property, 63,6% are on rent, and 34,5% are hosted; see pie charts of Figure 7. We can then statistically infer that 70% of the whole population of the disabled persons in West Athens has their private property ($p = 0.52 > 0.05$ for the corresponding binomial test in Table 7). It is interesting to mention here that the private property status is independent of employment status or non-employment categories ($p = 0.481 > 0.05$ for the chi-squared test between the property and employment status; see Table 7); see also left bar chart in Figure 7. On the other hand, considering the non-employment categories (pensioners, unemployed, domestic duties, and the ones that receive subsidy), there is a clear dependence on the property status ($p = 0.015 < 0.05$ for the chi-squared test; see sub-table of Table 7). For example, 30.3% of the non-

Table 6. Tests for the relation between the years in educational institutions and the employment/work categories

Hypothesis Test Summary

	Null Hypothesis	Test	Sig.	Decision
1	The distribution of Years of education in educ. institutions is the same across categories of Type of work - Mental/Muscle.	Independent-Samples Kruskal-Wallis Test	.014	Reject the null hypothesis.

Asymptotic significances are displayed. The significance level is .05.

Tests of Normality

	Type of work - Mental/Muscle	Kolmogorov-Smirnov[a] Statistic	df	Sig.	Shapiro-Wilk Statistic	df	Sig.
Education	Muscle work	.248	12	.039	.903	12	.175
	Mental work	.250	8	.150	.884	8	.208
	Other	.404	5	.008	.768	5	.044

a. Lilliefors Significance Correction

Group Statistics

	Type of work - Mental/Muscle	N	Mean	Std. Deviation	Std. Error Mean
Years of education in educ. institutions	Muscle work	12	11.75	3.079	.889
	Mental work	8	14.88	3.227	1.141

Independent Samples Test

		Levene's Test for Equality of Variances		t-test for Equality of Means					95% Confidence Interval of the Difference	
		F	Sig.	t	df	Sig. (2-tailed)	Mean Difference	Std. Error Difference	Lower	Upper
Years of education in educ. institutions	Equal variances assumed	.025	.876	-2.183	18	.043	-3.125	1.432	-6.133	-.117
	Equal variances not assumed			-2.161	14.643	.048	-3.125	1.446	-6.214	-.036

Hypothesis Test Summary

	Null Hypothesis	Test	Sig.	Decision
1	The distribution of Years of education in educ. institutions is the same across categories of Employment status.	Independent-Samples Mann-Whitney U Test	.001	Reject the null hypothesis.

Asymptotic significances are displayed. The significance level is .05.

Hypothesis Test Summary

	Null Hypothesis	Test	Sig.	Decision
1	The distribution of Years of education in educ. institutions is the same across categories of Non-employed status.	Independent-Samples Kruskal-Wallis Test	.001	Reject the null hypothesis.

Asymptotic significances are displayed. The significance level is .05.

employed in our sample are pensioners owning private property, while only 6.9% are pensioners without private property; see right bar chart in Figure 7.

The non-private property categories (rent, hosted or being house guests) are not affected by their employment status (working or not) or by their non-employed status, i.e. these categories show the same (statistically) behaviour for the disabled persons whether working or not ($p = 0.898 > 0.05$ for the corresponding chi-squared test in Table 7) or whether they are pensioners, unemployed, occupied with domestic duties, or receiving subsidy ($p = 0.27 > 0.05$ for the corresponding chi-squared test in Table 7).

Regarding their average personal monthly income (expressed in levels, from 1 for "no income", to 10 for income "> €3000"), it corresponds to the 3rd income level (mean value 3.85), i.e. from €301 to €500. In the bar chart of Figure 8, the distribution of the annual income is presented with the mean level (green bar) highlighted; see also the corresponding boxplot right to it (5% of our sample has a monthly income from less than €300 to €1200). Concerning employment and non-employment status, there is no difference between the distributions of the income level of the employed and the non-employed persons ($p = 0.062 > 0.05$ for the Mann-Whitney U test of the income level about their employed status, Table 8). The above result can be empowered since it can also be deduced via the corresponding chi-square test (p = 0.059 > 0.05; see the corresponding table in Table 8), i.e. the employed and non-employed status does not reflect the 'respondent's income level. This is not the case when we take into account

Figure 7. Private property ownership and non-private property categories

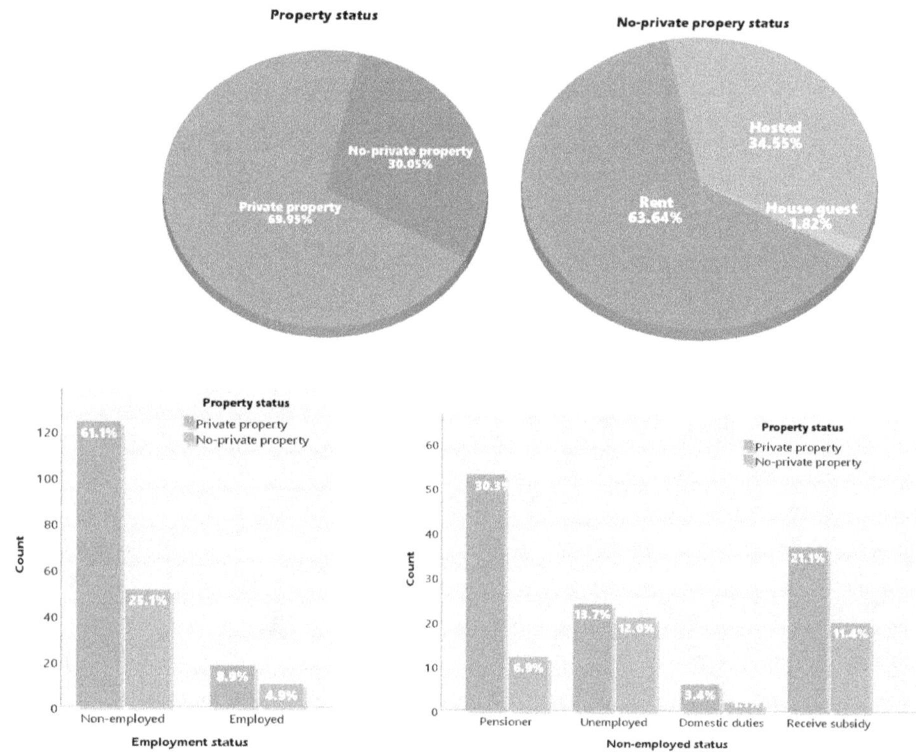

their past professional status ($p < 0.001$ for the Mann-Whitney U test for the income level the past employed status, Table 8), i.e. **the fact that the respondents were employed (or not) in the past is now strongly reflected their income level**. For a graphic illustration, see the corresponding clustered bar chart and boxplot of Figure 8b.

Regarding the non-employed categories, there is a clear difference in the income level between those categories ($p < 0.01$ for the Kruskal-Wallis test, in Table 8), with the pensioners acquiring the (on average) most. See also the bottom boxplot of Figure 8a,b.

The private property ownership seems to differentiate regarding personal income level ($p = 0.02 < 0.05$ for the Mann-Whitney U test in Table 9). However, the corresponding upper boxplot in Figure 9 provides some equality evidence for our sample. On the other hand, the personal income level does not affect the non-private property categories ($p = 0.18 > 0.05$ for the Kruskal-Wallis test, Table 9) and the mental/muscle type of work ($p = 0.181 > 0.05$ for the Kruskal-Wallis test, Table 9), i.e. the physical type work (muscle or mental) has no effect (on average) on their income. Moreover,_there is **no significant correlation between the personal income level and the years spent in educational institutes** ($p = 0.346 > 0.05$ for the Pearson correlation, Table 9). Note that, in the corresponding bar chart in Figure 9, 'there's some trend with the income level increasing as the years in Educ. Institutes is rising.

Table 7. Tests for private and on-private property ownership in relation to employment status

Binomial Test

		Category	N	Observed Prop.	Test Prop.	Exact Sig. (1-tailed)
Property status	Group 1	Private property	142	.7	.7	.520[a]
	Group 2	No-private property	61	.3		
	Total		203	1.0		

a. Alternative hypothesis states that the proportion of cases in the first group < .7.

Chi-Square Tests

	Value	df	Asymptotic Significance (2-sided)	Exact Sig. (2-sided)	Exact Sig. (1-sided)
Pearson Chi-Square	.496[a]	1	.481		
Continuity Correction[b]	.233	1	.630		
Likelihood Ratio	.483	1	.487		
Fisher's Exact Test				.509	.309
Linear-by-Linear Association	.493	1	.482		
N of Valid Cases	203				

a. 0 cells (0.0%) have expected count less than 5. The minimum expected count is 8.41.

b. Computed only for a 2x2 table

Chi-Square Tests

	Value	df	Asymptotic Significance (2-sided)
Pearson Chi-Square	10.427[a]	3	.015
Likelihood Ratio	10.638	3	.014
Linear-by-Linear Association	2.386	1	.122
N of Valid Cases	175		

a. 1 cells (12.5%) have expected count less than 5. The minimum expected count is 2.51.

Chi-Square Tests

	Value	df	Asymptotic Significance (2-sided)
Pearson Chi-Square	.216[a]	2	.898
Likelihood Ratio	.377	2	.828
Linear-by-Linear Association	.091	1	.763
N of Valid Cases	55		

a. 3 cells (50.0%) have expected count less than 5. The minimum expected count is .16.

Chi-Square Tests

	Value	df	Asymptotic Significance (2-sided)
Pearson Chi-Square	7.585[a]	6	.270
Likelihood Ratio	8.828	6	.184
Linear-by-Linear Association	.181	1	.671
N of Valid Cases	50		

a. 7 cells (58.3%) have expected count less than 5. The minimum expected count is .04.

Living Conditions

Figure 10 presents the housing living conditions of the respondents. We mention here that **the building status where the disabled persons live does not depend on their income** ($p = 0.059 > 0.05$ for the Kruskal-Wallis test, Table 10) or if the building is their private property or not ($p = 0.614 > 0.05$ for the corresponding chi-square test, Table 10), **but depends on their employment status**, employed or not ($p = 0.027 < 0.05$ for the chi-square test, Table 10) with a positive correlation/trend as we switch from the non-employed to employed status ($p = 0,008 < 0,05$ for the corresponding 'Pearson's correlation with value -0.185 which is negative since the categories range in reverse order, i.e. from 1for "Very good" to 5 for "Very bad"); see clustered bar chart of Figure 10.

Healthy Living and Lifestyle

Regarding their diet, the corresponding bar chart in Figure 11a,b presents the mean of the weekly consumption index (valued with 1: "Never", 2: "Rarely", 3: "1-2 times", 4: "2-3 times", 5: "3-4 times" and 6: "Daily") for various food types. Bread and oil are the most consumed products with an almost daily consumption (index of over 5), followed by dairy products (4.4), potatoes and fruits (4.0). Fast food and food complements have the lowest consumption (1.5 and 1.6 respectively).

Concerning their drinking habits, only 20.7% drinks alcohol; pie chart of Figure 11a,b. The bar chart of the mean (daily) alcohol consumption index (i.e. the amount of alcohol type consumed daily concerning the total amount of alcohol consumed) shows that wines (red or white) have the highest consumption

Figure 8a. Personal annual income distribution and its relation to the employment and non-employment status

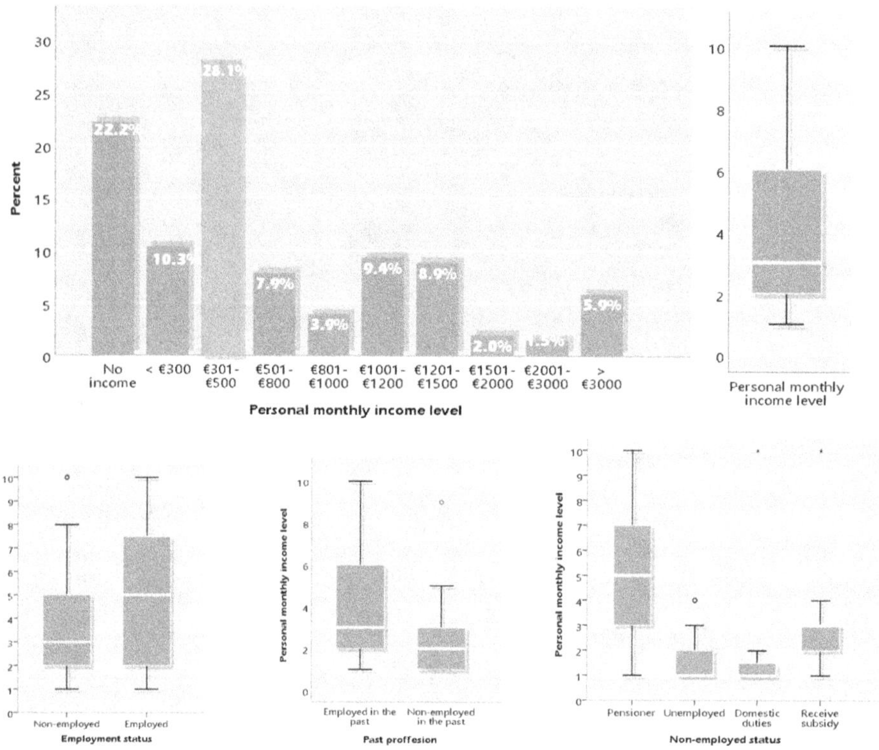

(0.52 and 0.55, i.e. over 50% of the 'alcohol's daily consumption is due to wine) and ouzo and whisky the lowest (0.14 and 0.17).

Coffee is a much more preferred choice for 64% of the respondents while drinking tea is preferred by only 32% of the respondents (I cup for the 80.7% of the ones that prefer this beverage).

Regarding their smoking habits, 24.6% are smoking (with an average of 17.8 cigarettes /day) while only 14.8% were smoking in the past; see a pie chart of Figure 12. The years of smoking to date are, on average, 17.1 while the years that the respondents were smoking in the past are 22.5. The ones that stopped smoking have quit smoking for, on average, 10.1 years. Figure 12a,b presents the corresponding histograms and boxplots for the distributions of their smoking years until now, smoking years in the past, and years that quit smoking. Only 2% of the respondents use drugs; see the corresponding pie chart and histogram of Figure 12. Note that their smoking habit today does not relate to their current employed status ($p = 0.072 > 0.05$ for the chi-square test), although **their smoking habits in the past were related to their past employed status** ($p = 0.038 < 0.05$ for the corresponding chi-square test). Indeed, the clustered bar chart in Figure 12a,b shows that the ones working in the past are the ones that smoked more.

Regarding the sleep habits, the disabled persons seem to change their sleep habits between weekdays and weekends ($p = 0.035 < 0.05$ for the Wilcoxon test, Table 11), although the sample means are pretty close (6.49 hours/day for the weekdays and 6.56 hours/day for the weekends); see the corresponding histograms for their sleep hours/day on weekdays and weekends (Figure 13). Moreover, the night sleep

Figure 8b. Personal annual income distribution and its relation to the employment and non-employment status

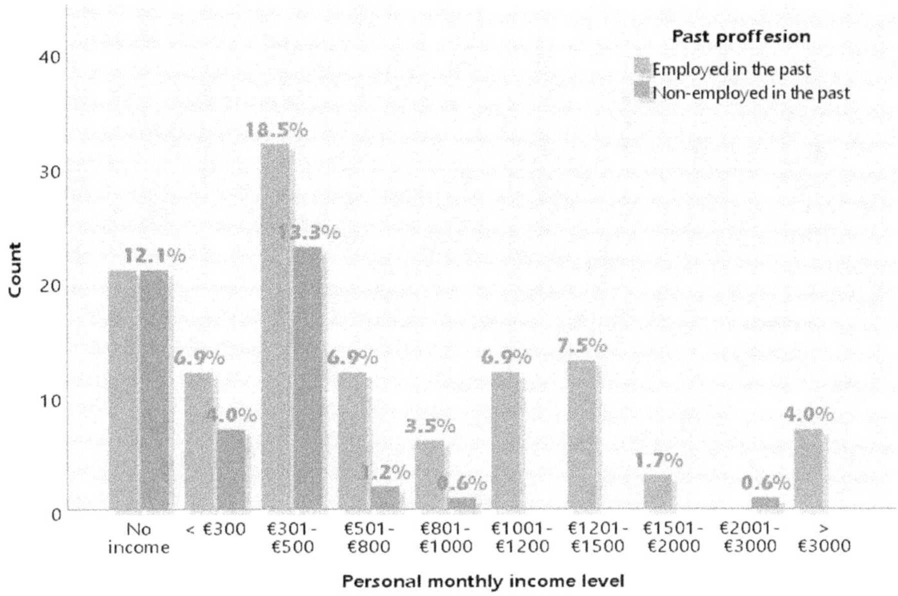

Table 8. Distribution tests for the relation of the income level and the employment status

Hypothesis Test Summary

	Null Hypothesis	Test	Sig.	Decision
1	The distribution of Personal annual income level is the same across categories of Employment status.	Independent-Samples Mann-Whitney U Test	.062	Retain the null hypothesis.

Asymptotic significances are displayed. The significance level is .05.

Hypothesis Test Summary

	Null Hypothesis	Test	Sig.	Decision
1	The distribution of Personal annual income level is the same across categories of Non-employed status.	Independent-Samples Kruskal-Wallis Test	.000	Reject the null hypothesis.

Asymptotic significances are displayed. The significance level is .05

Hypothesis Test Summary

	Null Hypothesis	Test	Sig.	Decision
1	The distribution of Personal monthly income level is the same across categories of Past proffesion	Independent-Samples Mann-Whitney U Test	.000	Reject the null hypothesis.

Asymptotic significances are displayed. The significance level is .05

Personal annual income level * Employment status Crosstabulation

Count

		Employment status		
		Non-employed	Employed	Total
Personal annual income level	No income	39	6	45
	< €300	19	2	21
	€301-€500	53	4	57
	€501-€800	15	1	16
	€801-€1000	6	2	8
	€1001-€1200	15	4	19
	€1201-€1500	16	2	18
	€1501-€2000	3	1	4
	€2001-€3000	1	2	3
	> €3000	8	4	12
Total		175	28	203

Chi-Square Tests

	Value	df	Asymptotic Significance (2-sided)
Pearson Chi-Square	16.421[a]	9	.059
Likelihood Ratio	13.180	9	.155
Linear-by-Linear Association	6.418	1	.011
N of Valid Cases	203		

a. 10 cells (50.0%) have expected count less than 5. The minimum expected count is .41.

Figure 9. Personal annual income distribution and its relation to other factors

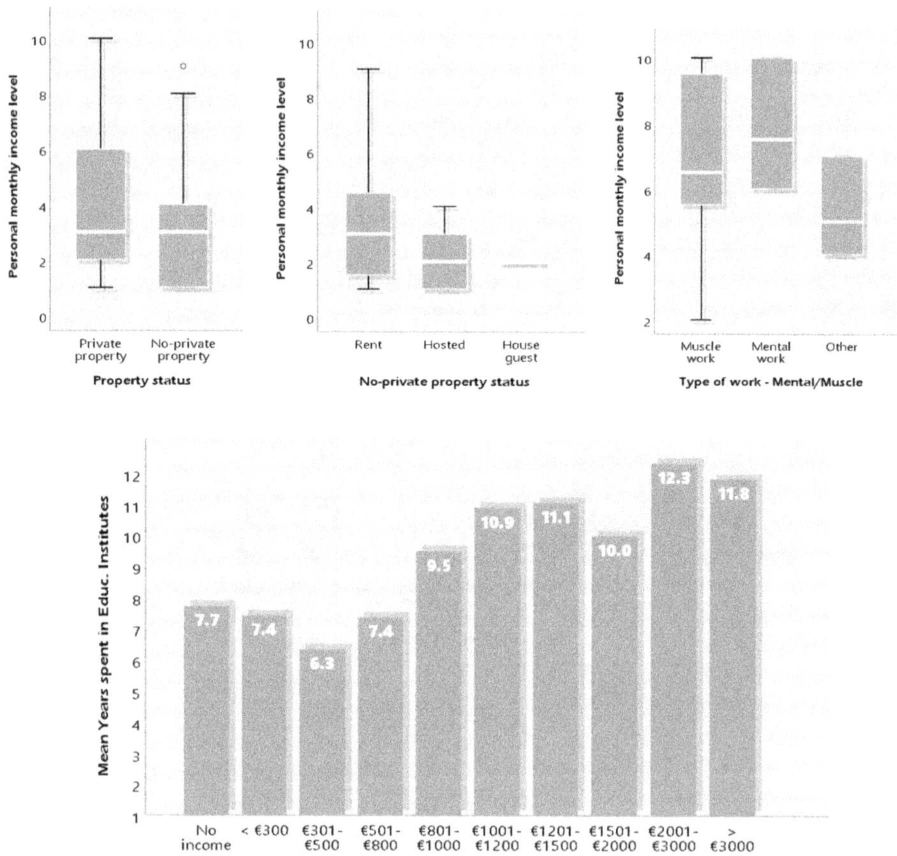

Table 9. Distribution tests for the relation of the income level and other factors

Hypothesis Test Summary

	Null Hypothesis	Test	Sig.	Decision
1	The distribution of Personal monthly income level is the same across categories of Property status.	Independent-Samples Mann-Whitney U Test	.002	Reject the null hypothesis.

Asymptotic significances are displayed. The significance level is .05.

Hypothesis Test Summary

	Null Hypothesis	Test	Sig.	Decision
1	The distribution of Personal monthly income level is the same across categories of No-private property status.	Independent-Samples Kruskal-Wallis Test	.180	Retain the null hypothesis.

Asymptotic significances are displayed. The significance level is .05.

Hypothesis Test Summary

	Null Hypothesis	Test	Sig.	Decision
1	The distribution of Personal monthly income level is the same across categories of Type of work - Mental/Muscle.	Independent-Samples Kruskal-Wallis Test	.181	Retain the null hypothesis.

Asymptotic significances are displayed. The significance level is .05.

Correlations

		Personal monthly income level	Years of education in educ. institutions
Personal monthly income level	Pearson Correlation	1	.346**
	Sig. (2-tailed)		.000
	N	203	203
Years of education in educ. institutions	Pearson Correlation	.346**	1
	Sig. (2-tailed)	.000	
	N	203	203

**. Correlation is significant at the 0.01 level (2-tailed).

Figure 10. Living conditions

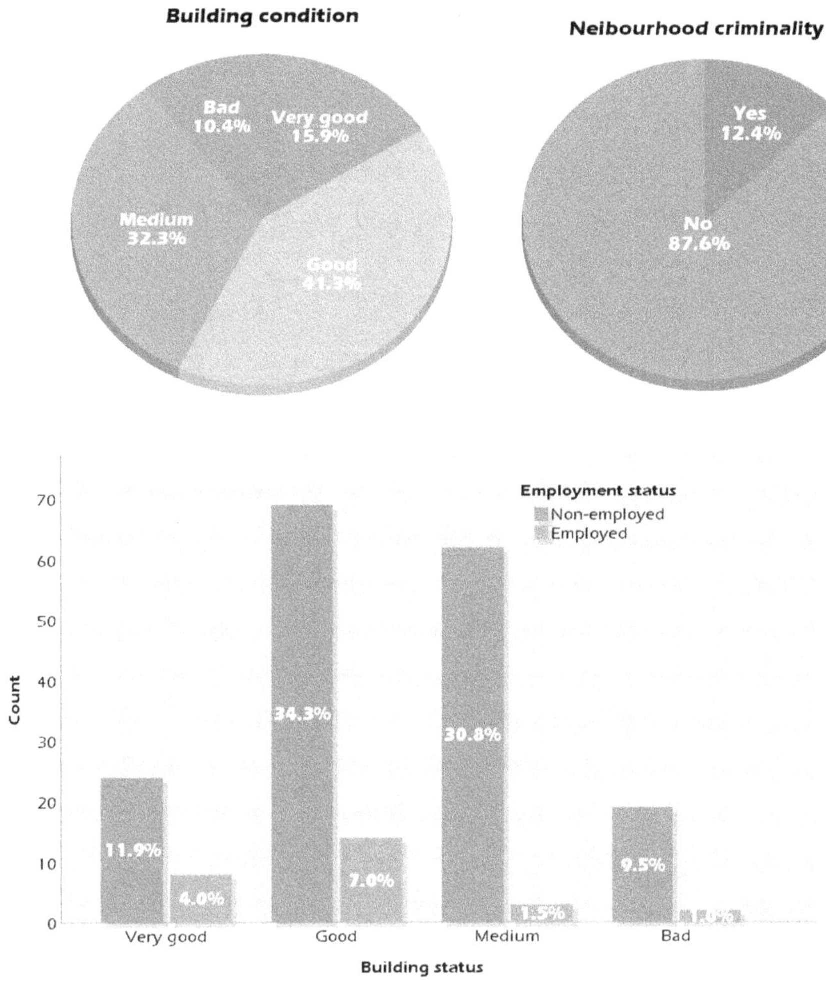

pattern for the weekdays of the disabled persons is the one that seems to alter regarding their marital status ($p = 0.042 > 0.05$ for the Kruskal-Wallis test, Table 11), with the other sleep patterns (noon sleep at weekdays and weekends) remain unchanged. Indeed, cohabit, and non-married appear to sleep the most at weeknights (7.2 and 7 hours/day respectively), while divorced sleep the least (5.7 hours/weeknight).

Health Status

Concerning the main diseases from which the respondents stated they are suffering or hospitalized, Figure 14a,b provides those diseases grouped into general categories. Heart diseases are the most common (29.4%) among the disabled persons in West Athens, followed by various types of cancer (17.9%), with respiratory (2.5%) and gastrointestinal (1.5%) diseases being the least common.

More than 91% of the respondents are stated that they hospitalized due to their disease. Histograms in Figure 14 depict, the hospitalisation dura distribution of the hospitalisation duration distribution for the whole sample (left, per month) and for the first three months (right, per week). For the five most

Table 10. Tests regarding the living conditions

Hypothesis Test Summary

	Null Hypothesis	Test	Sig.	Decision
1	The distribution of Personal monthly income level is the same across categories of Building status.	Independent-Samples Kruskal-Wallis Test	.059	Retain the null hypothesis.

Asymptotic significances are displayed. The significance level is .05.

Crosstab

Count

		Property status		
		Private property	No-private property	Total
Building status	Very good	25	7	32
	Good	58	25	83
	Medium	44	21	65
	Bad	13	8	21
Total		140	61	201

Chi-Square Tests

	Value	df	Asymptotic Significance (2-sided)
Pearson Chi-Square	1.803[a]	3	.614
Likelihood Ratio	1.847	3	.605
Linear-by-Linear Association	1.612	1	.204
N of Valid Cases	201		

a. 0 cells (0.0%) have expected count less than 5. The minimum expected count is 6.37.

Crosstab

Count

		Employment status		
		Non-employed	Employed	Total
Building status	Very good	24	8	32
	Good	69	14	83
	Medium	62	3	65
	Bad	19	2	21
Total		174	27	201

Chi-Square Tests

	Value	df	Asymptotic Significance (2-sided)
Pearson Chi-Square	9.146[a]	3	.027
Likelihood Ratio	9.763	3	.021
Linear-by-Linear Association	6.859	1	.009
N of Valid Cases	201		

a. 2 cells (25.0%) have expected count less than 5. The minimum expected count is 2.82.

Symmetric Measures

		Value	Asymptotic Standard Error[a]	Approximate T[b]	Approximate Significance
Interval by Interval	Pearson's R	-.185	.069	-2.658	.008[c]
Ordinal by Ordinal	Spearman Correlation	-.195	.067	-2.809	.005[c]
N of Valid Cases		201			

a. Not assuming the null hypothesis.

b. Using the asymptotic standard error assuming the null hypothesis.

c. Based on normal approximation.

common diseases, i.e., heart, cancer, neurological, psychiatric and musculoskeletal diseases, the average duration of hospitalization were the same ($p = 0.377 > 0.05$ for the Kruskal-Wallis test, Table 12a,b).

Most of the respondents, 88%, have taken medication, and only 12% have not. Statistically, 90% of the disabled persons in West Athens had taken medication ($p < 0.001$ for the binomial test, Table 12). It is interesting to mention here that the present and the past smoking habits of the disabled persons are not affected by the medication they have or had ($p = 0.127 > 0.05$ and $p = 0.115 > 0.05$ for the corresponding chi-square tests, Table 12a,b).

For the duration of the hospitalization, the majority of the respondents needed hospital care for a month, with 47.9 being the average hospitalization days with a maximum of 3 years (left histogram of Fig 14a,b), while when we are restricted to the first 30 months (90 days) of hospitalization the average drops to 35.1 days (right histogram of Figure 14) since we have a lot of dispersion. Regarding the disease category, the corresponding bar chart of Figure 14 shows the mean duration of the 'respondents' hospitalization per disease type. The disabled persons with neurological (72.8 days) and psychiatric

Figure 11a. Diet of the respondents

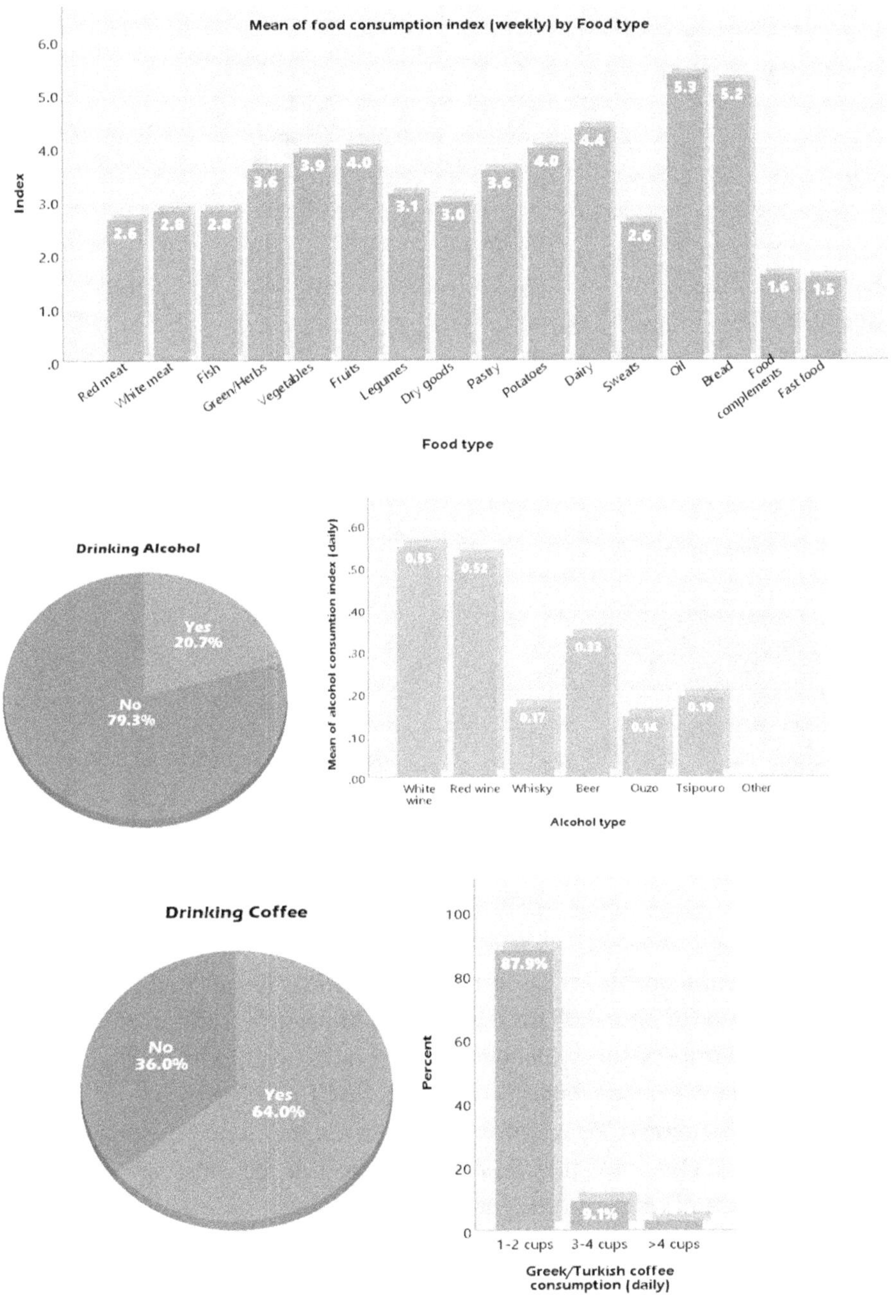

issues (63.7) have been hospitalized the most, followed by the ones with heart diseases (46.7 days) and cancer problems (48 days). However, due to the dispersion of the 'sample's hospitalization duration, the boxplots of Figures 14a,b are more informative regarding the hospitalization per disease type, where we observe that half of the respondents with heart, cancer, neurological, musculoskeletal and other issues were hospitalized for one month at most (median = 30). Finally, the respondents with heart problems

Figure 11b. Diet of the respondents

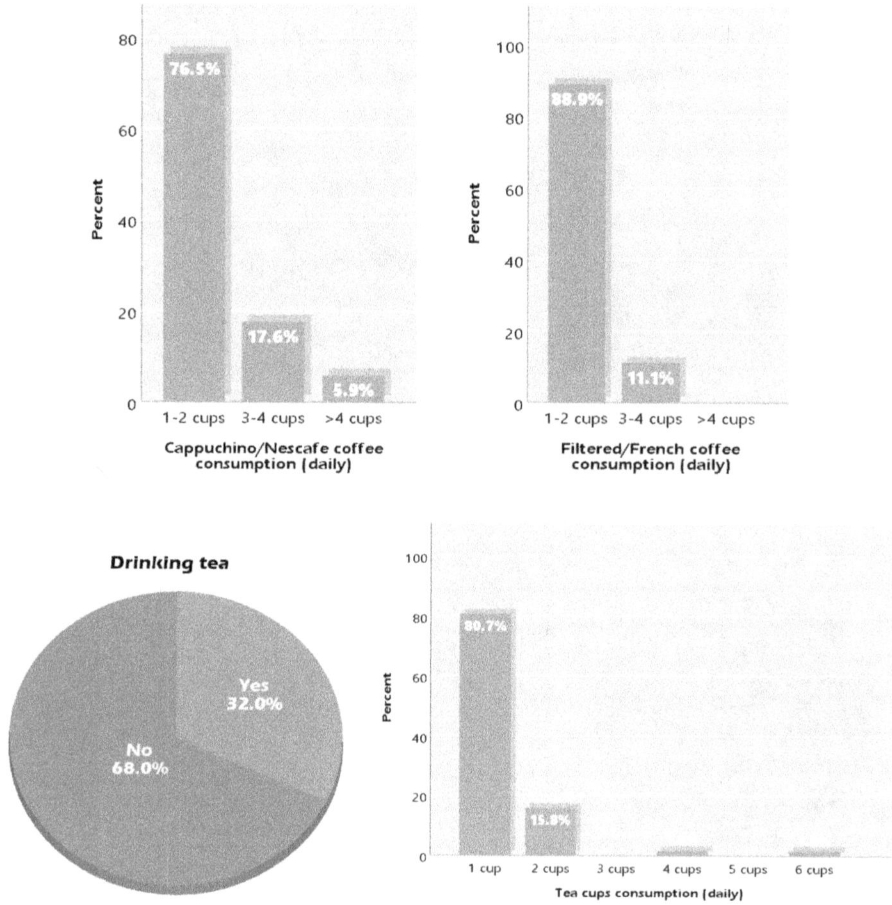

have the same "hospitalization profile" (i.e. the same distribution the hospitalization duration) with the ones with cancer issues ($p = 0.297 > 0.05$ for the Mann-Whitney test, sub-table Test Statistics of Table 12), as well as between the respondents with neurological and the ones with psychiatric issues ($p = 0.44 > 0.05$ for the Mann-Whitney test, sub-table Test Statistics of Table 12a,b).

As far as the stress is concerned, 80.2% of the respondents stated that they experience stress daily, while 19.2% have no stress. Over 50% they feel stressed most of the time, and almost 31% feel stress with very high intensity; see a pie chart of Figure 15a,b. Moreover, 68% experience social stress, with 32% are not. For the types of social stress, see the corresponding bar charts in Figure 15a,b.

Concerning the disease type that the respondents are/was facing, the most stressed are the ones with gastrointestinal issues (with an index of 1, from a binary index with 0 for "No stress" to 1 for "Stress") followed by the ones with psychiatric issues (0.92). Persons with psychiatric issues feel (on average) stress more frequently (1.3 from an index value 1 for "Most of the times", 2 for "Many times" and 3 for "Occasionally") followed by the ones that facing diabetes (1.4); see the corresponding bar chart of Figure 15a,b. In ones with psychiatric issues feel stress with the most intensity (4 on a scale ranging from 1 for "Low intense" to 4 for "Very high") followed by the ones with cancer problems (3.5). **The disabled persons feel the same stress, social or not, regardless of their marital status, income, employed**

Figure 12a. Smoking and drugs habits of the respondents

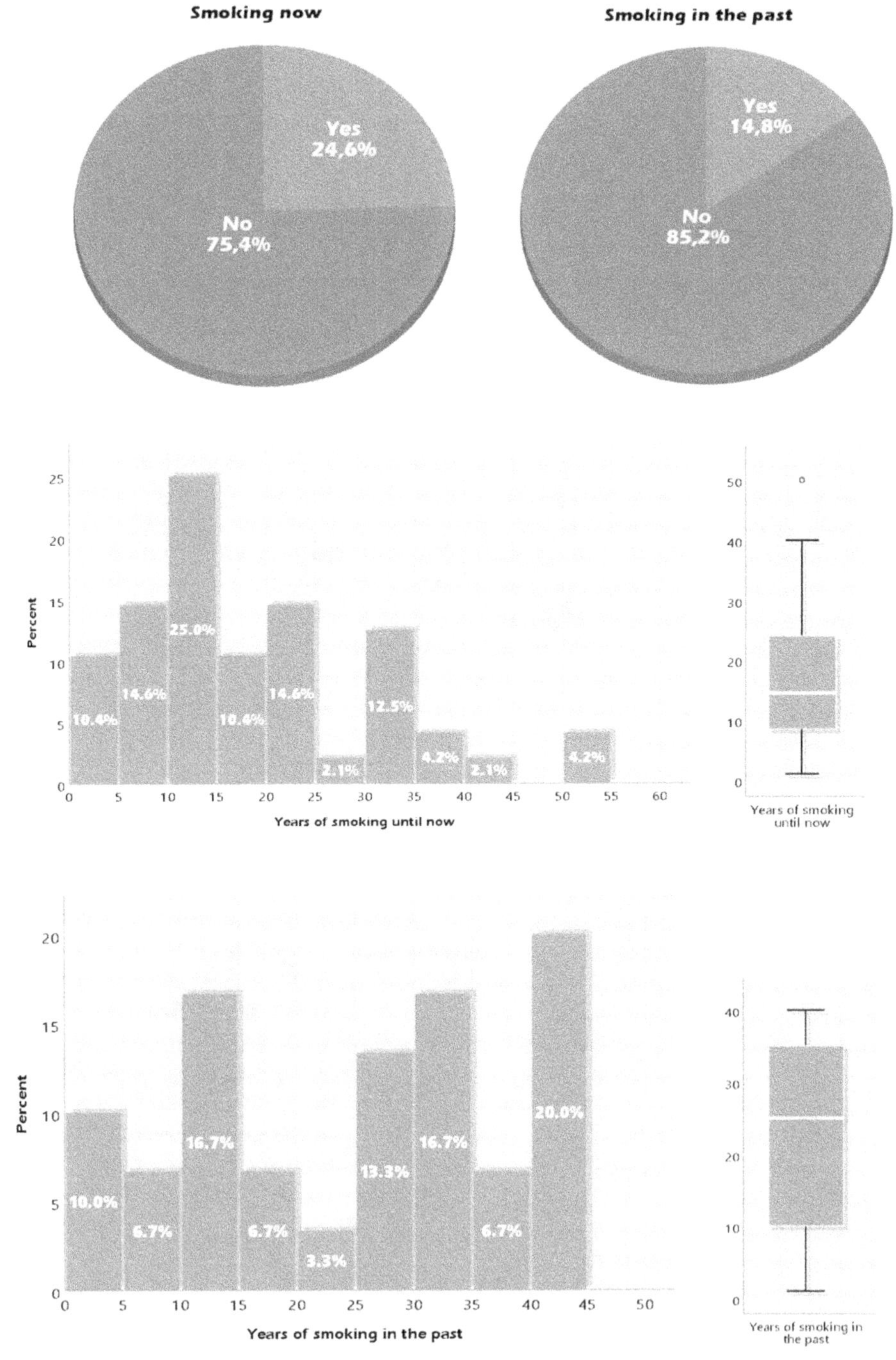

Figure 12b. Smoking and drugs habits of the respondents

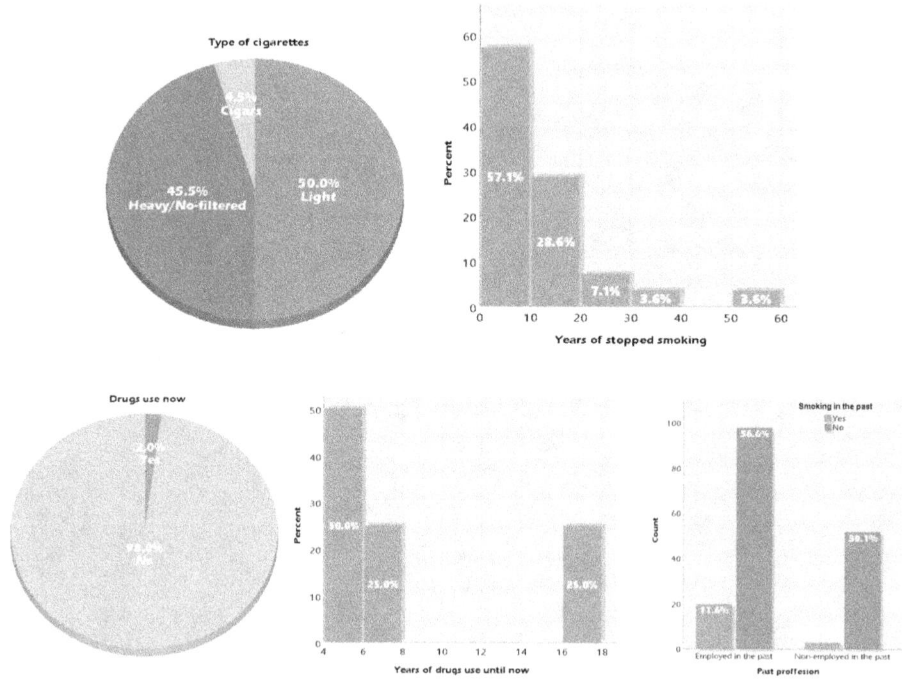

Figure 13. Sleep habits

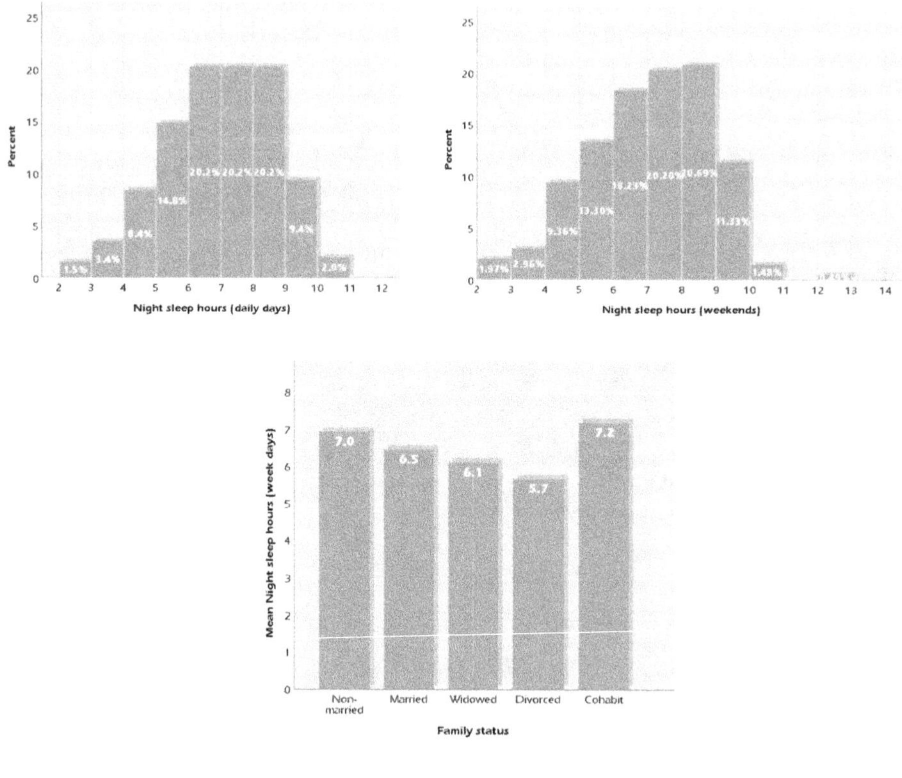

Table 11. Distribution tests regarding the sleep habits

Test Statistics[a]

	Night sleep hours (weekends) - Night sleep hours (week days)
Z	-2.111[b]
Asymp. Sig. (2-tailed)	.035

a. Wilcoxon Signed Ranks Test

b. Based on negative ranks.

Test Statistics[a,b]

	Night sleep hours (week days)	Night sleep hours (weekends)	Noon sleep hours (week days)	Noon sleep hours (weekends)
Kruskal-Wallis H	8.221	3.946	3.536	3.489
df	3	3	3	3
Asymp. Sig.	.042	.267	.316	.322

a. Kruskal Wallis Test

b. Grouping Variable: Family status

status, private property, and non-private property categories ($p > 0.05$ for all the corresponding tests, Table 13a,b), **while differs in the type of stress**; see the clustered bar charts at the end of Figure 15a,b.

Type of Support Received

Regarding the support (official and unofficial) that the respondents obtain, the pie charts of Figure 16a,b,c presents the portion of the respondents that receive official/government support (42.1%) and

Figure 14a. Main diseases and hospitalization

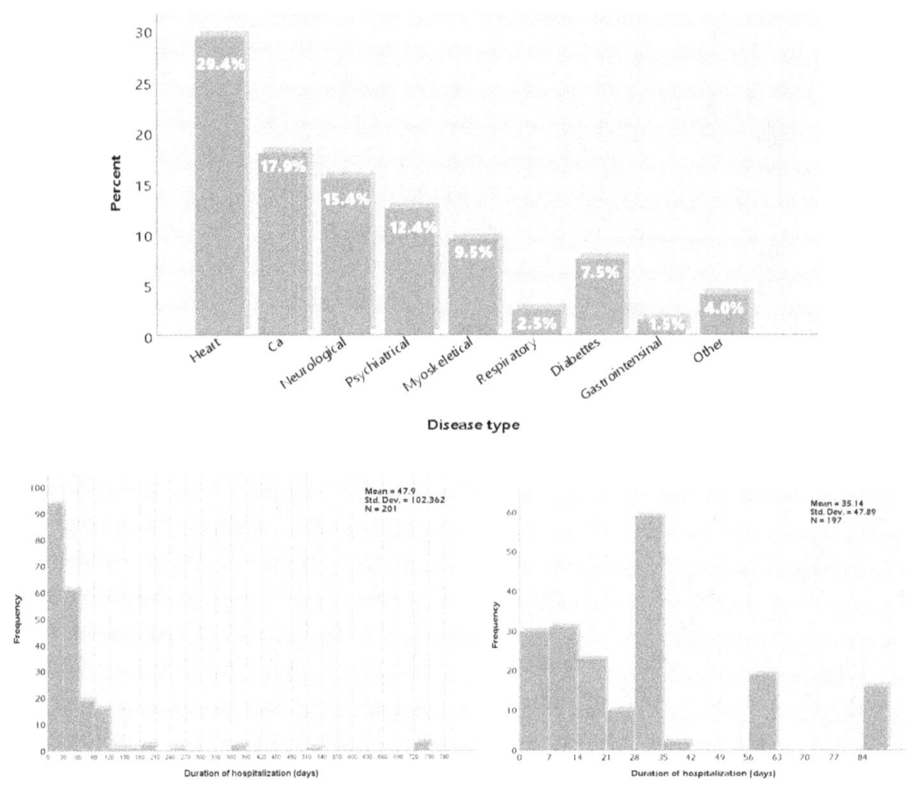

Figure 14b. Main diseases and hospitalization

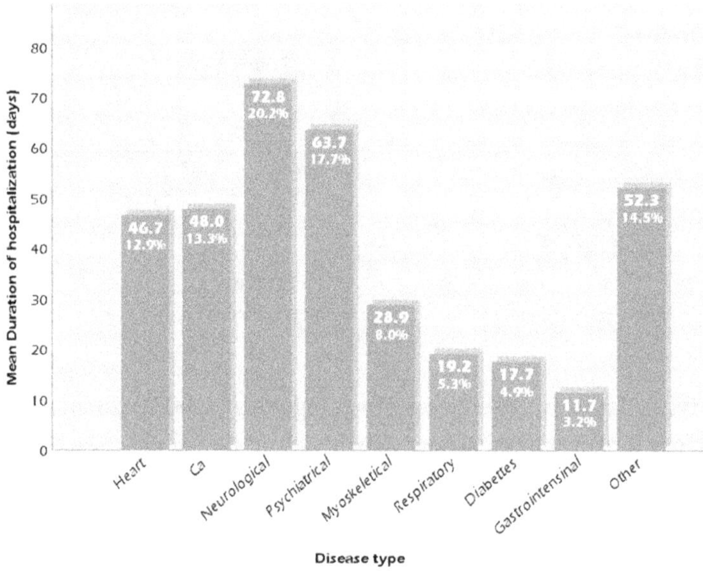

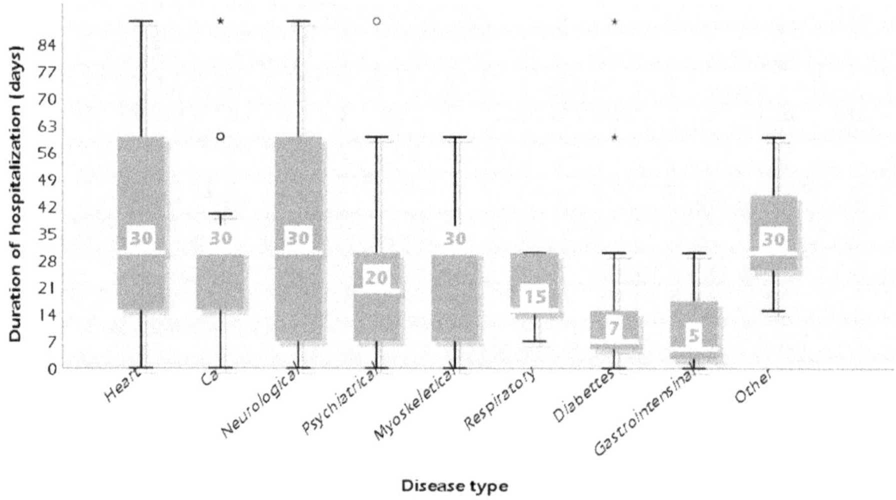

unofficial support (64.5%) as well as the corresponding types of support. The access to the health services was estimated as easy for only 37.2% of the respondents, with half of them stated that the access is as medium ease (Figure 16a,b,c). Two detailed clustered bar charts that concern the government and unofficial support per disease type is also provided in Figure 16a,b,c. It is statistically inferred that the official and unofficial support they receive is the same regardless of their disease category ($p = 0.17 > 0.05$ and $p = 0.209 > 0.05$ for the corresponding chi-squared tests, Table 14a,b). Government/official and unofficial support do not depend on the income level, i.e., the ones that receive government (as well as unofficial) support have the same distribution regarding their income as the ones that do not receive government (and unofficial resp.) support ($p = 0.917 > 0.05$ and $p = 0.424 > 0.05$ for the corresponding Mann-Whitney tests, Table 14a,b); see also the related clustered bar chart of the income level with

Table 12a. Distribution tests regarding main diseases

Binomial Test

		Category	N	Observed Prop.	Test Prop.	Exact Sig. (1-tailed)
Medication for disease	Group 1	No	24	.1	.9	.000[a]
	Group 2	Yes	177	.9		
	Total		201	1.0		

a. Alternative hypothesis states that the proportion of cases in the first group < .9.

Ranks

	Disease type	N	Mean Rank
Duration of hospitalization (days)	Cardiac	59	93.59
	Ca	36	84.51
	Neurological	31	84.10
	Psychiatrical	25	71.60
	Myoskeletical	18	77.97
	Total	169	

Test Statistics[a,b]

	Duration of hospitalization (days)
Kruskal-Wallis H	4.217
df	4
Asymp. Sig.	.377

a. Kruskal Wallis Test

b. Grouping Variable: Disease type

Crosstab

Count

		Smoking now: Yes	No	Total
Medication for disease	Yes	40	133	173
	No	9	15	24
Total		49	148	197

Chi-Square Tests

	Value	df	Asymptotic Significance (2-sided)	Exact Sig. (2-sided)	Exact Sig. (1-sided)
Pearson Chi-Square	2.332[a]	1	.127		
Continuity Correction[b]	1.626	1	.202		
Likelihood Ratio	2.158	1	.142		
Fisher's Exact Test				.136	.104
Linear-by-Linear Association	2.320	1	.128		
N of Valid Cases	197				

a. 0 cells (0.0%) have expected count less than 5. The minimum expected count is 5.97.

b. Computed only for a 2x2 table

regards to the government support in Figure 16a,b,c. Finally, the government (as well as the unofficial) support does not differentiate regarding the employed status of the marital/family status of the disabled persons ($p > 0.05$ for all the corresponding chi-square tests, four last lines of sub-tables of Table 14a,b); see the corresponding clustered bar charts in Figure 16a,b,c.

Table 12b. Distribution tests regarding main diseases

Crosstab

Count

		Smoking in the past: Yes	No	Total
Medication for disease	Yes	29	148	177
	No	1	23	24
Total		30	171	201

Chi-Square Tests

	Value	df	Asymptotic Significance (2-sided)	Exact Sig. (2-sided)	Exact Sig. (1-sided)
Pearson Chi-Square	2.484[a]	1	.115		
Continuity Correction[b]	1.615	1	.204		
Likelihood Ratio	3.215	1	.073		
Fisher's Exact Test				.138	.093
Linear-by-Linear Association	2.472	1	.116		
N of Valid Cases	201				

a. 1 cells (25.0%) have expected count less than 5. The minimum expected count is 3.58.

b. Computed only for a 2x2 table

Test Statistics[a]

	Duration of hospitalization (days)
Mann-Whitney U	929.000
Wilcoxon W	1595.000
Z	-1.042
Asymp. Sig. (2-tailed)	.297

a. Grouping Variable: Disease type

Test Statistics[a]

	Duration of hospitalization (days)
Mann-Whitney U	341.000
Wilcoxon W	666.000
Z	-.773
Asymp. Sig. (2-tailed)	.440

a. Grouping Variable: Disease type

Figure 15a. Feeling of stress

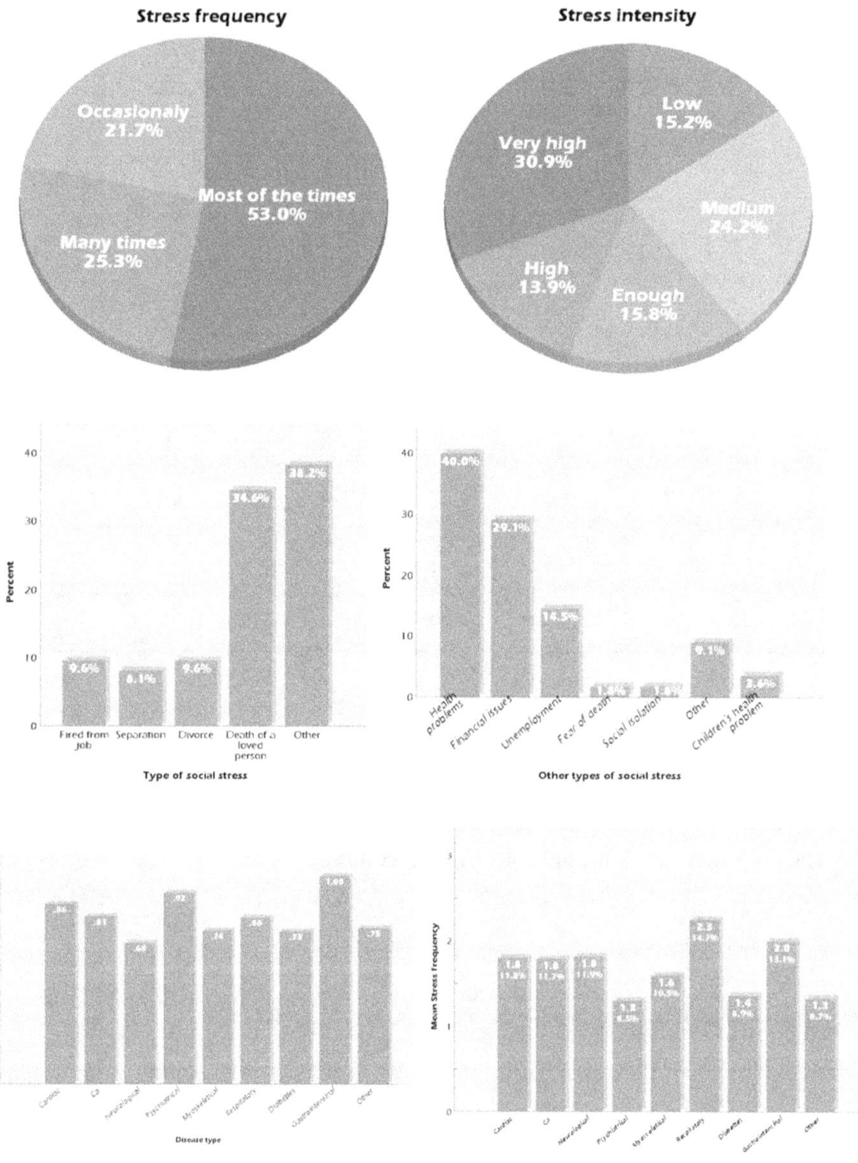

DISCUSSION AND CONCLUSION

The ageing of the population, along with the presence of disabilities in a large amount of population brings to the fore unprecedented needs that require special handling by those in charge and are closely related to humanitarian issues. To this end, there was a solid necessity to conduct the present study since there have not been any studies concerning the sociological and epidemiological analysis of West Athens's population health profile. The present study results are also consistent with that of previous studies, enriching the current academic literature and highlighting the impact of social factors on health and illness.

Figure 15b. Feeling of stress

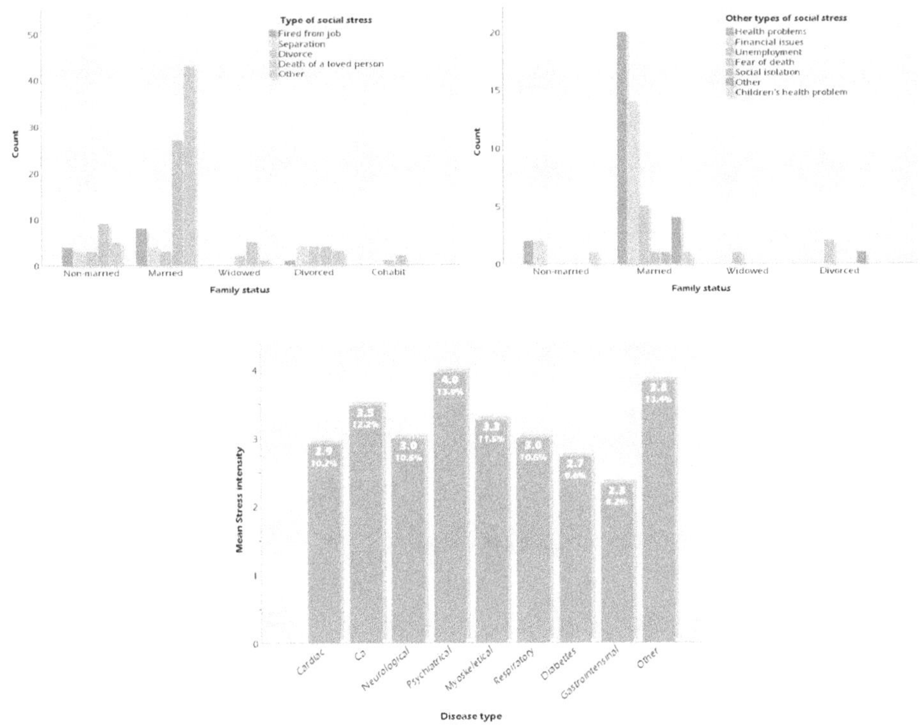

More specifically, the family was the most important social determinant of health and illness. The family seems to positively affect the target population's health or treatment, i.e., individuals living in West Athens in Greece who had been diagnosed with diseases and had a disability degree of over 67%. For those individuals, the family was beneficial to their physical, psychological and social health.

On the other hand, respondents' low education level (the majority had completed compulsory education) does not seem to positively affect their health. It can contribute neither to adopting a healthy lifestyle nor to managing their health issues.

Similarly, "working status" determinant is highly associated with the respondents' poor health. A substantial proportion of the participants were unemployed, while only a few of them stated that they were working (mostly doing manual-heavy work for many years). This finding comes to confirm the official unemployment indicators, which demonstrate that West Athens population has the highest unemployment rates at the national level, and it is consistent with existing literature about the negative effect of the "unemployment" determinant on individuals' health and quality of life. Additionally, no correlation was found between the working conditions and various types of professions.

On the contrary, a correlation was found between the respondents' total years of studies and various professions. Intellectual labor is associated with more years of study than other types of labor (manual labor, unemployment etc.); and it was also found that employed individuals invest many more years in their studies than unemployed individuals.

Similarly, a negative correlation was observed between respondents' economic status and health conditions (their maximum income consisted of the welfare allowance, a low-value pension or even no income). As existing literature has proved, 'individual's income level constitutes a severe determinant

Table 13a. Tests regarding stress

Crosstab

Count

		Stress		
		No	Yes	Total
Family status	Non-married	13	29	42
	Married	23	103	126
	Widowed	2	6	8
	Divorced	1	21	22
	Cohabit	0	5	5
Total		39	164	203

Chi-Square Tests

	Value	df	Asymptotic Significance (2-sided)
Pearson Chi-Square	8.215[a]	4	.084
Likelihood Ratio	9.786	4	.044
Linear-by-Linear Association	6.789	1	.009
N of Valid Cases	203		

a. 4 cells (40.0%) have expected count less than 5. The minimum expected count is .96.

Test Statistics[a]

	Personal monthly income level
Mann-Whitney U	4431.500
Wilcoxon W	6576.500
Z	-.140
Asymp. Sig. (2-tailed)	.889

a. Grouping Variable: Stress on social life

Test Statistics[a]

	Personal monthly income level
Mann-Whitney U	3145.500
Wilcoxon W	3925.500
Z	-.162
Asymp. Sig. (2-tailed)	.871

a. Grouping Variable: Stress

Crosstab

Count

		Stress		
		No	Yes	Total
Employment status	Non-employed	34	141	175
	Employed	5	23	28
Total		39	164	203

Chi-Square Tests

	Value	df	Asymptotic Significance (2-sided)	Exact Sig. (2-sided)	Exact Sig. (1-sided)
Pearson Chi-Square	.038[a]	1	.845		
Continuity Correction[b]	.000	1	1.000		
Likelihood Ratio	.039	1	.843		
Fisher's Exact Test				1.000	.540
Linear-by-Linear Association	.038	1	.845		
N of Valid Cases	203				

a. 0 cells (0.0%) have expected count less than 5. The minimum expected count is 5.38.

b. Computed only for a 2x2 table

of health. A high income is associated with good health and a low income with poor health, although many of the respondents live in their own house of several square meters and only a few of them live on rent or in another house as guests. Remarkably, no correlation was found between respondents' property status and working status. No correlation was also found between past working status and income, and as expected, between years of studies and income.

Moreover, the living conditions of the target population seemed to be adequate concerning fulfilling the basic requirements for good physical, psychological and social health, even though participants were considered to belong to socially disadvantaged groups. It is worth mentioning that no correlation was found between building conditions or house ownership and income, but it was observed that living conditions are associated with working status.

Concerning food habits, it seems that the majority of the respondents were following a relatively healthy lifestyle, consuming a regular or a small portion of food, which is a protective factor for their health or the progress of their disease.

Low alcohol consumption was found to positively affect the majority of the respondents' health, especially that of wine. Based on the literature, it has been proved that low to medium consumption of alcohol (1.5 to 30 grams every day), compared with no consumption is associated with lower mortality levels.

Similarly, daily medium coffee consumption seems to have a positive effect on the respondents' health. Many studies confirm this and especially those which support that medium consumption of caffeine (3

Table 13b. Tests regarding stress

Crosstab

Count

		Stress No	Stress Yes	Total
Property status	Private property	27	115	142
	No-private property	12	49	61
Total		39	164	203

Chi-Square Tests

	Value	df	Asymptotic Significance (2-sided)	Exact Sig. (2-sided)	Exact Sig. (1-sided)
Pearson Chi-Square	.012[a]	1	.913		
Continuity Correction[b]	.000	1	1.000		
Likelihood Ratio	.012	1	.913		
Fisher's Exact Test				1.000	.527
Linear-by-Linear Association	.012	1	.913		
N of Valid Cases	203				

a. 0 cells (0.0%) have expected count less than 5. The minimum expected count is 11.72.

b. Computed only for a 2x2 table

Crosstab

Count

		Stress No	Stress Yes	Total
No-private property status	Rent	11	24	35
	Hosted	1	18	19
	House guest	0	1	1
Total		12	43	55

Chi-Square Tests

	Value	df	Asymptotic Significance (2-sided)
Pearson Chi-Square	5.227[a]	2	.073
Likelihood Ratio	6.296	2	.043
Linear-by-Linear Association	4.929	1	.026
N of Valid Cases	55		

a. 3 cells (50.0%) have expected count less than 5. The minimum expected count is .22.

cups per day) may produce faster and, more detailed thoughts and better general body coordination, may also help maintain cognitive function in aging and may have protective effects against several diseases as some types of cancer, coronary heart disease, type 2 diabetes, Parkinson's disease and liver disease.

The minority of the respondents consume tea moderately, which seems to positively affect their health. According to several studies, moderate tea consumption decreases the likelihood of death caused by cardiovascular disease by 24% and increases physical activity.

On the other hand, cigarette smoking has been causally associated with many diseases (mainly chronic diseases, such as cardiovascular diseases, malignant neoplasms, lung diseases, etc.). However, it is worth mentioning that respondents' recent smoking habits do not seem to relate to their working status. Still, a correlation is observed between their smoking habits and their past working status.

As far as the sleep factor is concerned, there is a difference between the total number of sleep hours per weekdays versus the total number of sleep hours per weekend and between sleep hours per weekdays and family status. Generally, the respondents seem to ensure proper sleep in terms of duration as a prerequisite for their good physical, psychological and social well-being.

Concerning television consumption, the results indicate that the majority of the respondents, as well as smaller proportions of them, watch television for several hours daily and many more during the weekend, harming their health in that manner. Other studies indicate that people, even the youngest ones, who watch television for at least 3 hours daily, are twice as likely to die prematurely than people who watch less television (*to the detriment of physical activity*).

Figure 16a. Available support

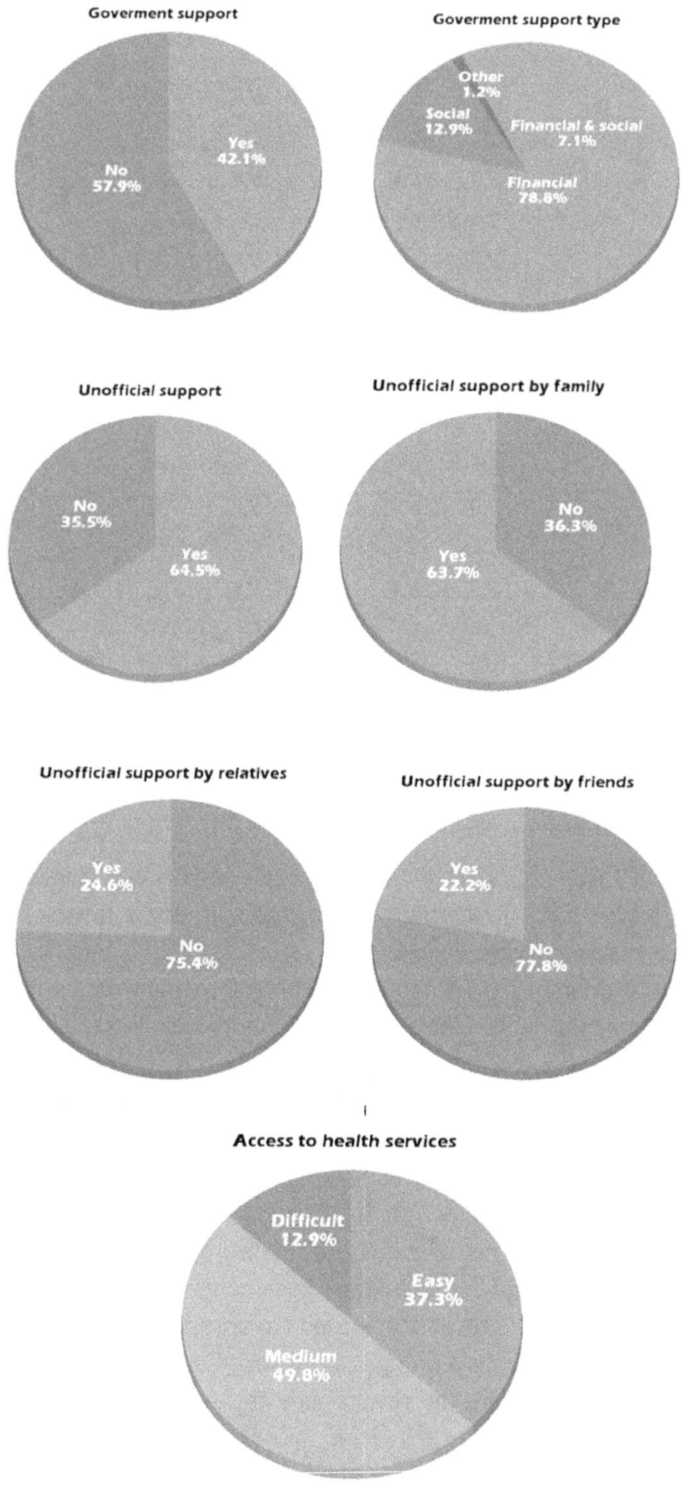

Figure 16b. Available support

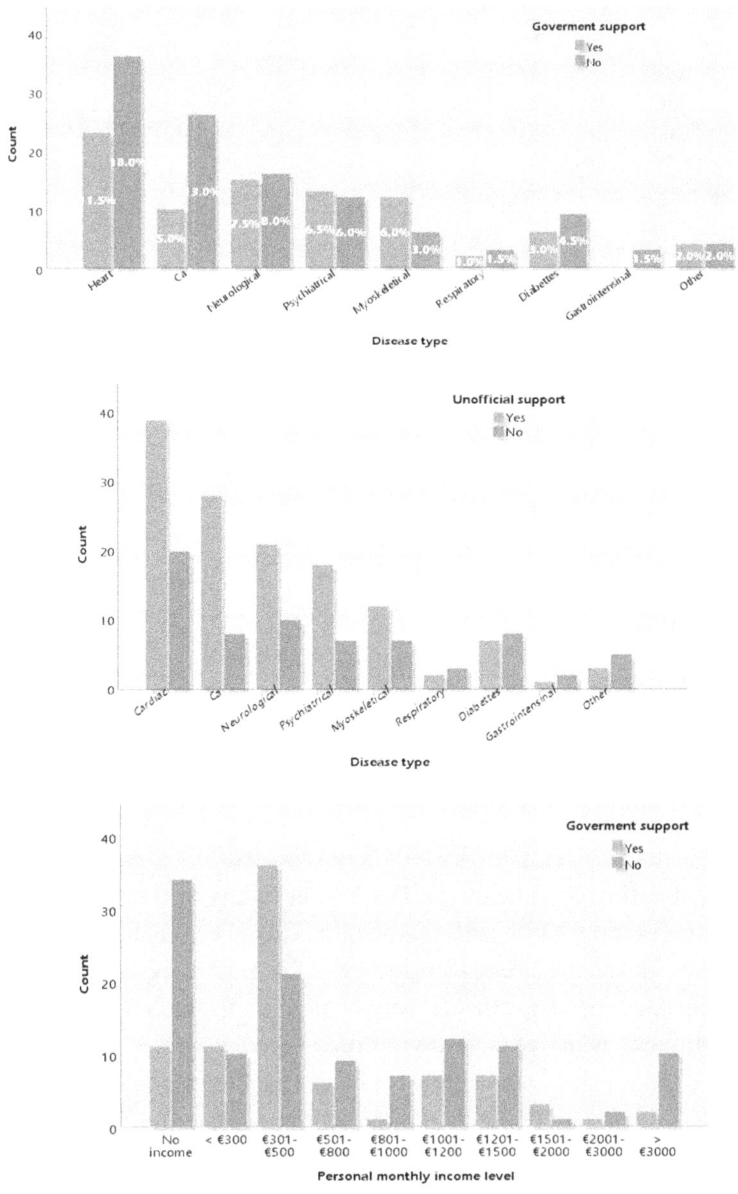

Outlining the epidemiological profile of the target population led to the observation that the majority of the diseases from which the target population suffered (cardiovascular diseases, several types of cancer, neurological and psychiatric diseases, diabetes, rheumatic and respiratory diseases, etc.) were chronic diseases and "diseases of civilization" with a strong social dimension and social severe effects on patients' daily life. The majority of the respondents had been hospitalized for a short time due to their disease, but there is no association between the length of hospital stay and health problems.

Concerning treatment which the majority of respondents received, it is worth noticing that present and past smoking habits do not seem to have affected their treatment.

Figure 16c. Available support

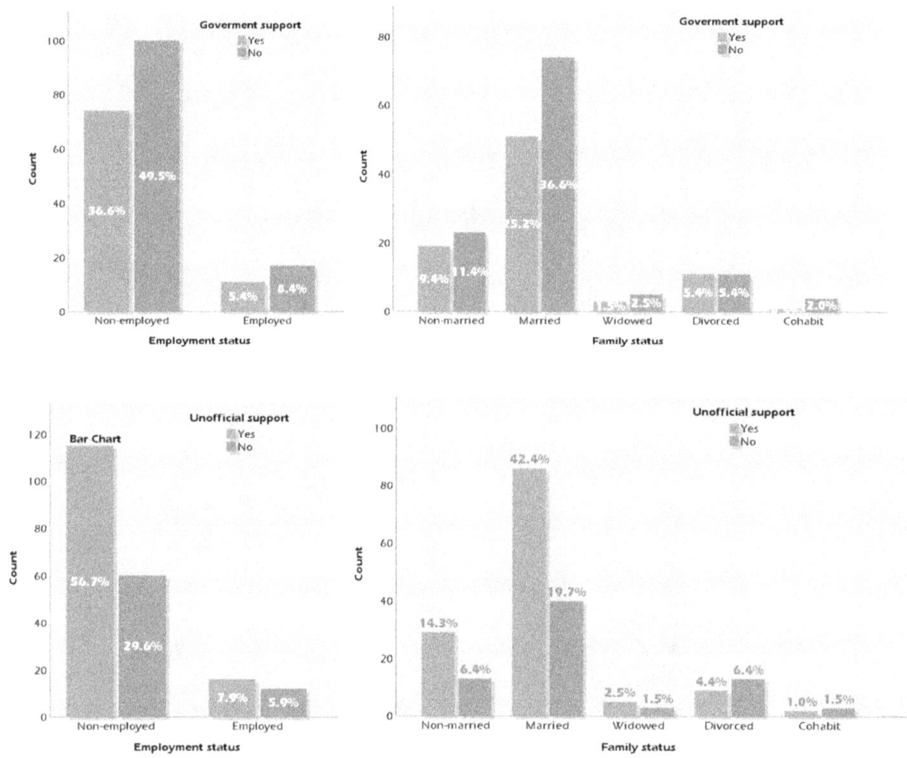

Moreover, "physical activity" factor, which did not represent the lifestyle of the majority of respondents, was negatively affecting good health and treatment of any of their diseases. The significance of this factor to health, disease prevention and treatment has also been documented by previous literature.

The factor of anxiety and social stress seems to exist for most respondents among different types of disease. However, it did not depend on other factors such as family, income, working and property status. In all respects, it constitutes a negative factor. According to many studies, stress seems to be a risk factor for the appearance and treatment of many physical health problems, including the aging process. Mental disorders, however, such as emotional disorders, seem to derive from "errors" of the stress process. Moreover, stress has been associated with tumor evolution and other health problems, such as diabetes. At the same time,, enough data confirm the essential role of stress in triggering cardiovascular diseases and the evolution and prognosis of a pre-existing CVD.

On the other hand, the criminal record of the majority of respondents seems to be clear, with only a tiny percentage of them being involved in illegal activity. This constitutes a positive health factor since, according to studies, people who repeatedly break the law and have delinquent and criminal behavior disrupt the community and disrupt the community and the community and their health. Although it is not so clear where this derives from, it might be explained by the risky behavior of those inclined to crime which leads more quickly to injuries and accidents.

Regarding social support (official or unofficial), it appears that the majority of the respondents do not accept any other than financial support from the official institutions, which mainly is the welfare

Table 14a. Tests regarding the available support

Employment status * Goverment support Crosstabulation

Count

		Goverment support		
		Yes	No	Total
Employment status	Non-employed	74	100	174
	Employed	11	17	28
Total		85	117	202

Chi-Square Tests

	Value	df	Asymptotic Significance (2-sided)	Exact Sig. (2-sided)	Exact Sig. (1-sided)
Pearson Chi-Square	.104[a]	1	.747		
Continuity Correction[b]	.014	1	.907		
Likelihood Ratio	.105	1	.746		
Fisher's Exact Test				.838	.457
Linear-by-Linear Association	.104	1	.748		
N of Valid Cases	202				

a. 0 cells (.0%) have expected count less than 5. The minimum expected count is 11.78.

b. Computed only for a 2x2 table

Crosstab

Count

		Goverment support		
		Yes	No	Total
Employment status	Non-employed	74	100	174
	Employed	11	17	28
Total		85	117	202

Chi-Square Tests

	Value	df	Asymptotic Significance (2-sided)	Exact Sig. (2-sided)	Exact Sig. (1-sided)
Pearson Chi-Square	.104[a]	1	.747		
Continuity Correction[b]	.014	1	.907		
Likelihood Ratio	.105	1	.746		
Fisher's Exact Test				.838	.457
Linear-by-Linear Association	.104	1	.748		
N of Valid Cases	202				

a. 0 cells (0.0%) have expected count less than 5. The minimum expected count is 11.78.

b. Computed only for a 2x2 table

Crosstab

Count

		Unofficial support		
		Yes	No	Total
Employment status	Non-employed	115	60	175
	Employed	16	12	28
Total		131	72	203

Chi-Square Tests

	Value	df	Asymptotic Significance (2-sided)	Exact Sig. (2-sided)	Exact Sig. (1-sided)
Pearson Chi-Square	.775[a]	1	.379		
Continuity Correction[b]	.446	1	.504		
Likelihood Ratio	.758	1	.384		
Fisher's Exact Test				.400	.250
Linear-by-Linear Association	.771	1	.380		
N of Valid Cases	203				

a. 0 cells (0.0%) have expected count less than 5. The minimum expected count is 9.93.

b. Computed only for a 2x2 table

benefit they receive because of their disability rate. On the contrary, significant receive social support from unofficial factors. The family holds the first place and relatives and friends follow. Based on relevant literature, it is observed that unofficial *non-institutionalized* activity protects health or even helps them to confront their disease. According to Durkheim, supportive social relationships are expected to positively impact an individual's physical and mental health. On the contrary, official institutions seem to adopt only a role of providing a small amount of money. In no case do they help prevent or engage disease physically, psychologically or socially.

Additionally, no correlation was found between official or unofficial social support and type of disease and between income levels and family status.

Finally, "access to health services" factor does not seem to protect the investigated population's health, leading to social exclusion and painful health consequences.

To summarize, the West Athens population, which has been investigated in this present study, seems to suffer from chronic diseases with a severe social impact for which they receive a disability rate over 67% by the primary Health Committee of West Athens. The social determinants which have been studied (family status, education level, working status, economic status, living conditions, food habits, alcohol-tea consumption, smoking habits - drugs use, sleep-television, diseases, physical activity, anxiety – social stress, delinquency, social support, access to health services) do not, most of them, appear to be protective in terms of prevention and treatment for any of their diseases.

Table 14b. Tests regarding the available support

Employment status * Goverment support Crosstabulation

Count

		Goverment support Yes	No	Total
Employment status	Non-employed	74	100	174
	Employed	11	17	28
Total		85	117	202

Chi-Square Tests

	Value	df	Asymptotic Significance (2-sided)	Exact Sig. (2-sided)	Exact Sig. (1-sided)
Pearson Chi-Square	.104ª	1	.747		
Continuity Correction[b]	.014	1	.907		
Likelihood Ratio	.105	1	.746		
Fisher's Exact Test				.838	.457
Linear-by-Linear Association	.104	1	.748		
N of Valid Cases	202				

a. 0 cells (.0%) have expected count less than 5. The minimum expected count is 11.78.

b. Computed only for a 2x2 table

Crosstab

Count

		Goverment support Yes	No	Total
Employment status	Non-employed	74	100	174
	Employed	11	17	28
Total		85	117	202

Chi-Square Tests

	Value	df	Asymptotic Significance (2-sided)	Exact Sig. (2-sided)	Exact Sig. (1-sided)
Pearson Chi-Square	.104ª	1	.747		
Continuity Correction[b]	.014	1	.907		
Likelihood Ratio	.105	1	.746		
Fisher's Exact Test				.838	.457
Linear-by-Linear Association	.104	1	.748		
N of Valid Cases	202				

a. 0 cells (0.0%) have expected count less than 5. The minimum expected count is 11.78.

b. Computed only for a 2x2 table

Crosstab

Count

		Unofficial support Yes	No	Total
Employment status	Non-employed	115	60	175
	Employed	16	12	28
Total		131	72	203

Chi-Square Tests

	Value	df	Asymptotic Significance (2-sided)	Exact Sig. (2-sided)	Exact Sig. (1-sided)
Pearson Chi-Square	.775ª	1	.379		
Continuity Correction[b]	.446	1	.504		
Likelihood Ratio	.758	1	.384		
Fisher's Exact Test				.400	.250
Linear-by-Linear Association	.771	1	.380		
N of Valid Cases	203				

a. 0 cells (0.0%) have expected count less than 5. The minimum expected count is 9.93.

b. Computed only for a 2x2 table

Crosstab

Count

		Goverment support Yes	No	Total
Family status	Non-married	19	23	42
	Married	51	74	125
	Widowed	3	5	8
	Divorced	11	11	22
	Cohabit	1	4	5
Total		85	117	202

Chi-Square Tests

	Value	df	Asymptotic Significance (2-sided)
Pearson Chi-Square	1.891ª	4	.756
Likelihood Ratio	1.980	4	.739
Linear-by-Linear Association	.052	1	.819
N of Valid Cases	202		

a. 4 cells (40.0%) have expected count less than 5. The minimum expected count is 2.10.

Crosstab

Count

		Unofficial support Yes	No	Total
Family status	Non-married	29	13	42
	Married	86	40	126
	Widowed	5	3	8
	Divorced	9	13	22
	Cohabit	2	3	5
Total		131	72	203

Chi-Square Tests

	Value	df	Asymptotic Significance (2-sided)
Pearson Chi-Square	7.830ª	4	.098
Likelihood Ratio	7.480	4	.113
Linear-by-Linear Association	6.289	1	.012
N of Valid Cases	203		

a. 3 cells (30.0%) have expected count less than 5. The minimum expected count is 1.77.

Another conclusion is the lack of an IT-based system that collects valuable data to be easily accessible and available to researchers and policymakers. Thus, no key indicators are available, and any future effort/study needs a similar approach to provide generalized conclusions. The current research stresses the need for an automated way to collect and process information to address the humanitarian needs associated with the disabled population. The use of appropriate classification or scales, like the International Classification of Functioning, Disability and Health – ICF (WHO, 2021a), or the World Health Organization Disability Assessment Schedule 2.0 - WHODAS 2.0 (WHO, 2021b) scale, could help in the way of collecting and processing data more homogeneously, to be easily comparable and helpful for researchers. The score of the WHODAS 2.0 scale could also be used for cross-cultural comparisons and has already been tested for its reliability and validity in Greece (Koumpouros et al., 2017; Koumpouros et al., 2018) as well as in many other countries, populations and pathologies (Baron et al., 2008; Silva et al., 2013; Garin et al., 2010).

On the other hand, one must consider the complexity of the ICF and the time required to classify and record people with disabilities in the system. Based on the previous analysis, the best solution seems to be the use of WHODAS 2.0 along with other socio-epidemiological details and information using an appropriate IT-based system. The key to effectively addressing humanitarian issues, like those related to disability and ageing of the population, is key performance indicators (KPIs) and related data provided in a digestible format for end users.

ACKNOWLEDGMENT

This research received no specific grant from any funding agency in the public, commercial, or not-for-profit sectors. The research was supported by the primary Health Committee of the West Athens in Greece.

REFERENCES

Aneshensel, C. (1992). Social Stress: Theory and Research. *Annual Review of Sociology*, *18*(1), 15–38. doi:10.1146/annurev.so.18.080192.000311

Avlund, K., Holstein, B., Osler, M., Damsqaard, M., Pedersen, P., & Rasmussen, M. (2003). Social position and health in old age: The relevance of different indicators of social position. *Scandinavian Journal of Public Health*, *31*(2), 126–136. doi:10.1080/14034940210134130 PMID:12745763

Bambra, C., & Eikemo, T. A. (2009). Welfare state regimes, unemployment and health: A comparative study of the relationship between unemployment and self reported health in 23 European countries. *Journal of Epidemiology and Community Health*, *63*(2), 92–98. doi:10.1136/jech.2008.077354 PMID:18930981

Baron, M., Schieir, O., Hudson, M., Steele, R., Kolahi, S., Berkson, L., Couture, F., Fitzcharles, M. A., Gagné, M., Garfield, B., Gutkowski, A., Kang, H., Kapusta, M., Ligier, S., Mathieu, J.-P., Ménard, H., Starr, M., Stein, M., & Zummer, M. (2008). The clinimetric properties of the world health organization disability assessment schedule II in early inflammatory arthritis. *Arthritis and Rheumatism*, *59*(3), 382–390. doi:10.1002/art.23314 PMID:18311751

Bleich, S. N., Jones-Smith, J., Wolfson, J. A., Zhu, X., & Story, M. (2015). The complex relationship between diet and health. *Health Affairs*, *34*(1), 1813–1820. doi:10.1377/hlthaff.2015.0606 PMID:26526238

Brandby, H. (2010). *Introduction to health and illness*. Pedio Editions.

Cockerham, W., Hambz, B., & Oates, G. (2017). *The social determinants of chronic disease*. PMC. doi:10.1016/j.amepre.2016.09.010

Dahlgren, G., & Whitehead, M. (1992). Policies and strategies for promotion equality in health (M. Gitona, Trans.). Ministry of Health, National School of Public Health.

Davou, B., & Christakis, N. (1994). *Children talk about health and illness: Social representations about health and illness of 10 year-old children*. Papazisis Editions.

Fakiolas, N. (2012). Social Networks Health impact. *Review of Social Research, 137-138,* 201-232.

Fellas, K. (2009). *Society and Health*. Kritiki Editions.

Fletcher, G., Landolfo, C., Niebauer, J., Ozemek, C., Arena, R., & Lavie, C. (2018). Promoting Physical Activity and Exercise. *Journal of the American College of Cardiology*, *72*(14), 1622–1639. doi:10.1016/j.jacc.2018.08.2141 PMID:30261965

Frank, J., Abel, T., Campostrini, S., Cook, S., Lin, V., & McQueen, D. (2020). The Social Determinants of Health: Time to Re-Think? *International Journal of Environmental Research and Public Health*, *17*(16), 5856. doi:10.3390/ijerph17165856 PMID:32806743

Garin, O., Ayuso-Mateos, J., Almansa, J., Nieto, M., Chatterji, S., Vilagut, G., Alonso, J., Cieza, A., Svetskova, O., Burger, H., Racca, V., Francescutti, C., Vieta, E., Kostanjsek, N., Raggi, A., Leonardi, M., & Ferrer, M. (2010). Validation of the "World Health Organization Disability Assessment Schedule, WHODAS-2" in patients with chronic diseases. *Health and Quality of Life Outcomes*, *8*(1), 51. doi:10.1186/1477-7525-8-51 PMID:20482853

Huttunen, R., Heineken, T., & Syrjänen, J. (2011). Smoking and the outcome of infection. *Journal of Internal Medicine*, *269*(3), 258–269. doi:10.1111/j.1365-2796.2010.02332.x PMID:21175903

Ioannidi, E., & Manti, P. (1999). Health services/hospital, special issues and challenges. In Health: limits and prospects. Patra: Hellenic Open University.

Irwin, A., & Solar, O. (2010). A conceptual framework for action on the social determinants of health. Discussion paper for the Commission on Social Determinants of Health, World Health Organization.

Kanuganti, S., Sarkar, A., Singh, A., & Arkatkar, S. (2015). Quantification of accessibility to health facilities in rural areas. *Case Studies On Transport Policy*, *3*(3), 311–320. doi:10.1016/j.cstp.2014.08.004

Koumpouros, Y., Papageorgiou, E., Sakellari, E., Prapas, X., Perifanou, D., & Lagiou, A. (2018). Adaptation and psychometric properties evaluation of the Greek version of WHODAS 2.0. Pilot application in Greek elderly population. *Health Services and Outcomes Research Methodology*, *18*(1), 63–74. doi:10.100710742-017-0176-x

Koumpouros, Y., Sakellari, E., Papageorgiou, E., & Lagiou, A. (2017). *Measuring health and disability. Manual for WHO Disability Assessment Schedule, WHODAS 2.0*. WHO and University of West Attica.

Kyriopoulos, J., Kalandidi, A., & Trichopoulos, D. (1983). Influence of socioeconomic factors on cause specific mortality in Greece. *Materia Medica Greca*, 11.

Link, B., & Phelan, J. (1995). Social conditions as fundamental causes of disease. *Journal of Health and Social Behavior*, *35*(Spec), 80–94. doi:10.2307/2626958 PMID:7560851

Lu, Z. N., Chen, H., Hao, Y., Wang, J., Song, X., & Mok, T. M. (2017). The dynamic relationship between environmental pollution, economic development and public health: Evidence from China. *Journal of Cleaner Production*, *166*, 134–147. doi:10.1016/j.jclepro.2017.08.010

Nettleton, S. (2020). *The sociology of health and illness* (4th ed.). Wiley.

Peirce, R. S., Frone, M. R., Russell, M., Cooper, M. L., & Mudar, P. (2000). A longitudinal model of social contact, social support, depression, and alcohol use. *Health Psychology*, *19*(1), 28–38. doi:10.1037/0278-6133.19.1.28 PMID:10711585

Petridou, E. (1988). *Social factors in health and illness*. Notes of social medicine. Department of Health Visiting of Technological and Educational Institute of Athens.

Pontisidis, G., & Mpilali, Th. (2015). Social Support and its impact on facing chronic diseases. *Hellenic Journal of Nursing Science*, *8*(4).

Price, A. (2004). Heart disease and work. *BMJ (Clinical Research Ed.)*, *2003*, 029298. doi:10.1136 / hrt PMID:15310715

Raphael, D. (2016). Social structure, living conditions, and health: Canadian. *Social Determinants of Health: Canadian Perspectives,* 31-56.

Roumelioti, A. (2010). *Principles of Epidemiology in Public Health practice*. Educational notes, National School of Public Health (NSPH).

Samuels-Kalow, M., Ciccolo, G., Lin, M., Schoenfeld, E., & Camargo, C. Jr. (2020). The terminology of social emergency medicine: Measuring social determinants of health, social risk, and social need. *Journal Of The American College Of Emergency Physicians Open*, *1*(5), 852–856. doi:10.1002/emp2.12191 PMID:33145531

Sarris, M. (2001). *Sociology of health and quality of life*. Papazisis publications.

Silva, C., Coleta, I., Silva, A., Amaro, A., Alvarelhão, J., Queirós, A., & Rocha, N. (2013). Adaptation and validation of WHODAS 2.0 in patients with musculoskeletal pain. *Revista de Saude Publica*, *47*(4), 1–6. PMID:24346666

Tountas, Y. (2000). Society and Health. Athens: Odisseas.

Tountas, Y. (2012). *Live without cigarettes*. Tipothito Publications.

Tselepi, C. (2000). *Social and cultural factors in the engagement and in the experience of illness. Sociological and psychological approach of hospitals services of health. Social cultural aspects of health and illness.* Hellenic Open University Publications, Patra.

Tzoufras, I., & Piperoglou, A. (2009). Introduction to Biostatistics and Epidemiology. Academic Press.

Uckino, B. N. (2006). Social Support and health. A review of physiological process potentially underlying links to disease outcomes. *Journal of Behavioral Medicine*, *29*(4), 377–387. doi:10.100710865-006-9056-5 PMID:16758315

Wang, J., Lesage, A., Schmitz, N., & Drapeau, A. (2008). The relationship between work stress and mental disorders in men and women: Findings from a population-based study. *Journal of Epidemiology and Community Health*, *62*(1), 42–47. doi:10.1136/jech.2006.050591 PMID:18079332

WHO. (2021a). *International Classification of Functioning, Disability and Health (ICF).* https://www.who.int/standards/classifications/international-classification-of-functioning-disability-and-health

WHO. (2021b). *WHO Disability Assessment Schedule (WHODAS 2.0).* https://www.who.int/standards/classifications/international-classification-of-functioning-disability-and-health/who-disability-assessment-schedule

Compilation of References

Abdullahi, Y., Badamasi, A., Galadimawa, N., & Abubakar, G. (2016). *Community based solid waste management strategy: a case study of Kaduna metropolis*. doi:10.2495/SDP160641

Abdullah, T. O., Ali, S. S., Al-Ansari, N. A., & Knutsson, S. (2018). Possibility of Groundwater Pollution in Halabja Saidsadiq Hydrogeological Basin, Iraq Using Modified DRASTIC Model Based on AHP and Tritium Isotopes. *Geosciences*, *8*(7). Advance online publication. doi:10.3390/geosciences8070236

Abrar, C. R. (2001, January). On the Margin: Refugees, Migrants and Minorities, Refugee and Migratory Movement Research Unit. *Refugee Survey Quarterly*, *20*(1).

ACAPS. (2017). *Nigeria: Humanitarian Needs of Returnees from Cameroon, Chad & Niger* (ACAPS Thematic report: Returning refugees in Borno State, Nigeria, Issue). ACAPS. https://www.acaps.org/sites/acaps/files/products/files/20170731_acaps_thematic_report_returnign_refugees_in_borno_statenigeria.pdf

ACAPS. (2018). *Rohingya Influx Overview Key Changes During 2018 Monsoon Season*. Retrieved 12 August 2021 from https://reliefweb.int/report/bangladesh/rohingya-influx-overview-key-changes-during-2018-monsoon-season

Agenda for Change. (2020). *Strengthening WASH Systems: Concepts, Examples and Experiences*. Agenda for Change. https://washagendaforchange.org/wp-content/uploads/2020/04/20200227_agenda_for_change_systems_strengthening_experiences_final.pdf

Aguilar, G., Rozgic, V., Wang, W., & Wang, C. (2019). Multimodal and Multi-view Models for Emotion Recognition. *Proceedings of the 57th Annual Meeting of the Association for Computational Linguistics*, 991–1002. doi:10.18653/v1/P19-1095

Ahmed, N., Michelin, R. A., Xue, W., Ruj, S., Malaney, R., Kanhere, S. S., Seneviratne, A., Hu, W., Janicke, H., & Jha, S. K. (2020). A Survey of COVID-19 Contact Tracing Apps. *IEEE Access: Practical Innovations, Open Solutions*, *8*, 134577–134601. doi:10.1109/ACCESS.2020.3010226

Ahmed, R. (2003a). Climate of Bangladesh. In S. Islam (Ed.), *Banglapedia* (Vol. 3, pp. 45–48). Asiatic Society of Bangladesh.

Ahmed, R. (2003b). Climate of Banglaesh. In R. Ahmed (Ed.), *Abhaoa O Jalabayu Bijnan* [Meteorology and Climatology] (3rd ed.). University of Rajshahi.

Ainul, S., Ehsan, I., Haque, E.F., Amin, S., Rob, U., Melnikas, A.J., & Falcone, J. (2018). *Marriage and Sexual and Reproductive Health of Rohingya Adolescents and Youth in Bangladesh: A Qualitative Study*. Population Council. UNFPA.

Akpan, V. I., Igwe, U. A., Ikechukwu, B., Mpamah, I., & Okoro, C. O. (2020). Social constructivism: Implications on teaching and learning. *Brock Journal of Education*, *8*(8), 49–56.

Alanazi, A. (2016). A critical review of constructivist theory and the emergence of constructionism. *American Research Journal of Humanities and Social Sciences*, *2*, 1–8.

Alberro, A., Iribarren-Lopez, A., Sáenz-Cuesta, M., Matheu, A., Vergara, I., & Otaegui, D. (2021). Inflammaging markers characteristic of advanced age show similar levels with frailty and dependency. *Scientific Reports*, *11*(1), 4358. https://doi.org/10.1038/s41598-021-83991-7

Alcántara-Ayala, I., & Goudie, A. S. (2011). *Geomorphological Hazards and Disaster Prevention*. Cambridge University Press.

Alexandre, K. (2020). Frontiers in Climate Change Adaptation Science: Advancing Guidelines to Design Adaptation Pathways. *Current Climate Change Reports*, *6*(4), 166–177. doi:10.100740641-020-00166-8

Alexopoulos, G. S., & Kiosses, D. (2011). Geriatric Psychiatry, An Issue of Psychiatric Clinics (Illustrated edition). Saunders.

Ali, A. (1999). Climate change impacts and adaptation assessment in Bangladesh. *Climate Research*, *12*, 109–116. doi:10.3354/cr012109

Ali, S. S. (2019). Problem based learning: A student-cemtred approach. *English Language Teaching*, *12*(5), 73–78. doi:10.5539/elt.v12n5p73

Almeida, M., Martins, M. A., Soares, A. M. V., Cuesta, A., & Oliveira, M. (2019). Polystyrene nanoplastics alter the cytotoxicity of human pharmaceuticals on marine fish cell lines. *Environmental Toxicology and Pharmacology*, *169*, 56–65. doi:10.1016/j.etap.2019.03.019 PMID:30953935

Aloi, G., Caliciuri, G., Fortino, G., Gravina, R., Pace, P., Russo, W., & Savaglio, C. (2016). A Mobile Multi-Technology Gateway to Enable IoT Interoperability. *2016 IEEE First International Conference on Internet-of-Things Design and Implementation (IoTDI)*, 259–264. doi:10.1109/IoTDI.2015.29

Alves, B., Angnuureng, D. B., Morand, P., & Almar, R. (2020). A review on coastal erosion and flooding risks and best management practices in West Africa: What has been done and should be done. *Journal of Coastal Conservation*, *24*(3), 38. Advance online publication. doi:10.100711852-020-00755-7

Amankwa, M. O., Tetteh, E. K., Mohale, G. T., Dagba, G., & Opoku, P. (2021). The production of valuable products and fuel from plastic waste in Africa. *Discover Sustainability*, *2*(1), 31. Advance online publication. doi:10.100743621-021-00040-z

Anderson, C. C., & Renaud, F. G. (2021). A review of public acceptance of nature-based solutions: The 'why', 'when', and 'how' of success for disaster risk reduction measures. *Ambio*, *50*(8), 1552–1573. Advance online publication. doi:10.100713280-021-01502-4 PMID:33606249

Andriopoulou, F., Dagiuklas, T., & Orphanoudakis, T. (2016). In G. Keramidas, N. Voros, & M. Hübner (Eds.), *Integrating IoT and Fog Computing for Healthcare Service Delivery* (pp. 213–232). Springer. https://link.springer.com/chapter/10.1007%2F978-3-319-42304-3_11

Aneshensel, C. (1992). Social Stress: Theory and Research. *Annual Review of Sociology*, *18*(1), 15–38. doi:10.1146/annurev.so.18.080192.000311

Anguera, J. A., & Gazzaley, A. (2015). Video games, cognitive exercises, and the enhancement of cognitive abilities. *Current Opinion in Behavioral Sciences*, *4*, 160–165. https://doi.org/10.1016/j.cobeha.2015.06.002

Anh & Ngoc. (2019). Public Participatory Role in Urban Flood Risk Management of Ho Chi Minh City - Vietnam : From Awareness to Action. *South Asian Journal of Social Studies and Economics, 4*(4), 1–10. doi:10.9734/sajsse/2019/v4i430133

Anh. (2017a). How public Participation Contributed to the Infrastructure Projects in Vietnam: The case of MRT 2 project in Ho Chi Minh City. *Imperial Journal of Interdisciplinary Research, 3*(4).

Anholt, R., & Sinatti, G. (2020). Under the Guise of Resilience: The EU Approach to Migration and Forced Displacement in Jordan and Lebanon. *Contemporary Security Policy*, *41*(2), 311–335. doi:10.1080/13523260.2019.1698182

Anh, P. T. (2017b). Public Awareness and participation in Canal Environmental Protection: Case Studies in Ho Chi Minh City, Vietnam. *Journal of Shipping and Ocean Engineering*, *7*(3), 121–126. doi:10.17265/2159-5879/2017.03.005

Anh, P. T., Bush, S. R., Mol, A. P. J., & Kroeze, C. (2011). The multi-level environmental governance of Vietnamese aquaculture: Global certification, national standards, local cooperatives. *Journal of Environmental Policy and Planning*, *13*(4), 373–397. doi:10.1080/1523908X.2011.633701

Anh, P. T., & Ngoc, N. T. B. (2021). Promoting Public Participation in Environmental Protection and Flood Risk Management in Ho Chi Minh City - Vietnam. In *Insights into Economics and Management* (Vol. 11). Issue SE-Chapters. doi:10.9734/bpi/ieam/v11/2508E

Annele, U., Satu, K. J., & Timo, E. S. (2019). Definitions of successful ageing: A brief review of a multidimensional concept. *Acta Bio Medica : Atenei Parmensis*, *90*(2), 359–363. https://doi.org/10.23750/abm.v90i2.8376

Antony, C., Unnikrishnan, A. S., & Woodworth, P. L. (2016). Evolution of extreme high waters along the east coast of India and at the head of the Bay of Bengal. *Global and Planetary Change*, *140*, 59–67. doi:10.1016/j.gloplacha.2016.03.008

Arias, Bellouin, Coppola, Jones, Krinner, Marotzke, & Naik. (2021). *Climate Change 2021: The Physical Science Basis.* Contribution of Working Group14 I to the Sixth Assessment Report of the Intergovernmental Panel on Climate Change Technical Summary.

Arnstein, S. (1969). A Ladder of citizen participation. *Journal of the American Institute of Planners*, *35*(4), 216–224. doi:10.1080/01944366908977225

Arroio, A. (2010). Context based learning: A role for cinema in science education. *Science Education International*, *21*(3), 131–143.

Arthur, C., Baker, J., & Bamford, H. (2009). *Proceedings of the International Research Workshop on the Occurrence, Effects, and Fate of Microplastic Marine Debris.* NOAA Technical Memorandum NOS-OR&R-30. https://marinedebris.noaa.gov/proceedings-international-research-workshop-microplastic-marine-debris

Asim, M., Batool, S. A., & Chaudhry, M. N. (2012). Scavengers and their role in the recycling of waste in Southwestern Lahore. *Resources, Conservation and Recycling*, *58*, 152–162. doi:10.1016/J.RESCONREC.2011.10.013

Atiyeh, B., & Gunn, S. (2017). *Refugee camps, fire disasters and burn injuries. Ann Burns Fire Disaster.* https://www.ncbi.nlm.nih.gov/pmc/articles/PMC5946753/

Avlund, K., Holstein, B., Osler, M., Damsqaard, M., Pedersen, P., & Rasmussen, M. (2003). Social position and health in old age: The relevance of different indicators of social position. *Scandinavian Journal of Public Health*, *31*(2), 126–136. doi:10.1080/14034940210134130 PMID:12745763

Babu, K. E. K. (2020). Impacts and Challenges to Host Country Bangladesh Due to Sheltering the Rohingya Refugees. *Cogent Social Sciences*, *6*, 1770943. doi:10.1080/23311886.2020.1770943

Baby, B. (2018, January 1). Rohingya crisis: A postscript. *The Daily Star*. Retrieved 1 January 2018 from https://www.thedailystar.net/supplements/unpacking-2017/rohingya-crisis-postscript-1513045

Baltes, P. B., & Baltes, M. M. (1990). Psychological perspectives on successful aging: The model of selective optimization with compensation. In Successful aging: Perspectives from the behavioral sciences (pp. 1–34). Cambridge University Press. https://doi.org/10.1017/CBO9780511665684.003.

Bambra, C., & Eikemo, T. A. (2009). Welfare state regimes, unemployment and health: A comparative study of the relationship between unemployment and self reported health in 23 European countries. *Journal of Epidemiology and Community Health*, *63*(2), 92–98. doi:10.1136/jech.2008.077354 PMID:18930981

Banglapedia. (2021, June 18). *National Encyclopedia of Bangladesh*. Retrieved 21 September 2021 from https://en.banglapedia.org/index.php?title=Bangladesh_Geography

Barbosa, F., Adeyemi, J. A., Bocato, M. Z., Comas, A., & Campiglia, A. (2020). A critical viewpoint on current issues, limitations, and future research needs on micro- and nanoplastic studies: From the detection to the toxicological assessment. *Environmental Research*, *182*, 109089. Advance online publication. doi:10.1016/j.envres.2019.109089 PMID:32069751

Barnett, M. N. (2013). *Empire of Humanity: A History of Humanitarianism*. Cornell University Press. doi:10.7591/9780801460616

Baron, M., Schieir, O., Hudson, M., Steele, R., Kolahi, S., Berkson, L., Couture, F., Fitzcharles, M. A., Gagné, M., Garfield, B., Gutkowski, A., Kang, H., Kapusta, M., Ligier, S., Mathieu, J.-P., Ménard, H., Starr, M., Stein, M., & Zummer, M. (2008). The clinimetric properties of the world health organization disability assessment schedule II in early inflammatory arthritis. *Arthritis and Rheumatism*, *59*(3), 382–390. doi:10.1002/art.23314 PMID:18311751

Baumgärtner, L., Dmitrienko, A., Freisleben, B., Gruler, A., Höchst, J., & Kühlberg, J. Mezini. M., Mitev, R., Miettinen. M., Muhamedagic. A., Nguyen, T. D., Penning, A., Pustelnik, D. F., Roos, F., Sadeghi, A., Schwarz. M., & Uhl, C. (2020). *Mind the GAP: Security & Privacy Risks of Contact Tracing Apps*. https://arxiv.org/abs/2006.05914

BBC. (2016, September 21). Refugee crisis: Plan to create 100,000 jobs in Ethiopia. *BBC News*. Retrieved from https://www.bbc.com/news/world-africa-37433085

BBC. (2020). *Lesbos: Hundreds test positive for Covid-19 after migrant camp fire*. https://www.bbc.co.uk/news/world-europe-54239446

Becker, M., Papa, F., Karpytchev, M., Delebecque, C., Krien, Y., Khan, J. U., Ballu, V., Durand, F., Le Cozannet, G., Islam, A. K. M. S., Calmant, S., & Shum, C. K. (2020). Water level changes, subsidence, and sea level rise in the Ganges–Brahmaputra–Meghna delta. *Proceedings of the National Academy of Sciences of the United States of America*, *117*(4), 1867–1876. doi:10.1073/pnas.1912921117 PMID:31907308

Beierle, T. C. (1998). *Public participation in Environmental DEcisions: An Evaluation Framework Using Social Goals*. Resources for the Future.

Bell, S. (2010). Project-based learning for the 21st century: Skills for the future. *The Clearing House: A Journal of Educational Strategies, Issues and Ideas*, *83*(2), 39–43. doi:10.1080/00098650903505415

Benalcázar, M. E., Jaramillo, A. G., Zea, A., Páez, A., & Andaluz, V. H. (2017). Hand gesture recognition using machine learning and the Myo armband. *2017 25th European Signal Processing Conference (EUSIPCO)*, 1040–1044. doi:10.23919/EUSIPCO.2017.8081366

Bergami, E., Krupinski Emerenciano, A., González-Aravena, M., Cárdenas, C. A., Hernández, P., Silva, J. R. M. C., & Corsi, I. (2019). Polystyrene nanoparticles affect the innate immune system of the Antarctic sea urchin Sterechinus neumayeri. *Polar Biology*, *42*(4), 743–757. doi:10.100700300-019-02468-6

Besseling, E., Quik, J. T. K., Sun, M., & Koelmans, A. A. (2017). Fate of nano- and microplastic in freshwater systems: A modeling study. *Environmental Pollution*, *220*, 540–548. doi:10.1016/j.envpol.2016.10.001 PMID:27743792

Better Shelter. (2020). *Designed with and for Refugees.* Product: Better Shelter. https://bettershelter.org/product/

Betts, A., Bloom, L., Kaplan, J. D., & Omata, N. (2017). *Refugee Economies: Forced Displacement and Development*. Oxford University Press.

Betts, A., & Collier, P. (2015). Help Refugees Help Themselves: Let Displaced Syrians Join the Labor Market. *Foreign Affairs*, *94*(6), 84–92.

Betts, A., & Collier, P. (2017). *Refuge: Transforming a Broken Refugee System.* Penguin.

Bhandari, M., & Raj, S. (2017). Practical approach to green chemistry. *International Journal of Pharmacy and Pharmaceutical Sciences*, *9*(4), 10–26. doi:10.22159/ijpps.2017v9i4.15640

Bhatia, A., Mahmud, A., Fuller, A., Shin, R., Rahman, A., Shatil, T., Sultana, M., Morshed, K.A.M., Leaning, J., & Balsari, S. (2018). The Rohingya in Cox's Bazar: When the Stateless Seek Refugee. *Health and Human Rights Journal, 20*(2).

Bilzen, V. G. (2015). *The Development of Aid.* Cambridge Scholars Publishing.

Bleich, S. N., Jones-Smith, J., Wolfson, J. A., Zhu, X., & Story, M. (2015). The complex relationship between diet and health. *Health Affairs*, *34*(1), 1813–1820. doi:10.1377/hlthaff.2015.0606 PMID:26526238

Block, I. (2019). *Refugee Shelters Could Be Built from Concrete Fabric in 24 Hours.* Dezeen. Available at: https://www.dezeen.com/2019/09/17/cortex-shelter-cutwork-refugee-architecture/

Bogner, J. E., Spokas, K. A., & Chanton, J. P. (2011). Seasonal Greenhouse Gas Emissions (Methane, Carbon Dioxide, Nitrous Oxide) from Engineered Landfills: Daily, Intermediate, and Final California Cover Soils. *Journal of Environmental Quality*, *40*(3), 1010–1020. doi:10.2134/jeq2010.0407 PMID:21546687

Bowman, B. (2020). *As Coronavirus Spreads, Experts Warn Refugees Are Particularly Vulnerable.* The Globe Post. Available at: https://theglobepost.com/2020/02/28/coronavirus-outbreak-refugees

Brandby, H. (2010). *Introduction to health and illness*. Pedio Editions.

Bremner, L. (2020). Sedimentary Logics and the Rohingya Refugee Camps in Bangladesh. *Political Geography*, *77*, 102109. doi:10.1016/j.polgeo.2019.102109

Bruun, O. (2012). Sending the right bill to the right people: Climate change, environmental degradation, and social vulnerabilities in central Vietnam. *Weather, Climate, and Society*, *4*(4), 250–262. doi:10.1175/WCAS-D-11-00040.1

Bui, V. T., & Ly, H. M. (2016). Detection of shoreline changes: A geospatial data analysis in Vung Tau City, Southern Vietnam. *Science and Technology Development Journal*, *19*(1), 131–139. doi:10.32508tdj.v19i1.567

Bulley, D. (2014). Inside the Tent: Community and Government in Refugee Camps. *Security Dialogue*, *45*(1), 63–80. doi:10.1177/0967010613514788

Burmeister, M., Rauch, F., & Eilks, I. (2012). Education for sustainable development and chemistry education. *Chemistry Education Research and Practice*, *13*(2), 59–68. doi:10.1039/C1RP90060A

Burntley, S. J. (2007). A review of municipal solid waste composition in the United Kingdom. *Journal of Waste Management, 27*(10), 1274–1285. PMID:17011771

Byler, R., Gelaw, F., & Khoshnood, K. (2015). G: Considerations for Altering the Standard Refugee Camp Design for Improved Health Outcomes. *2015 IEEE Global Humanitarian Technology Conference (GHTC)*. 10.1109/GHTC.2015.7343964

Caleda & Anh. (2014). *Report on public consultation for Ho Chi Minh city urban mass rapid transit line 2*. Academic Press.

Cammack, P. (2015). What International Organizations Do, and Why They Do It. *Spectrum: Journal of Global Studies, 7*(1), 62–77.

Campanale, C., Massarelli, C., Savino, I., Locaputo, V., & Uricchio, V. F. (2020). A Detailed Review Study on Potential Effects of Microplastics and Additives of Concern on Human Health. *International Journal of Environmental Research and Public Health, 17*(4), 1212. Advance online publication. doi:10.3390/ijerph17041212 PMID:32069998

Campbell, R. C., & Wilson, D. (2011). The unique value of huanitarian enginering. In *2011 ASEE Annual Conference & Exposition* (p. 12). ASEE.

Carpenter, E. J., & Smith, K. L. Jr. (1972). Plastics on the Sargasso sea surface. *Science, 175*(4027), 1240–1241. doi:10.1126cience.175.4027.1240 PMID:5061243

Cavedon, S. (2019). *The Failures of Refugee Camps in Europe*. Columbia Political Review. http://www.cpreview.org/blog/2019/11/the-failures-of-refugee-camps-in-europe

Cerghit, I. (2006). *Metode de invatamant*. Polirom.

Chamma, N. & Mendoza, C.. (2016). *Rethinking Refugee Camp Design: From 'Temporary' Camps To Sustainable Settlements*. https://www.researchgate.net/publication/317694602_RETHINKING_REFUGEE_CAMP_DESIGN_FROM_'TEMPORARY'_CAMPS_TO_SUSTAINABLE_SETTLEMENTS/stats

Chang, H.-K., Chen, W.-W., Tsai, F.-C., Liou, J.-C., & Kuo, S.-G. (2018). Annual and Seasonal Shoreline Evolution of Tainan Gold Coast Using Multiband Satellite Images. *IACSIT International Journal of Engineering and Technology, 10*(1), 52–56. doi:10.7763/IJET.2018.V10.1034

Chatham House. (2015). *The role of Local Government in Addressing the Impact of Syrian Refugees: Jordan Case Study*. https://www.chathamhouse.org/sites/default/files/events/02-06-15-jordan-workshop-summary-final2.pdf

Chatzistefanou, V., & Limniotis, K. (2017). On the (non-)anonymity of anonymous social networks. *e-Democracy – Privacy-Preserving, Secure, Intelligent E-Government Services Communications in Computer and Information Science, 792*, 153–168. doi:10.1007/978-3-319-71117-1_11

Cheng, X., Yang, Y., Schwebel, D. C., Liu, Z., Li, L., Cheng, P., Ning, P., & Hu, G. (2020). Population ageing and mortality during 1990–2017: A global decomposition analysis. *PLoS Medicine, 17*(6), e1003138. https://doi.org/10.1371/journal.pmed.1003138

Chen, H., Qin, Y., Huang, H., & Xu, W. (2020). A regional difference analysis of microplastic pollution in global freshwater bodies based on a regression model. *Water (Basel), 12*(7), 1889. Advance online publication. doi:10.3390/w12071889

Chen, Y. C., & Shih, C. H. (2019). Sustainable management of coastal wetlands in Taiwan: A review for invasion, conservation, and removal of mangroves. *Sustainability (Switzerland), 11*(16), 4305. Advance online publication. doi:10.3390u11164305

Chin, T., Buckee, C. O., & Mahmud, A. S. (2020). Quantifying the Success of Measles Vaccination Campaigns in the Rohingya Refugee Camps. *Epidemics, 30*, 100385. doi:10.1016/j.epidem.2020.100385 PMID:31951876

Chiu, S., & Small, C. (2016). Observations of cyclone-induced storm surge in coastal Bangladesh. *Journal of Coastal Research, 32*(5), 1149–1161. doi:10.2112/JCOASTRES-D-15-00030.1

Chowdhary, N. (2020). Environmental Refugees: A Humanitarian Crisis in India and Bangladesh. *Refugee Crises and Third-World Economies,* 37-43. doi:10.1108/978-1-83982-190-520201007

Cockerham, W., Hambz, B., & Oates, G. (2017). *The social determinants of chronic disease.* PMC. doi:10.1016/j.amepre.2016.09.010

Cohen-Shacham, E., Andrade, A., Dalton, J., Dudley, N., Jones, M., Kumar, C., & Walters, G. (2019). Core principles for successfully implementing and upscaling Nature-based Solutions. *Environmental Science and Policy, 98*(June), 20–29. doi:10.1016/j.envsci.2019.04.014

Conceição, P. (2019). *Human Development Report 2019: beyond income, beyond averages, beyond today: Inequalities in human development in the 21st century.* United Nations Development Program. http://hdr.undp.org/sites/default/files/hdr2019.pdf

Corsellis, T., & Vitale, A. (2005). *Transitional Settlement: Displaced Populations.* Oxfam GB.

Cortese, A. D. (2003). The critical role of higher education in creating a sustainable future. *Planning for Higher Education*, 15–22.

Cotten, S. R., Ford, G., Ford, S., & Hale, T. M. (2014). Internet Use and Depression Among Retired Older Adults in the United States: A Longitudinal Analysis. *The Journals of Gerontology: Series B, 69*(5), 763–771. doi:10.1093/geronb/gbu018

Crawley, E. E., Malmqvist, J., Ostlund, S., Brodeur, D. R., & Edstrom, K. (2014). Introduction and motivation. In E. F. Crawley (Ed.), *Rethinking engineering education* (pp. 1–9). Springer. doi:10.1007/978-3-319-05561-9_1

Cristea, M., Noja, G. G., Stefea, P., & Sala, A. L. (2020). The Impact of Population Aging and Public Health Support on EU Labor Markets. *International Journal of Environmental Research and Public Health, 17*(4), E1439. https://doi.org/10.3390/ijerph17041439

Cuong, N. Q., & Cu, N. V. (2014). Integrated coastal management in Vietnam: Current situation and orientation. *Journal of Marine Science and Technology, 14*(1), 89–96. https://www.vjs.ac.vn/index.php/jmst

Dabaieh, M., Emami, N., Heinonen, J., & Marteinsson, B. (1970). *A life cycle assessment of a 'minus carbon' refugee house: Global warming potential and sensitivity analysis: Semantic Scholar.* https://www.semanticscholar.org/paper/A-life-cycle-assessment-of-a-'minus-carbon'-refugee-Dabaieh-Emami/7e7136057c4617b313a560d7c76b0845ed6835a4

Dahlgren, G., & Whitehead, M. (1992). Policies and strategies for promotion equality in health (M. Gitona, Trans.). Ministry of Health, National School of Public Health.

Dasgupta, S., Huq, M., Khan, Z. H., Ahmed, M. M., Mukherjee, N., Khan, M. F., & Pandey, K. (2010). *Vulnerability of Bangladesh to Cyclones in a Changing Climate:Potential Damages and Adaptation Cost.* World Bank.

Davou, B., & Christakis, N. (1994). *Children talk about health and illness: Social representations about health and illness of 10 year-old children.* Papazisis Editions.

De Jong, B., Keijsers, J. G. S., Riksen, M. J. P. M., Krol, J., & Slim, P. A. (2014). Soft engineering vs. a dynamic approach in coastal dune management: A case study on the North Sea barrier Island of Ameland, the Netherlands. *Journal of Coastal Research, 30*(4), 670–684. doi:10.2112/JCOASTRES-D-13-00125.1

de Krom, M. P. M. M., Oosterveer, P., & Mol, A. P. J. (2013). Public Interests and Values in Multi-Level Food Risk Governance: European Responses to Avian Influenza. *Journal of Environmental Policy and Planning, 15*(2), 161–177. doi:10.1080/1523908X.2012.746867

De Vriend, H. J., & Van Koningsveld, M. (2012). *Building with Nature: Thinking, acting and interacting differently.* Ecoshape. doi:10.1080/02513625.2014.925714

Dehghani, M., Hashemi, H., Abedi, T., Shamsedini, N., Khodabakhshi, A., & Ghasemi, R. (2014). The survey of knowledge, attitude, and Behavior of students in Shiraz University of Medical Sciences about the recycling of solid waste in 2012. *Health System Research, 10*, 821–829.

DeJarnette, N. K. (2018). Impleenting STEM in the early childhood classroom. *European Journal of STEM Education, 3*(3), 1–9. doi:10.20897/ejsteme/3878

Demiris, G., Thompson, H. J., Reeder, B., Wilamowska, K., & Zaslavsky, O. (2013). Using informatics to capture older adults' wellness. *International Journal of Medical Informatics, 82*(11), e232–e241. https://doi.org/10.1016/j.ijmedinf.2011.03.004

Denselow, J. (2016). *The end of the refugee camp?* https://www.aljazeera.com/opinions/2016/5/20/the-end-of-the-refugee-camp

Department for Environment, Food and Rural Affairs. (2009). *Good practice and regulatory guidance on composting and odour control for local authorities.* Author.

Department for International Development. (2011a). *Humanitarian Emergency Response Review.* Retrieved from https://assets.publishing.service.gov.uk/government/uploads/system/uploads/attachment_data/file/67579/HERR.pdf

Department for International Development. (2011b). *Department for International Development Annual Report and Resource Accounts 2009-10: Third Report of Session 2010-11.* London: Stationery Office.

Department for International Development. (2016). *Jordan Compact Economic Opportunities Program: Summary Sheet.* Retrieved from http://iati.dfid.gov.uk/iati_documents/5519561.odt

Dolui, K., & Datta, S. K. (2017). *Comparison of edge computing implementations: Fog computing, cloudlet and mobile edge computing.* doi:10.1109/GIOTS.2017.8016213

Donadio, R. (2019). *Welcome to Europe. Now Go Home.* https://www.theatlantic.com/international/archive/2019/11/greeces-moria-refugee-camp-a-european-failure/601132/

DONRE. (2018). *Annual report on solid waste management program in Ho Chi Minh City.* Author.

Dreze, J. (1990). Famine Prevention in India. In J. Dreze & A. K. Sen (Eds.), *The Political Economy of Hunger: Famine Prevention* (pp. 13–122). Clarendon Press.

EMERGENCY UK. (2020). *Italy: The Makeshift Camps of San Ferdinando.* EMERGENCY UK. https://emergencyuk.org/2018/03/italy-the-makeshift-camps-of-san-ferdinando/

Enfrin, M., Dumée, L. F., & Lee, J. (2019). Nano/microplastics in water and wastewater treatment processes – Origin, impact and potential solutions. *Water Research, 161*, 621–638. doi:10.1016/j.watres.2019.06.049 PMID:31254888

Enria, N., & Gerwins, S. (2020). *Greek-Turkish border crisis: Refugees are paying the price for the EU's failure to reform its asylum system.* https://blogs.lse.ac.uk/europpblog/2020/03/25/greek-turkish-border-crisis-refugees-are-paying-the-price-for-the-eus-failure-to-reform-its-asylum-system/

Euromonitor International. (2017). *Global trends in food and drinks packaging: a report compiled for auspack 2017.* https://go.euromonitor.com/00-EV-APAC2017-AUSPACK_Thank-You.html?aliId=eyJpIjoidWZmM1dPUzBiblV3RDVRVSIsInQiOiJXdDBjRVN1NXR2SG5aRjRKa1plWEVnPT0ifQ%253D%253D

European Data Protection Board. (2020). *Guidelines 04/2020 on the use of location data and contact tracing tools in the context of the COVID-19 outbreak.* https://edpb.europa.eu/sites/default/files/files/file1/edpb_guidelines_20200420_contact_tracing_covid_with_annex_en.pdf

European Environment Agency. (2019). *Industrial wastewater treatment – pressures on Europe's environment (23/2018).* https://www.eea.europa.eu/publications/industrial-waste-water-treatment-pressures

European Union Agency for Cybersecurity. (2017). *Privacy and data protection in mobile applications - A study on the app development ecosystem and the technical implementation of GDPR.* https://www.enisa.europa.eu/publications/privacy-and-data-protection-in-mobile-applications

EWARS. (2018). *Epidemiological bulletin week 25.* Retrieved 21 June 2018 from http://www.searo.who.int/bangladesh/ewarsw252018.pdf

Exodus Privacy. (2021). *What Exodus Privacy Does.* https://exodus-privacy.eu.org/en/page/what/

Ezeah, C., Fazakerley, J. A., & Roberts, C. L. (2013). Emerging trends in informal sector recycling in developing and transition countries. *Waste Management (New York, N.Y.)*, *33*(11), 2509–2519. doi:10.1016/J.WASMAN.2013.06.020 PMID:23886489

Fahad, M. G., Islam, A. S., Nazari, R., Hasan, M. A., Islam, G. M. T., & Bala, S. K. (2018). Regional changes of precipitation and temperature over Bangladesh using bias-corrected multi-model ensemble projections considering high emission pathways. *International Journal of Climatology*, *38*(4), 1634–1648. doi:10.1002/joc.5284

Fakiolas, N. (2012). Social Networks Health impact. *Review of Social Research, 137-138,* 201-232.

Fellas, K. (2009). *Society and Health.* Kritiki Editions.

Finck, M., & Pallas, F. (2020). They who must not be identified—Distinguishing personal from non-personal data under the GDPR. *International Data Privacy Law, 10*(1), 11–36. doi:10.1093/idpl/ipz026

Fletcher, G., Landolfo, C., Niebauer, J., Ozemek, C., Arena, R., & Lavie, C. (2018). Promoting Physical Activity and Exercise. *Journal of the American College of Cardiology*, *72*(14), 1622–1639. doi:10.1016/j.jacc.2018.08.2141 PMID:30261965

Fogel, A. L., & Kvedar, J. C. (2018). Artificial intelligence powers digital medicine. *NPJ Digital Medicine*, *1*(1), 1–4. https://doi.org/10.1038/s41746-017-0012-2

Folke, L., Riisgaard, L., Ponte, S., Hartwich, F., & Kormawa, P. (2010). Agro-Food Value Chain Interventions in Asia: A Review and Analysis of Case Studies. Vienna: United Nations Industrial Development Organization (UNIDO).

Fonseca, C. (2015). *Reality check for the life-cycle cost approach in the context of refugee camps.* https://www.ircwash.org/blog/reality-check-life-cycle-cost-approach-context-refugee-camps

Frank, J., Abel, T., Campostrini, S., Cook, S., Lin, V., & McQueen, D. (2020). The Social Determinants of Health: Time to Re-Think? *International Journal of Environmental Research and Public Health*, *17*(16), 5856. doi:10.3390/ijerph17165856 PMID:32806743

Friedrich, M. J. (2019, May 21). Water and Sanitation Deficits Take a Toll in Armed Conflict Regions. *Journal of the American Medical Association*, *321*(19), 1861. doi:10.1001/jama.2019.6186 PMID:31112265

Furukawa, M., & Hanh, P. T. T. (2007). Impact of sea level rise on coastal zone of Vietnam. *Proceedings of the 10th International River Symposium and Environmental Flows Conference*, *59*(84), 1–9.

Garin, O., Ayuso-Mateos, J., Almansa, J., Nieto, M., Chatterji, S., Vilagut, G., Alonso, J., Cieza, A., Svetskova, O., Burger, H., Racca, V., Francescutti, C., Vieta, E., Kostanjsek, N., Raggi, A., Leonardi, M., & Ferrer, M. (2010). Validation of the "World Health Organization Disability Assessment Schedule, WHODAS-2" in patients with chronic diseases. *Health and Quality of Life Outcomes*, *8*(1), 51. doi:10.1186/1477-7525-8-51 PMID:20482853

Gensch, R., Jennings, A., Renggli, S., & Reymond, P. (2018). *Compendium of Sanitation Technologies in Emergencies*. https://www.washnet.de/wp-content/uploads/emergency-sanitation-compendium.pdf

Gesing, F. (2019b). Towards a more-than-human political ecology of coastal protection: Coast Care practices in Aotearoa New Zealand. *Environment and Planning E: Nature and Space*. doi:10.1177/2514848619860751

Gesing, F. (2019a). The politics of artificial dunes: Sustainable coastal protection measures and contested socio-natural objects. *Die Erde*, *150*(3), 145–157. doi:10.12854/erde-2019-423

Geyer, R., Jambeck, J. R., & Law, K. L. (2017). Production, use, and fate of all plastics ever made. *Science Advances*, *3*(7), 25–29. doi:10.1126ciadv.1700782 PMID:28776036

Ghattasheh, E. (2019). *Managing Syrian Refugee Camps Using ArcGIS*. https://www.esri.com/about/newsroom/arcuser/managing-syrian-refugee-camps-using-arcgis/

Ghosh, D., & Dinda, S. (2020). Determinants of the Quality of Life Among Elderly: Comparison Between China and India. *The International Journal of Community and Social Development*, *2*(1), 71–98. https://doi.org/10.1177/2516602620911835

Giri, C., Ochieng, E., Tieszen, L. L., Zhu, Z., Singh, A., Loveland, T., & Duke, N. (2011). Status and distribution of mangrove forests of the world using earth observation satellite data. *Global Ecology and Biogeography*, *20*(1), 154–159. doi:10.1111/j.1466-8238.2010.00584.x

GIS. (2020). *What is GIS?* Geographic Information System Mapping Technology. https://www.esri.com/en-us/what-is-gis/overview

Giuli, C., Papa, R., Lattanzio, F., & Postacchini, D. (2016). The Effects of Cognitive Training for Elderly: Results from My Mind Project. *Rejuvenation Research*, *19*(6), 485–494. https://doi.org/10.1089/rej.2015.1791

Global WASH Cluster. (2020). *Delivering Humanitarian WASH at scale, Anywhere and Any Time: Road Map for 2020-2025*. U. Global WASH Cluster (GWC). https://washcluster.net/sites/default/files/inline-files/RoadMap_2020-2025.pdf

Gold, M. R., Stevenson, D., & Fryback, D. G. (2002). HALYS and QALYS and DALYS, Oh My: Similarities and differences in summary measures of population Health. *Annual Review of Public Health*, *23*, 115–134. https://doi.org/10.1146/annurev.publhealth.23.100901.140513

Gomez, O. A., & Kawaguchi, C. (2018). A Theory for the Continuum: Multiple Approaches to Humanitarian Crises Management. In Crisis Management Beyond the Humanitarian-Development Nexus (pp. 15–36). Routledge.

Gonzalez, A. (2016). Making Global Value-Chains Work for Economic Development and Shared Prosperity. In *Speech at Global Value Chain Development Report 2016 Background Paper Conference, Beijing*. Retrieved from https://www.worldbank.org/en/news/speech/2016/03/17/making-global-value-chains-work-for-economic-development-and-shared-prosperity

Gopinath, P. M., Saranya, V., Vijayakumar, S., Mythili Meera, M., Ruprekha, S., Kunal, R., Pranay, A., Thomas, J., Mukherjee, A., & Chandrasekaran, N. (2019). Assessment on interactive prospectives of nanoplastics with plasma proteins and the toxicological impacts of virgin, coronated and environmentally released-nanoplastics. *Scientific Reports*, *9*(1), 1–15. doi:10.103841598-019-45139-6 PMID:31222081

Gowing, J. W., Tuong, T. P., Hoanh, C. T., & Khiem, N. T. (2006). Social and environmental impact of rapid change in the coastal zone of Vietnam: An assessment of sustainability issues. *Environment and Livelihoods in Tropical Coastal Zones: Managing Agriculture-Fishery-Aquaculture Conflicts*, 48–60. doi:10.1079/9781845931070.0048

Gracia, A., Rangel-Buitrago, N., Oakley, J. A., & Williams, A. T. (2018). Use of ecosystems in coastal erosion management. *Ocean and Coastal Management*, *156*, 277–289. doi:10.1016/j.ocecoaman.2017.07.009

Grant, H. (2020). *'Moria is a Hell': New Arrivals Describe Life in a Greek Refugee Camp.* The Guardian. https://www.theguardian.com/global-development/2020/jan/17/moria-is-a-hell-new-arrivals-describe-life-in-a-greek-refugee-camp

Gülseven, Y. (2020). Reconsidering Development Aid: A Systemic Analysis. *World Review of Political Economy*, *11*(2), 232–255. doi:10.13169/worlrevipoliecon.11.2.0232

Guo, H. J., & Sapra, A. (2021). *Instrumental Activity of Daily Living*. StatPearls Publishing. https://www.ncbi.nlm.nih.gov/books/NBK553126/

Gvili, Y. (2020). *Security Analysis of the COVID-19 Contact Tracing Specifications by Apple Inc. and Google Inc.* Cryptology ePrint Archive, Report 2020/428. https://eprint.iacr.org/2020/428

Gworek, B., Dmuchowski, W., Koda, E., Marecka, M., Baczewska, A. H., Bragoszewska, P., Sieczka, A., & Osiński, P. (2016). Impact of the municipal solid waste lubna landfill on environmental pollution by heavy metals. *Water (Switzerland)*, *8*(10). Advance online publication. doi:10.3390/w8100470

Han, Q., Liang, S., & Zhang, H. (2015). Mobile Cloud Sensing, Big Data, and 5G Networks Make an Intelligent and Smart World. *IEEE Network*, *29*, 40–45. https://doi.org/10.1109/MNET.2015.7064901

Hanson, R. (2014). The impact and challenges of integrating micro chemistry experiments into e-learning. *International Journal for Cross-Disciplinary Subjects in Education*, *4*(1), 1884–1892.

Hanson, R. (2017). Enhancing students' performance in organic chemistry through context-based learning and microactivities: A casestudy. *International Journal of Research and Reflection in Educational Sciences*, *5*(6), 7–20.

Hanson, R., & Acquah, S. (2014). Enhancing concept understanding through the use of micro chemistry equipment and collaborative activities. *Journal of Education and Practice*, *5*(12), 120–130.

Hargroves, K. (2011). *Teaching sustainability in high schools.* Retrieved February 14, 2021, from Resources for rethinking: https://resources4rethinking.ca/en/resource/chemistry-innovations-in-sustainable-development

Harmer, A., & Macrae, J. (Eds.). (2004). *Beyond the Continuum: An Overview of the Changing Role of Aid Policy in Protracted Crises*. Humanitarian Policy Group Research Briefing. Retrieved from https://cdn.odi.org/media/documents/341.pdf

Hartmann, N. B., Hüffer, T., Thompson, R. C., Hassellöv, M., Verschoor, A., Daugaard, A. E., Rist, S., Karlsson, T., Brennholt, N., Cole, M., Herrling, M. P., Hess, M. C., Ivleva, N. P., Lusher, A. L., & Wagner, M. (2019). Are We Speaking the Same Language? Recommendations for a Definition and Categorization Framework for Plastic Debris. *Environmental Science & Technology*, *53*(3), 1039–1047. doi:10.1021/acs.est.8b05297 PMID:30608663

Hendriksen, A., Tukahirwa, J., Oosterveer, P. J. M., & Mol, A. P. J. (2012). Participatory Decision Making for Sanitation Improvements in Unplanned Urban Settlements in East Africa. *Journal of Environment & Development*, *21*(1), 98–119. doi:10.1177/1070496511426778

Hoang, V. C. (2020). Long-Term Evolution of Morphology at Loc An Estuary, Vung Tau, Vietnam. *Marine Geodesy*, *43*(2), 163–188. doi:10.1080/01490419.2019.1606125

Hodson, M. E., Duffus-Hodson, C. A., Clark, A., Prendergast-Miller, M. T., & Thorpe, K. L. (2017). Plastic Bag Derived-Microplastics as a Vector for Metal Exposure in Terrestrial Invertebrates. *Environmental Science & Technology*, *51*(8), 4714–4721. doi:10.1021/acs.est.7b00635 PMID:28355064

Hoepman, J.-H. (2014). Privacy Design Strategies. In N. Cuppens-Boulahia, F. Cuppens, S. Jajodia, A. Abou El Kalam, & T. Sans (Eds.), ICT Systems Security and Privacy Protection (pp. 446–459). Springer. https://doi.org/10.1007/978-3-642-55415-5_38.

Hsu, T. W., Tseng, I. F., Lin, T. Y., Shin, C. Y., & Ou, S. H. (2008). Review of countermeasures against beach erosion on the Taiwanese coast. *Coastal Management*, *36*(3), 274–293. doi:10.1080/08920750801968678

Hsu, T., Lin, T., & Tseng, I. (2007). Human Impact on Coastal Erosion in Taiwan. *Journal of Coastal Research*, *234*, 961–973. doi:10.2112/04-0353R.1

Humanitarian Practice Network. (2017). *Special feature- The Lake Chad Basin: An Overlooked Crisis?* https://odihpn.org/wp-content/uploads/2017/10/he-70-web.pdf

Humphrey, J., & Navas-Alemán, L. (2010). Value Chains, Donor Interventions and Poverty Reduction: A Review of Donor Practice. *IDS Research Reports*, *2010*(63), 1–106. doi:10.1111/j.2040-0217.2010.00063_2.x

Hung, J.-Y., & Lu, K.-S. (2014). Research on the Healthy Lifestyle Model, Active Ageing, and Loneliness of Senior Learners. *Educational Gerontology*, *40*(5), 353–362. https://doi.org/10.1080/03601277.2013.822200

Hung, N. T., Vinh, B. T., Nam, S. Y., & Lee, J. L. (2017). Cause Analysis of Erosion-Induced Resort Washout on Cua Dai Beach, Vietnam. *Journal of Coastal Research*, *79*, 214–218. doi:10.2112/SI79-044.1

Hutton, G., & Chase, C. (2017). Water Supply, Sanitation, and Hygiene. In C. N. Mock, R. Nugent, O. Kobusingye, & K. R. Smith (Eds.), Injury Prevention and Environmental Health. The World Bank. doi:10.1596/978-1-4648-0522-6_ch9

Huttunen, R., Heineken, T., & Syrjänen, J. (2011). Smoking and the outcome of infection. *Journal of Internal Medicine*, *269*(3), 258–269. doi:10.1111/j.1365-2796.2010.02332.x PMID:21175903

Idrus, M. M., Mohd Baharudin, R., Shahabudin, M., & Rashidah, A. (2008). *An overview of landfill management and technologies: A Malaysian case study at Ampar Tenang.* 1st National Seminar on Environment, Development and Sustainability, Selangor, Malaysia.

International Water Management Institute. (2020, October 20). *Annual Report 2020.* Retrieved 2021, from Summary of sustainable development goals: https://www.iwmi.cgiar.org

Ioannidi, E., & Manti, P. (1999). Health services/hospital, special issues and challenges. In Health: limits and prospects. Patra: Hellenic Open University.

IOM. (2019). In North-East Nigeria, Access to Water Paves the Way for a Brighter Future. International Organization of Migration (IOM).

IPCC. (2014). *Impacts, adaptation, and vulnerability: Summary for policy maker.* Author.

IPCC. (2019). *Climate change and land - IPCC, Summary for policymakers.* doi:10.4337/9781784710644

IRC. (2020). *COVID-19 Threatens Refugee Camps.* International Rescue Committee (IRC). https://www.rescue-uk.org/coronavirus?utm_source=www.rescue.org&utm_medium=georedirect

Irwin, A., & Solar, O. (2010). A conceptual framework for action on the social determinants of health. Discussion paper for the Commission on Social Determinants of Health, World Health Organization.

ISCG. (2017a, November 2). *ISCG Situation Update: Rohingya Refugee Crisis, Cox's Bazar*. Retrieved 21 September 2021 from https://reliefweb.int/report/bangladesh/iscg-situation-update-rohingya-refugee-crisis-cox-s-bazar-2-november-2017

ISCG. (2017b, October 15). *Bangladesh Humanitarian Situation Report No.6 (Rohingya influx)*. Retrieved 12 August 2021 from https://reliefweb.int/report/bangladesh/iscg-situation-report-cox-s-bazar-influx-15-october-2017

ISCG. (2019a). *Government of Bangladesh Humanitarian Stakeholders*. Retrieved 23 September 2021 from https://www.humanitarianresponse.info/en/operations/bangladesh/infographic/iscg-coordination-structure-2019

ISCG. (2019b). *Monsoon Response in Rohingya Refugee Camps Season Update | 21 April to 24 October 2019 Cox's Bazar, Bangladesh*. Retrieved 12 August 2021 from https://reliefweb.int/report/bangladesh/monsoon-response-rohingya-refugee-camps-season-update-21-april-24-october-2019

ISCG. (2020). *Monsoon Response in Rohingya Refugee Camps | 12 June to 15 October 2020 Cox's Bazar, Bangladesh*. Retrieved 12 August 2021 from https://reliefweb.int/report/bangladesh/monsoon-response-rohingya-refugee-camps-season-update-12-june-15-october-2020

Ismail, H. (2015). Gait and Postural Sway Analysis, A Multi-Modal System. *Proceedings of the 2015 ACM on International Conference on Multimodal Interaction*, 629–633. doi:10.1145/2818346.2823310

Jahre, M., Kembro, J., Adjahossou, A., & Altay, N. (2018). Approaches to the design of Refugee Camps. *Journal of Humanitarian Logistics and Supply Chain Management*, *8*(3), 323–345. doi:10.1108/JHLSCM-07-2017-0034

Jonsson, A., Andersson, L., Alkan-Olsson, J., & Arnheimer, B. (2007). How participatory can participatory planning be? Degrees of influence of stakeholder and expert perspectives in six dimensions of participatory modeling. *Water Science and Technology*, *56*(1), 207–214. doi:10.2166/wst.2007.453 PMID:17711017

Kaczan, D. J., & Orgill-Meyer, J. (2019). The impact of climate change on migration: A synthesis of recent empirical insights. *Climatic Change*, *158*(3-4), 281–300. doi:10.100710584-019-02560-0

Kaga, M., & Nakache, D. (2019). *Protection and the Humanitarian-Development Nexus: A Literature Review* (Working paper No. 1). Retrieved from https://carleton.ca/lerrn/?s=Protection

Kankanige, D., & Babel, S. (2021). Contamination by ≥6.5 μm-sized microplastics and their removability in a conventional water treatment plant (WTP) in Thailand. *Journal of Water Process Engineering*, *40*, 101765. doi:10.1016/j.jwpe.2020.101765

Kanuganti, S., Sarkar, A., Singh, A., & Arkatkar, S. (2015). Quantification of accessibility to health facilities in rural areas. *Case Studies On Transport Policy*, *3*(3), 311–320. doi:10.1016/j.cstp.2014.08.004

Kaplan, B., & Harris-Salamone, K. D. (2009). Health IT Success and Failure: Recommendations from Literature and an AMIA Workshop. *Journal of the American Medical Informatics Association*, *16*(3), 291–299. https://doi.org/10.1197/jamia.M2997

Karbalaei, S., Hanachi, P., Walker, T. R., & Cole, M. (2018). Occurrence, sources, human health impacts and mitigation of microplastic pollution. *Environmental Science and Pollution Research International*, *25*(36), 36046–36063. doi:10.100711356-018-3508-7 PMID:30382517

Karim, M. F., & Mimura, N. (2008). Impacts of climate change and sea-level rise on cyclonic storm surge floods in Bangladesh. *Global Environmental Change*, *18*(3), 490–500. doi:10.1016/j.gloenvcha.2008.05.002

Karpudewan, M., Ismail, Z., & Roth, W.-M. (2012a). Ensuring sustainability of tomorrow through green chemistry integrated with sustainable development concepts (SDCs). *Chemistry Education Research and Practice*, *13*(2), 120–127. doi:10.1039/C1RP90066H

Karpudewan, M., Ismail, Z., & Roth, W.-M. (2012b). Fostering pre-service teachers' self-determined environmental motivation through green chemistry experiments. *Journal of Science Teacher Education*, *23*(6), 673–693. doi:10.100710972-012-9298-8

Kastner, M., Hayden, L., Wong, G., Lai, Y., Makarski, J., Treister, V., Chan, J., Lee, J. H., Ivers, N. M., Holroyd-Leduc, J., & Straus, S. E. (2019). Underlying mechanisms of complex interventions addressing the care of older adults with multimorbidity: A realist review. *BMJ Open*, *9*(4), e025009. https://doi.org/10.1136/bmjopen-2018-025009

Kennedy, J. (2005). Challenging Camp Guidelines. *Forced Migration Review*, (23), 46–47.

Khalili, P., Vazirinejad, R., Jamalizadeh, A., Nakhaee, P., Fatehi, F., & Fallah, A. (2014). The impact of healthy lifestyle education and integrated care on the quality of life of the elderly. *Journal of Occupational Health and Epidemiology*, *3*, 81–87. https://doi.org/10.18869/acadpub.johe.3.2.81

Khan, M.J.U., Islam, A.S., Bala, S.K., & Islam, G.M.T (2020). Changes of climate extremes over Bangladesh at 1.5°C, 2°C and 4°C of global warming. *Theoretical and Applied Climatology*.

Khatmullina, L., & Isachenko, I. (2017). Settling velocity of microplastic particles of regular shapes. *Marine Pollution Bulletin*, *114*(2), 871–880. doi:10.1016/j.marpolbul.2016.11.024 PMID:27863879

Khatun, F. (2017). *Addressing Rohingya Crisis: Options for Bangladesh*. Center for Policy Dialogue.

Kingsley, P. (2020). Fire Destroys Most of Europe's Largest Refugee Camp, on Greek Island of Lesbos. *The New York Times*. https://www.nytimes.com/2020/09/09/world/europe/fire-refugee-camp-lesbos-moria.html

Kitaoka, S. (2018). Foreword. In Crisis Management Beyond the Humanitarian-Development Nexus (pp. xiv-xv). Routledge.

Kon, B., Lam, A., & Chan, J. (2017). *Evolution of Smart Homes for the Elderly*. doi:10.1145/3041021.3054928

Koumpouros, Y., Papageorgiou, E., Sakellari, E., Prapas, X., Perifanou, D., & Lagiou, A. (2018). Adaptation and psychometric properties evaluation of the Greek version of WHODAS 2.0. Pilot application in Greek elderly population. *Health Services and Outcomes Research Methodology*, *18*(1), 63–74. doi:10.100710742-017-0176-x

Koumpouros, Y., Sakellari, E., Papageorgiou, E., & Lagiou, A. (2017). *Measuring health and disability. Manual for WHO Disability Assessment Schedule, WHODAS 2.0.* WHO and University of West Attica.

Kyriopoulos, J., Kalandidi, A., & Trichopoulos, D. (1983). Influence of socioeconomic factors on cause specific mortality in Greece. *Materia Medica Greca*, 11.

Labbe, J. A. (2010). Health-Adjusted Life Expectancy: Concepts and Estimates. In V. R. Preedy & R. R. Watson (Eds.), Handbook of Disease Burdens and Quality of Life Measures (pp. 417–424). Springer. https://doi.org/10.1007/978-0-387-78665-0_23.

Laville, S., & Taylor, M. (2017, June 28). A million bottles a minute: world's plastic binge "as dangerous as climate change." *The Guardian*. https://www.theguardian.com/environment/2017/jun/28/a-million-a-minute-worlds-plastic-bottle-binge-as-dangerous-as-climate-change

Lawlor, R. (2016). Engineering in society - Beyond the technical...what every engineering student should know. Royal Academy of Engineering.

LeBaron, G. (2015). It's Time to Get Serious about Forced Labour in Supply Chains. *Forced Labour in the Global Economy, Beyond Trafficking and Slavery Series*. https://cdn-prod.opendemocracy.net/media/documents/BTS-2-Global-Economy.pdf

LeBaron, G. (2016). Slavery, Human Trafficking, and Forced Labour: Implications for international development. In The Palgrave Handbook of International Development (pp. 381-398). Palgrave Macmillan.

LeBaron, G., Howard, N., Thibos, C., & Kyristis, P. (2018). *Confronting Root Causes: Forced Labour in Global Supply Chains*. https://eprints.whiterose.ac.uk/126167/1/Confronting_Root_Causes_Forced_Labour_In_Global_Supply_Chains.pdf

Lee, M., Kim, M. J., Suh, D., Kim, J., Jo, E., & Yoon, B. (2016). Feasibility of a Self-Determination Theory-Based Exercise Program in Community-Dwelling South Korean Older Adults: Experiences from a 13-Month Trial. *Journal of Aging and Physical Activity*, *24*(1), 8–21. https://doi.org/10.1123/japa.2014-0056

Leidman, E., Humphreys, A., & Greene, C. B. (2018, April 10). Acute Malnutrition and Anemia Among Rohingya Children in Kutupalong Camp, Bangladesh. *Journal of the American Medical Association*, *319*(14), 1505–1506. doi:10.1001/jama.2018.2405 PMID:29634821

Leith, D. J., & Farrell, S. (2021). Contact Tracing App Privacy: What Data Is Shared By Europe's GAEN Contact Tracing Apps. In *IEEE INFOCOM 2021 - IEEE Conference on Computer Communications* (pp. 1-10). IEEE.

Levine, S., & Mosel, I. (2014). *Remaking the Case for Linking Relief, Rehabilitation, and Development: How LRRD can become a practically useful concept for assistance in difficult places*. Humanitarian Policy Group Overseas Development Institute Report. Retrieved from https://odi.org/en/publications/remaking-the-case-for-linking-relief-rehabilitation-and-development/

Li, C., Busquets, R., & Campos, L. C. (2020). Assessment of microplastics in freshwater systems: A review. *The Science of the Total Environment*, *707*, 135578. doi:10.1016/j.scitotenv.2019.135578 PMID:31784176

Lin, T.- Y., Onselen van, V. M., & Vo, L. P. (2021). Coastal erosion in Vietnam: Case studies and implication for integrated coastal zone management in the Vietnamese south-central coastline. *IOP Conference Series: Earth and Environmental Science, 652*(1), 0–11. doi:10.1088/1755-1315/652/1/012009

Lin, T.-Y., Vo, L. P., Thao, N. D., & Bao, N. G. (2017). Case Studies of Coastal Erosion in Central Vietnam and Their Application in Hazard Management. *Proceedings of the 39th Ocean Engineering Conference in Taiwan*, 431–436.

Lin, J. C. (1996). Coastal modification due to human influence in south-western Taiwan. *Quaternary Science Reviews*, *15*(8–9), 895–900. doi:10.1016/S0277-3791(96)00060-1

Link, B., & Phelan, J. (1995). Social conditions as fundamental causes of disease. *Journal of Health and Social Behavior*, *35*(Spec), 80–94. doi:10.2307/2626958 PMID:7560851

Linn, M. C., Bell, P., & Davis, E. A. (2004). Specific design principles: Elaborating the scaffolded knowledge integration framework. In M. C. Linn, E. A. Davis, & P. Bell (Eds.), *Internet environments for science education* (pp. 315–339). Lawrence Erlbaum Associates Publishers.

Lin, T.-Y. (2016). Morphological response to the installation of detached breakwaters along the Cigu coast of Tainan, Taiwan. *Journal of Coastal Research*, *1*(75), 887–891. doi:10.2112/SI75-178.1

Li, R., Cheng, X., Schwebel, D. C., Yang, Y., Ning, P., Cheng, P., & Hu, G. (2021). Disability-adjusted life years associated with population ageing in China, 1990-2017. *BMC Geriatrics*, *21*(1), 369. https://doi.org/10.1186/s12877-021-02322-7

Li, T. H. Y. Y., Thomas Ng, S., & Skitmore, M. (2012). Public participation in infrastructure and construction projects in China : From an EIA-based to a whole-cycle process. *Habitat International*, *36*(1), 47–56. doi:10.1016/j.habitatint.2011.05.006

Li, W., Liu, J., & Li, D. (2012). Getting their voices heard : Three cases of public participation in environmental protection in China. *Journal of Environmental Management*, *98*, 65–72. doi:10.1016/j.jenvman.2011.12.019 PMID:22245866

Li, Z.-W., Zeng, G.-M., Zhang, H., Yang, B., & Jiao, S. (2007). The integrated eco-environment assessment of the red soil hilly region based on GIS—A case study in Changsha City, China. *Ecological Modelling*, *202*(3), 540–546.

Löder, M. G. J., & Gerdts, G. (2015). *Methodology Used for the Detection and Identification of Microplastics—A Critical Appraisal. In Marine Anthropogenic Litter*. Springer International Publishing., doi:10.1007/978-3-319-16510-3_8

Long, B. T., Truong, C. D., & Loi, L. L. (2006). Building up a tool for integrated coastal management in Da Nang City. *International Symposium on Geoinformatics for Spatial Infrastructure Development in Earth and Allied Sciences.*

Longobucco, Y., Benedetti, C., Tagliaferri, S., Angileri, V., Adorni, E., Pessina, M., Zerbinati, L., Cicala, L., Pelà, G., Giacomini, V., Barbolini, M., Lauretani, F., & Maggio, M. (2019). Proactive interception and care of Frailty and Multimorbidity in older persons: The experience of the European Innovation Partnership on Active and Healthy Ageing and the response of Parma Local Health Trust and Lab through European Projects. *Acta Bio-Medica: Atenei Parmensis*, *90*, 364–374. https://doi.org/10.23750/abm.v90i2.8419

Losaria & Anh. (2015). *Reports on public consultation of Ho Chi Minh City urban mass rapid transit line 2*. Academic Press.

Lunenfeld, B., & Stratton, P. (2013). The clinical consequences of an ageing world and preventive strategies. *Best Practice & Research. Clinical Obstetrics & Gynaecology*, *27*(5), 643–659. https://doi.org/10.1016/j.bpobgyn.2013.02.005

Lu, Z. N., Chen, H., Hao, Y., Wang, J., Song, X., & Mok, T. M. (2017). The dynamic relationship between environmental pollution, economic development and public health: Evidence from China. *Journal of Cleaner Production*, *166*, 134–147. doi:10.1016/j.jclepro.2017.08.010

Ma, B., Xue, W., Ding, Y., Hu, C., Liu, H., & Qu, J. (2019). Removal characteristics of microplastics by Fe-based coagulants during drinking water treatment. *Journal of Environmental Sciences (China)*, *78*, 267–275. doi:10.1016/j.jes.2018.10.006 PMID:30665645

Mafumiko, F. M. (2012). The potential of micro-scale chemistry experimentation in enhancing teaching and learning of secondary chemistry: Experiences from Tanzanian classrooms. *NUE Journal of International Cooperation*, *3*, 63–79.

Mahmood, S. S., Wroe, E., Fuller, A., & Leaning, J. (2016, December 1). The Rohingya People of Myanmar: Health, Human Rights, and Identity. *Lancet*. PMID:27916235

Major, D. C. (2010). *Adaptation assessment guidebook*. Academic Press.

Malik, N. K. A., Abdullah, S. H., & Manaf, L. A. (2015). Community Participation on Solid Waste Segregation Through Recycling Programmes in Putrajaya. *Procedia Environmental Sciences*, *30*, 10–14. doi:10.1016/J.PROENV.2015.10.002

Mallick, J. (2021). Municipal solid waste landfill site selection based on fuzzy-ahp and geoinformation techniques in ASIR region Saudi Arabia. *Sustainability (Switzerland)*, *13*(3), 1–33. doi:10.3390u13031538

Martin, D., Mica, C., & Katz, I. (2019). Rethinking the Camp: On Spatial Technologies of Power and Resistance. *Progress in Human Geography*, *44*(4), 743–768. doi:10.1177/0309132519856702

Martinez-Martin, P., Prieto-Flores, M.-E., Forjaz, M. J., Fernandez-Mayoralas, G., Rojo-Perez, F., Rojo, J.-M., & Ayala, A. (2012). Components and determinants of quality of life in community-dwelling older adults. *European Journal of Ageing*, *9*(3), 255–263. https://doi.org/10.1007/s10433-012-0232-x

Mary Zarate, J., Tian, X., Woods, K. J. P., & Poeppel, D. (2015). Multiple levels of linguistic and paralinguistic features contribute to voice recognition. *Scientific Reports*, *5*(1), 11475. https://doi.org/10.1038/srep11475

Mason, S. A., Welch, V. G., & Neratko, J. (2018). Synthetic Polymer Contamination in Bottled Water. *Frontiers in Chemistry*, *6*, 407. doi:10.3389/fchem.2018.00407 PMID:30255015

Matlin, S. A., Krief, A., Hopf, H., & Mehta, G. (2021). *Re-imagining priorities for chemistry: A central science for 'freedom from fear and want* (International Edition). Angewandte Chemie. doi:10.1002/anie.2021

Matter, A., Dietschi, M., & Zurbrügg, C. (2013). Improving the informal recycling sector through segregation of waste in the household. The case of Dhaka Bangladesh. *Habitat International*, *38*, 150–156.

Ma, X., Wang, Z., Zhou, S., Wen, H., & Zhang, Y. (2018). Intelligent Healthcare Systems Assisted by Data Analytics and Mobile Computing. *Wireless Communications and Mobile Computing*, *2018*, e3928080. https://doi.org/10.1155/2018/3928080

Mc Donnell, C., O'Connor, C., & Seery, M. K. (2007). Developing practical chemistry skills by means of student-driven problem-based learning mini-projects. *Chemistry Education Research and Practice*, *8*(2), 130–139. doi:10.1039/B6RP90026G

McCormick, A. R., Hoellein, T. J., London, M. G., Hittie, J., Scott, J. W., & Kelly, J. J. (2016). Microplastic in surface waters of urban rivers: Concentration, sources, and associated bacterial assemblages. *Ecosphere*, *7*(11), e01556. Advance online publication. doi:10.1002/ecs2.1556

McCullough, J. E., & Hayles, C. S. (2009). Education for sustainable development- Interim paper for COST C25. Sustainability of Construction-Integrated Approach to Lifetime Structural Engineering, 290-297.

McMorrow, K., & Roeger, W. (n.d.). *The economic consequences of ageing populations.* European Commission. Retrieved September 23, 2021, from https://ec.europa.eu/economy_finance/publications/pages/publication_summary11156_en.htm

Meekes, W., & Stanmore, E. K. (2017). Motivational Determinants of Exergame Participation for Older People in Assisted Living Facilities: Mixed-Methods Study. *Journal of Medical Internet Research*, *19*(7), e238. https://doi.org/10.2196/jmir.6841

Meidiana, C., Yakin, H., & Wijayanti, W. (2017). Household's Willingness to Accept Waste Separation for Improvement of Rural Waste Bank's Effectivity. doi:10.5772/intechopen.69428

Meijer, L. J. J., van Emmerik, T., van der Ent, R., Schmidt, C., & Lebreton, L. (2021). More than 1000 rivers account for 80% of global riverine plastic emissions into the ocean. *Science Advances*, *7*(18), 1–14. doi:10.1126ciadv.aaz5803 PMID:33931460

Mekonen, S. (2017). Roles of Traditional Ecological Knowledge for Biodiversity Conservation. *Journal of Natural Sciences Research*, *7*.

Meng, Y., Kelly, F. J., & Wright, S. L. (2020). Advances and challenges of microplastic pollution in freshwater ecosystems: A UK perspective. *Environmental Pollution*, *256*, 113445. doi:10.1016/j.envpol.2019.113445 PMID:31733965

Ministry of Environment and Forestry. (2020). *National Plastic Waste Reduction Strategic Actions for Indonesia, Republic of Indonesia.* https://wedocs.unep.org/bitstream/handle/20.500.11822/32898/NPWRSI.pdf?sequence=1&isAllowed=y

Ministry of Health and Sports—MoHS/Myanmar and ICF. Myanmar Demographic and Health Survey 2015–16. (2017). Myanmar: MoHS and ICF. Retrieved 20 August 2021 from https:// dhsprogram.com/publications/publication-fr324-dhs-final-reports.cfm

Mintenig, S. M., Löder, M. G. J., Primpke, S., & Gerdts, G. (2019). Low numbers of microplastics detected in drinking water from ground water sources. *The Science of the Total Environment*, *648*, 631–635. doi:10.1016/j.scitotenv.2018.08.178 PMID:30121540

Miranda, I. T. P., Fidelis, R., Fidelis, D. A. de S., Pilatti, L. A., & Picinin, C. T. (2020). The integration of recycling cooperatives in the formal management of municipal solid waste as a strategy for the circular economy—The case of Londrina, Brazil. *Sustainability (Switzerland)*, *12*(24), 1–22. doi:10.3390u122410513

Mitcham, C., & Munoz, D. R. (2010). *Humanitarian engineering*. Morgan & Claypool Publishers. doi:10.2200/S00248ED1V01Y201006ETS012

MoEF. (2021). *Performance Report of 2020.* Ministry of Environment and Forestry.

Mohammed, K., Islam, A. S., Islam, G. T., Alfieri, L., Bala, S. K., & Khan, M. J. (2018). Extreme flows and water availability of the Brahmaputra River under 1.5 and 2 °C global warming scenarios. *Climatic Change.*

Mongkolnchaiarunya, J. (2005). Promoting a community-based solid-waste management initiative in local government: Yala municipality, Thailand. *Habitat International*, *29*, 27–40. doi:10.1016/S0197-3975(03)00060-2

Monogios, S., Limniotis, K., Kolokotronis, N., & Shiaeles, S. (2019). A case study of intra-library privacy issues on Android GPS navigation apps. *e-Democracy 2019: Safeguarding Democracy and Human Rights in the Digital Age Communications in Computer and Information Science*, *1111*, 34–48. doi:10.1007/978-3-030-37545-4_3

Monwaza, M. (2020). Politics of Nationalism of Myanmar an Analysis from the Perspective of Rohingya Crisis. *Journal of International Affairs*, *22*(1 & 2), 1–24.

Moran, A., Gu, D., Zhao, D., Coxson, P., Wang, Y. C., Chen, C.-S., Liu, J., Cheng, J., Bibbins-Domingo, K., Shen, Y.-M., He, J., & Goldman, L. (2010). Future cardiovascular disease in china: Markov model and risk factor scenario projections from the coronary heart disease policy model-china. *Circulation: Cardiovascular Quality and Outcomes*, *3*(3), 243–252. https://doi.org/10.1161/CIRCOUTCOMES.109.910711

Mozos, O. M., Sandulescu, V., Andrews, S., Ellis, D., Bellotto, N., Dobrescu, R., & Ferrandez, J. M. (2017). Stress Detection Using Wearable Physiological and Sociometric Sensors. *International Journal of Neural Systems*, *27*(2), 1650041. https://doi.org/10.1142/S0129065716500416

MSF. (2015). *Greece: Testimonies from migrants and asylum seekers in Lesbos.* MSF. https://www.msf.org/greece-testimonies-migrants-and-asylum-seekers-lesbos

MSF. (2018). *Parched Pulka Searches for Water.* Medicine San Frontier (MSF). Retrieved November from https://www.msf.org/nigeria-parched-pulka-searches-water

MSF. (2021). *Pulka, Where Water is the Source of Life… and Disease.* https://www.msf.org/no-clean-water-pulka-nigeria-leads-disease

Murphy, T. (2015). *Swiss Ditch Ikea Refugee Shelters Over Fire Concerns.* https://www.humanosphere.org/basics/2015/12/ikea-shelters-fail/

MVA Systra Group. (2013). *Sustainable Urban Transport for Ho Chi Minh City MRT line 2 Project Integrated MRT2 Station Designs- Feasibility Study Report.* Academic Press.

Nabben, K., Gardner-Stephen, & Poblet, M. (2020). Identifying and Mitigating Humanitarian Challenges to COVID-19 Contact Tracing. In *2020 IEEE Global Humanitarian Technology Conference (GHTC)* (pp. 1-8), IEEE. 10.1109/GHTC46280.2020.9342852

Nare, H. (2021). Political Restructuring, Bureaucratization and Minority Groups in Myanmar: The Case of Rohingya from 2011 to 2018. *Asian Journal of Social Science*, *49*, 16–20.

Nechita, C., Alfayo, L., & Yakubu, D. (2018). *Trip Report- Field Visit to Pulka Community, Gwoza LGA*. https://www.humanitarianresponse.info/sites/www.humanitarianresponse.info/files/documents/files/acf_trip_report_-_pulka_12122018.pdf

Nehren, U., Ho, D. T. H., Trung, N. D., Raedig, C., & Alfonso, S. (2017). *Sand Dunes and Mangroves for Disaster Risk Reduction and Climate Change Adaptation in the Coastal Zone of Quang Nam Province*. Water Resources Development and Management. doi:10.1007/978-981-10-2624-9_13

Nettleton, S. (2020). *The sociology of health and illness* (4th ed.). Wiley.

Nguyen, H. (2019). *Vietnam's Tourism Sector: Opportunities for Investors in 2020*. Retrieved from: https://www.vietnam-briefing.com/

Nguyen, K., Liou, Y., & Terry, J. P. (2019). Science of the Total Environment Vulnerability of Vietnam to typhoons: A spatial assessment based on hazards, exposure and adaptive capacity. *The Science of the Total Environment*, *682*, 31–46. doi:10.1016/j.scitotenv.2019.04.069 PMID:31121354

Nolet, V. (2009). Preparing sustainability-literat teachers. *Teachers College Record*, *111*(2), 409–442. doi:10.1177/016146810911100207

Noll, B., Filatova, T., Need, A., & Taberna, A. (2021). *Contextualizing cross-national patterns in household climate change adaptation*. Nature Climate Change Articles. doi:10.103841558-021-01222-3

Nooke, G. (2019). *Migrant Cities instead of Refugee Camps*. SDZ Alliance. Retrieved from https://sdzalliance.org/migrant-cities-instead-of-refugee-camps

Noshi, Y., Uda, T., Kobayashi, A., Miyahara, S., & Serizawa, M. (2015). Beach changes observed in Phan Rang City in southeast Vietnam. *Procedia Engineering*, *116*(1), 163–170. doi:10.1016/j.proeng.2015.08.278

Noura, M., Atiquzzaman, M., & Gaedke, M. (2019). Interoperability in Internet of Things: Taxonomies and Open Challenges. *Mob. Networks Appl.* doi:10.1007/S11036-018-1089-9

Novotna, K., Cermakova, L., Pivokonska, L., Cajthaml, T., & Pivokonsky, M. (2019). Microplastics in drinking water treatment – Current knowledge and research needs. *The Science of the Total Environment*, *667*, 730–740. doi:10.1016/j.scitotenv.2019.02.431 PMID:30851606

NRC. (2021). *Humanitarian needs in Iran rise as 300,000 Afghans arrive since Taliban takeover*. https://www.nrc.no/news/2021/november/humanitarian-needs-in-iran-rise-as-300000-afghans-arrive-since-taliban-takeover/

Nutz, N. (2017). *A Guide to Market-Based Livelihood Interventions for Refugees*. Project documentation. UNHCR. Retrieved from https://www.unhcr.org/594b7d7f7.pdf

OCHA. (2020). *Fact Sheet: Pulka, Gwoza Local Government Area Borno State, North-east Nigeria*. N. UN Office for the Coordination of Humanitarian Affairs (OCHA). https://www.humanitarianresponse.info/sites/www.humanitarianresponse.info/files/documents/files/ocha_nga_pulka_factsheet_01052020.pdf

OCHA. (2021). *Periodic Monitoring Report- Nigeria-January-June 2021*. N. UN Office for the Coordination of Humanitarian Affairs (OCHA). https://reliefweb.int/sites/reliefweb.int/files/resources/20210906_nga_par_january_june.pdf_0.pdf

OECD. (2012). *Towards Better Humanitarian Donorship: 12 Lessons from DAC Peer Reviews*. Retrieved from https://www.oecd.org/dac/peer-reviews/12lessons.pdf

OECD. (2016). *OECD Resilience Systems Analysis: Turning Strategy into Action*. Retrieved from https://www.oecd.org/dac/conflict-fragility-resilience/publications/Resiliance%20brief.pdf

OECD. (2017a). *Assessing the Contribution of Refugees to the Development of Their Host Countries*. OECD Development Center. Retrieved from https://www.oecd.org/officialdocuments/publicdisplaydocumentpdf/?cote=DEV/DOC(2017)1&docLanguage=En

OECD. (2017b). *Addressing Forced Displacement through Development Planning and Cooperation Guidance for Donor Policy Makers and Practitioners*. OECD Publishing.

OECD. (2021). *What is ODA?* Retrieved from https://www.oecd.org/dac/financing-sustainable-development/development-finance-standards/What-is-ODA.pdf

Okanlawon, A. E. (2018). Reorienting chemistry teacher education towards integration of sustainability into classroom instructions. In G. Bello, M. T. Mustapha, & M. M. Osokoya (Eds.), *Contemporary isues in science,technology, engineering, arts and mathematics education in Nigeria* (pp. 226–236). Department of Science Education.

Oladosun, M., & Njobdi, S. (2018). *Report on End of Project Evaluation Project Title: SHO Four Famine Appeal 2017 - Nigeria*. https://www.researchgate.net/publication/337828350_End_of_Project_Evaluation_Project_Title_SHO_Four_Famine_Appeal_2017_-_Nigeria

Ortiz, D. (2020). *Geographic Information Systems (GIS) in Humanitarian Assistance: A Meta-Analysis*. https://repository.upenn.edu/cgi/viewcontent.cgi?article=1023&context=pathways_journal

Osa, Y., & Hanatani, A. (2018). Introduction: Addressing the Humanitarian-Development Nexus since the Cold War. In Crisis Management Beyond the Humanitarian-Development Nexus (pp. 3–14). Routledge.

Osiewicz, P. (2017). *Europe's Islamophobia and the Refugee Crisis*. https://www.mei.edu/publications/europes-islamophobia-and-refugee-crisis

Ostrom, E. (2009). A General Framework for Analyzing Sustainability of Social-Ecological Systems. *Science*, *35*(5939), 419–422. doi:10.1126cience.1172133 PMID:19628857

Palmiotto, M., Fattore, E., Paiano, V., Celeste, G., Colombo, A., & Davoli, E. (2014). Influence of a municipal solid waste landfill in the surrounding environment: Toxicological risk and odor nuisance effects. *Environment International*, *68*, 16–24. doi:10.1016/J.ENVINT.2014.03.004 PMID:24685488

Papantonis, A., & Becatoros, E. (2020). *Greek Minister Says 5,000 Move into New Refugee Camp on Lesbos After Fire*. https://www.stripes.com/news/europe/greek-minister-says-5-000-move-into-new-refugee-camp-on-lesbos-after-fire-1.645470

Pariatamby, A., & Tanaka, M. (2013). *Municipal Solid Waste Management in Asia and the Pacific Islands: Challenges and Strategic Solutions*. Springer Singapore. https://books.google.co.id/books?id=ZfsUswEACAAJ

Parmar, P. K., Leigh, J., Venters, H., & Nelson, T. (2019). Violence and Mortality in the Northern Rakhine State of Myanmar, 2017: Results of a Quantitative Survey of Surviving Community Leaders in Bangladesh. *The Lancet. Planetary Health*, *3*(3), e144–e153. doi:10.1016/S2542-5196(19)30037-3 PMID:30904113

Patinkin, J. (2018, May 4). Rohingya Refugees Face Malnutrition. *VOA*. Retrieved 20 August 2021 from https://www.voanews.com/a/rohingya-refugees-facing-malnutrition/4377429.html

Pearson, J., Bruce, T., Allsop, W., Kortenhaus, A., & van der Meer, J. (2004). Effectiveness of recurve walls in reducing wave overtopping on seawalls and breakwaters. *Coastal Engineering*, *148*, 4404–4416.

Peirce, R. S., Frone, M. R., Russell, M., Cooper, M. L., & Mudar, P. (2000). A longitudinal model of social contact, social support, depression, and alcohol use. *Health Psychology*, *19*(1), 28–38. doi:10.1037/0278-6133.19.1.28 PMID:10711585

Petridou, E. (1988). *Social factors in health and illness*. Notes of social medicine. Department of Health Visiting of Technological and Educational Institute of Athens.

Phillips, D. (2007). The good, the bad, and the ugly: The many faces of constructivism. *Educational Researcher*, *24*(7), 5–12. doi:10.3102/0013189X024007005

Pivokonsky, M., Cermakova, L., Novotna, K., Peer, P., Cajthaml, T., & Janda, V. (2018). Occurrence of microplastics in raw and treated drinking water. *The Science of the Total Environment*, *643*, 1644–1651. doi:10.1016/j.scitotenv.2018.08.102 PMID:30104017

Pivokonský, M., Pivokonská, L., Novotná, K., Čermáková, L., & Klimtová, M. (2020). Occurrence and fate of microplastics at two different drinking water treatment plants within a river catchment. *The Science of the Total Environment*, *741*, 140236. doi:10.1016/j.scitotenv.2020.140236 PMID:32603938

Plastics Europe. (2020). *Plastics – the Facts 2020: An analysis of European plastics production, demand and waste data*. https://plasticseurope.org/knowledge-hub/plastics-the-facts-2020/

Pontisidis, G., & Mpilali, Th. (2015). Social Support and its impact on facing chronic diseases. *Hellenic Journal of Nursing Science*, *8*(4).

Pooi, C. K., & Ng, H. Y. (2018). Review of low-cost point-of-use water treatment systems for developing communities. *Npj Clean Water*, *1*(1), 11. Advance online publication. doi:10.103841545-018-0011-0

Popoola, K., Talabi, A. O., Afolagboye, L. O., Oyedele, A. A., & Ojo, A. A. (2020). 2020/03/01). Groundwater potential evaluation in the basement complex terrain of Ekiti East Local Government Area, Southwestern Nigeria. *International Journal of Energy and Water Resources*, *4*(1), 81–90. doi:10.100742108-019-00053-5

Powell, N., Osbeck, M., Bach Tan, S., & Vu Canh, T. (2007). *Mangrove restoration and rehabilitation for climate change adaptation in Vietnam*. World Resources Report. Retrieved from http://www.worldresourcesreport.org

Price, A. (2004). Heart disease and work. *BMJ (Clinical Research Ed.)*, *2003*, 029298. doi:10.1136/hrt PMID:15310715

Prince, M. J., Wu, F., Guo, Y., Gutierrez Robledo, L. M., O'Donnell, M., Sullivan, R., & Yusuf, S. (2015). The burden of disease in older people and implications for health policy and practice. *Lancet*, *385*(9967), 549–562. https://doi.org/10.1016/S0140-6736(14)61347-7

Privacy International. (2018). *Investigating Apps interactions with Facebook on Android*. https://privacyinternational.org/taxonomy/term/552

Purwaningsih, A. (2015). Analysis of customer mindset change and accounting practice of garbage bank as medium of edupreneurship. Int. J. Monetary Economics and Finance, 8(4).

Raggi, A., Corso, B., Minicuci, N., Quintas, R., Sattin, D., De Torres, L., Chatterji, S., Frisoni, G. B., Haro, J. M., Koskinen, S., Martinuzzi, A., Miret, M., Tobiasz-Adamczyk, B., & Leonardi, M. (2016). Determinants of Quality of Life in Ageing Populations: Results from a Cross-Sectional Study in Finland, Poland and Spain. *PLoS One*, *11*(7), e0159293. https://doi.org/10.1371/journal.pone.0159293

Ragusa, A., Svelato, A., Santacroce, C., Catalano, P., Notarstefano, V., Carnevali, O., Papa, F., Rongioletti, M. C. A., Baiocco, F., Draghi, S., D'Amore, E., Rinaldo, D., Matta, M., & Giorgini, E. (2021). Plasticenta: First evidence of microplastics in human placenta. *Environment International*, *146*, 106274. doi:10.1016/j.envint.2020.106274 PMID:33395930

Rahman, K., Ensor, J., & Berger, R. (2009). River erosion and flooding in northern Bangladesh. *Understanding Climate Change Adaptation: Lessons from Community-Based Approaches*, 39-54.

Rahman, S., Islam, A. S., Tazkia, A. R., Shaha, P. K., Krien, Y., Durand, F., Testut, L., Islam, G. M. T., & Bala, S. K. (2019). Projected changes of inundation of cyclonic storms in the Ganges–Brahmaputra–Meghna delta of Bangladesh due to SLR by 2100. *Journal of Earth System Science*, *128*(6), 145. doi:10.100712040-019-1184-8

Rakhine State. (2021). Retrieved 12 July 2021 from https://en.wikipedia.org/wiki/Rakhine_State

Raphael, D. (2016). Social structure, living conditions, and health: Canadian. *Social Determinants of Health: Canadian Perspectives,* 31-56.

Rawlani, A. K., & Sovacool, B. K. (2011). Building responsiveness to climate change through community-based adaptation in Bangladesh. *Mitigation and Adaptation Strategies for Global Change*, *16*(8), 845–863. doi:10.100711027-011-9298-6

Reed, B., & Reed, B. (2013). *How Much Water is Needed in Emergencies* (Technical Notes on Drinking-Water, Sanitation and Hygiene in Emergencies). https://www.who.int/water_sanitation_health/publications/2011/WHO_TN_09_How_much_water_is_needed.pdf

Regencia, T. (2021). *Myanmar Coup Displaces Thousands as Global Refugee Numbers Rise.* https://www.aljazeera.com/news/2021/6/18/unhcr-urges-action-as-refugees-hit-record-high-of-82-4-million

Reichert, L., Brack, S., & Scheuermann, B. (2020). *A Survey of Automatic Contact Tracing Approaches Using Bluetooth Low Energy.* Cryptology ePrint Archive, Report 2020/672. https://eprint.iacr.org/2020/672

Resources for Thinking. (2021). *Sustainability Education for high schools*. Retrieved February 2, 2021, from https://resources 4 rethinking.ca/en/sustainable development

Ronteltap, A., & van Trijp, H. (2007). Consumer acceptance of personalised nutrition. *Genes & Nutrition*, *2*(1), 85–87. https://doi.org/10.1007/s12263-007-0003-z

Rosenzweig, C., Solecki, W. D., Blake, R., Bowman, M., Faris, C., Gornitz, V., Horton, R., Jacob, K., LeBlanc, A., Leichenko, R., Linkin, M., Major, D., O'Grady, M., Patrick, L., Sussman, E., Yohe, G., & Zimmerman, R. (2011). Developing Coastal Adaptation to Climate Change in the New York City Infrastructure-Shed: Process, Approach, Tools, and Strategies. *Climatic Change*, *106*(1), 93–127. doi:10.100710584-010-0002-8

Roumelioti, A. (2010). *Principles of Epidemiology in Public Health practice.* Educational notes, National School of Public Health (NSPH).

Roy, B., Khan, M., Mostafa, S., Saiful Islam, A. K. M., Khan, M., Uddin, J., & Mohammed, K. (2021). Integrated flood risk assessment of the Arial Khan River under changing climate using IPCC AR5 risk framework. *Journal of Water and Climate Change*, *12*(7), 3421–3447. doi:10.2166/wcc.2021.341

Rücker, J. (2018). *Sustainable Development Zones: Tools for Economic Development for Communities in Mass Displacement Scenarios*. Politas Consulting. Retrieved from https://refugeecities.files.wordpress. com/2018/07/sdzconcept-proposal-071118-final. pdf

S, O., & A, L. (2005). Evaluation of stakeholder influence in the implementation of construction projects. *International Journal of Project Management, 23*(4), 321–328.

Saengsupavanich, C. (2020). Deconstructing a jetty to rectify the downdrift erosion. *Journal of Sustainability Science and Management, 15*(2), 79–88.

Safdari, M., Mirzaei Alavijeh, M., Ehrampoush, M. H., Qhaneyan, M. T., & Morowatisharifabad, M. A. (2013). Knowledge, attitude and performance students of shahid sadoughi university of medical ciences-yazd about recycling solid material: A short report. *Majallah-i Ilmi-i Danishgah-i Ulum-i Pizishki-i Rafsanjan, 12*, 157–164.

Samuels-Kalow, M., Ciccolo, G., Lin, M., Schoenfeld, E., & Camargo, C. Jr. (2020). The terminology of social emergency medicine: Measuring social determinants of health, social risk, and social need. *Journal Of The American College Of Emergency Physicians Open, 1*(5), 852–856. doi:10.1002/emp2.12191 PMID:33145531

Sanò, M., Jiménez, J. A., Medina, R., Stanica, A., Sanchez-Arcilla, A., & Trumbic, I. (2011). The role of coastal setbacks in the context of coastal erosion and climate change. *Ocean and Coastal Management, 54*(12), 943–950. doi:10.1016/j.ocecoaman.2011.06.008

Sarris, M. (2001). *Sociology of health and quality of life*. Papazisis publications.

Schmitz, F. (2020). *Europe's Largest Refugee Camp Braces for COVID-19 Outbreak*. https://www.dw.com/en/europes-largest-refugee-camp-braces-for-covid-19-outbreak/a-54640747

Schoene, D., Heller, C., Aung, Y. N., Sieber, C. C., Kemmler, W., & Freiberger, E. (2019). <p>A systematic review on the influence of fear of falling on quality of life in older people: Is there a role for falls?</p>. *Clinical Interventions in Aging, 14*, 701–719. https://doi.org/10.2147/CIA.S197857

Schülke-Gill, B. (2020). *MSF: Moria Refugee Camp 'An ideal Breeding Ground for a Rapid Spread' of Corona Virus*. https://www.dw.com/en/msf-moria-refugee-camp-an-ideal-breeding-ground-for-a-rapid-spread-of-coronavirus/a-52824830

Schwarzer, R., & Luszczynska, A. (2008). How to overcome health-compromising behaviors: The health action process approach. *European Psychologist, 13*(2), 141–151. https://doi.org/10.1027/1016-9040.13.2.141

Schymanski, D., Goldbeck, C., Humpf, H. U., & Fürst, P. (2018). Analysis of microplastics in water by micro-Raman spectroscopy: Release of plastic particles from different packaging into mineral water. *Water Research, 129*, 154–162. doi:10.1016/j.watres.2017.11.011 PMID:29145085

Seddon, N., Smith, A., Smith, P., Key, I., Chausson, A., Girardin, C., & Turner, B. (2021). Getting the message right on nature-based solutions to climate change. *Global Change Biology, 27*(8), 1518–1546. doi:10.1111/gcb.15513 PMID:33522071

Sekito, T., Prayogo, T. B., Dote, Y., Yoshitake, T., & Bagus, I. (2013). Influence of a community-based waste management system on people's behavior and waste reduction. *Resources, Conservation and Recycling, 72*, 84–90. doi:10.1016/J.RESCONREC.2013.01.001

Selwyn, B. (2018). *The Struggle for Development*. Polity Press.

Shen, H.-W., & Yu, Y.-H. (1997). Social and Economic Factors in the Spread of the NIMBY Syndrome against Waste Disposal Sites in Taiwan. *Journal of Environmental Planning and Management, 40*(2), 273–282. doi:10.1080/09640569712227

Shih, Y. C. (2016). Coastal Management and Implementation in Taiwan. *Journal of Coastal Zone Management, 19*(4). Advance online publication. doi:10.4172/2473-3350.1000437

Shim, W. J., Hong, S. H., & Eo, S. E. (2017). Identification methods in microplastic analysis: A review. *Analytical Methods, 9*(9), 1384–1391. doi:10.1039/C6AY02558G

Shoeb, M. (2020). *Project Report on Resilience in Cox's Bazar and Environmental Impact of Rohingya Influx*. Ministry of Science and Technology Government of Bangladesh.

Shusterman, J. (2020). Gap or Prehistoric monster? A history of the Humanitarian–Development Nexus at UNICEF. *Disasters, 45*(2), 355–377. doi:10.1111/disa.12427 PMID:31799696

Silva, C., Coleta, I., Silva, A., Amaro, A., Alvarelhão, J., Queirós, A., & Rocha, N. (2013). Adaptation and validation of WHODAS 2.0 in patients with musculoskeletal pain. *Revista de Saude Publica, 47*(4), 1–6. PMID:24346666

Sims, T., Reed, A. E., & Carr, D. C. (2017). Information and Communication Technology Use Is Related to Higher Well-Being Among the Oldest-Old. *The Journals of Gerontology: Series B, 72*(5), 761–770. doi:10.1093/geronb/gbw130

Singh, Iyer, New, Few, Kuchimanchi, Segnon, & Morchain. (2021). Interrogating 'effectiveness' in climate change adaptation: 11 guiding principles for adaptation research and practice. *Climate and Development*. doi:10.1080/17565529.2021.1964937

Singhirunnusorn, W., Donlakorn, K., & Kaewhanin, W. (2017). Household Recycling Behaviours and Attitudes toward Waste Bank Project: Mahasarakham Municipality. *Journal of ASIAN Behavioural Studies, 2*(5), 17–26. doi:10.21834/jabs.v2i5.215

Slater, J. (2014). *Urban Systems of the Refugee Camp*. https://surface.syr.edu/cgi/viewcontent.cgi?article=1279&context=architecture_tpreps

SOMERSET Educational Products and Systems. (2017). RADMASTE Microchemistry experiments. University of Witwatersrand.

Sphere Association. (2018). *The Sphere Handbook: Humanitarian Charter and Minimum Standards in Humanitarian Response*. S. Association. www.spherestandards.org/handbook

Srivastava, V., Ismail, S. A., Singh, P., & Singh, R. P. (2015). Urban solid waste management in the developing world with emphasis on India: Challenges and opportunities. *Reviews in Environmental Science and Biotechnology, 14*(2), 317–337. doi:10.100711157-014-9352-4

Statista. (2021). *Android - Statistics & Facts*. https://www.statista.com/topics/876/android/

Statista. (2021). *Distribution of global plastic materials production in 2019, by region*. https://www.statista.com/statistics/281126/global-plastics-production-share-of-various-countries-and-regions/

Stirling, A. E., & Kerr, G. A. (2015). Creating meaningful co-curricular experiences in higher education. *Journal of Education & Social Policy, 2*(6), 1–7.

Strategic Executive Group. (2020). *Joint Response Plan Rohingya Humanitarian Crisis*. United Nations (UN). https://reporting.unhcr.org/sites/default/files/2021%20JRP.pdf

Strohfeldt, K., & Khutoryanskaya, O. (2015). Using problem-based learning in chemistry practical class for pharmacy students and engaging them with feedback. *American Journal of Pharmaceutical Education, 79*(9), 1–11. doi:10.5688/ajpe799141 PMID:26839430

Subbaraman, N. (2020). *'Distancing is Impossible': Refugee Camps Race to Avert Corona Virus Catastrophe.* https://www.nature.com/articles/d41586-020-01219-6

Suttibak, S., & Nitivattananon, V. (2008). Assessment of factors influencing the performance of solid waste recycling programs. *Resources, Conservation and Recycling*, *53*, 45–56. doi:10.1016/j.resconrec.2008.09.004

Szalacsi, A. (2015). Paradigm shift in 21st century: A call for action. In *Culture and Education, ISCHE 37* (pp. 1-2). Istanbul, Turkey: *Annual Conference of the Association of Caribbean Higher Education Administrators.*

Szitar, M.-A. (2013). Learning about sustainability community development. *Procedia-Social and Behavioural Sciences, 116*(2014), 3462-3466.

Takagi, H., Nguyen, D. T., Miquel, E., Tran, T. T., Hanne, L. K., Takahito, M., & Mitsuko, L. (2012). *Assessment of Disaster Vulnerability of Coastal Areas in Vietnam and Basic Study for Disaster Mitigation.* JST FY2011 SATREPS. doi:10.13140/RG.2.2.26388.30088

Talanquer, V. (2019). Some insights into assessing chemical systems thinking. *Journal of Chemical Education*, *96*(12), 2918–2925. doi:10.1021/acs.jchemed.9b00218

Talarska, D., Tobis, S., Kotkowiak, M., Strugała, M., Stanisławska, J., & Wieczorowska-Tobis, K. (2018). Determinants of Quality of Life and the Need for Support for the Elderly with Good Physical and Mental Functioning. *Medical Science Monitor*, *24*, 1604–1613. https://doi.org/10.12659/MSM.907032

Taoyuan Office of Coast Administration Construction. (2021). *Caota sand dunes geopark.* https://www.tyoca.gov.tw/wp-content/uploads/2021/02/%E8%8D%89%E6%BC%AF%E6%B2%99%E4%B8%98%E6%89%8B%E5%86%8A.pdf

Taylor, V. F., Beresford, A. R., & Martinovic, I. (2017). *Intra-Library Collusion: A Potential Privacy Nightmare on Smartphones.* https://arxiv.org/pdf/1708.03520.pdf

Tazzioli, M. (2020). *Confine to Protect: Greek Hotspots and the Hygienic-Sanitary Borders of Covid-19.* https://www.law.ox.ac.uk/research-subject-groups/centre-criminology/centreborder-criminologies/blog/2020/09/confine-protect

Thang, H. C., Misdorp, R., Laboyrie, H., Pos, H., van Zetten, R., & Huan, N. N. (2011). *Vietnam: a decade of Coastal Cooperation.* Climate of Coastal Cooperation.

Thapa, R. B., & Murayama, Y. (2008). Land evaluation for peri-urban agriculture using analytical hierarchical process and geographic information system techniques: A case study of Hanoi. *Land Use Policy*, *25*(2), 225–239.

The Agreement Establishing the European Union Regional Trust Fund in Response to the Syrian Crisis. (2016). http//ec.europa.eu/neighbourhood-enlargement/sites/near/files/neighbourhood/countries/syria/madad/20160526-revised-madad-fund-constitutive-agreement.pdf

Thomas Ng, S., Li, T. H., & Wong, J. M. W. (2012). Rethinking public participation in infrastructure projects. *Proceedings of the Institution of Civil Engineers: Municipal Engineer,* 101 - 113.

Timlett, R. E., & Williams, I. D. (2001). Public participation and recycling performance in England: A comparison of tools for behaviour change. *Resources, Conservation and Recycling*, *31*, 149–170.

Tolba, A., Al-Makhadmeh, Z., & Hussein, A. (2018). Montreal Imaging Stress Task Based Human Stress and Strain Analysis Using Image Processing Techniques. *Journal of Medical Imaging and Health Informatics*, *8*(4), 775–780. https://doi.org/10.1166/jmihi.2018.2349

Tountas, Y. (2000). Society and Health. Athens: Odisseas.

Tountas, Y. (2012). *Live without cigarettes.* Tipothito Publications.

Tran, K. C. (2006). Public Perception of Development Issues: Public Awareness can Contribute to Sustainable Development of a Small Island. *Ocean and Coastal Management*, *49*(3), 67–83. doi:10.1016/j.ocecoaman.2006.02.005

Truelove, S., Abrahim, O., Altare, C., Lauer, S. A., Grantz, K. H., Azman, A. S., & Spiegel, P. (2020). The potential impact of COVID-19 in refugee camps in Bangladesh and beyond: A modeling study. *PLoS Medicine*, *17*(6), e1003144–e1003144. doi:10.1371/journal.pmed.1003144 PMID:32544156

Tsaparlis, G., & Gorezi, M. (2007). Addition of a project-based component to a conventional expository physical chemistry laboratory. *Journal of Chemical Education*, *84*(4), 668–670. doi:10.1021/ed084p668

Tselepi, C. (2000). *Social and cultural factors in the engagement and in the experience of illness. Sociological and psychological approach of hospitals services of health. Social cultural aspects of health and illness.* Hellenic Open University Publications, Patra.

Tzoufras, I., & Piperoglou, A. (2009). Introduction to Biostatistics and Epidemiology. Academic Press.

Uckino, B. N. (2006). Social Support and health. A review of physiological process potentially underlying links to disease outcomes. *Journal of Behavioral Medicine*, *29*(4), 377–387. doi:10.100710865-006-9056-5 PMID:16758315

Uda, T., & Yamamoto, K. (1994). Beach erosion around a sand spit -An Example of Mihono-Matsubara Sand Spit-. *Coastal Engineering*.

UN High Commissioner for Refugees (UNHCR). (2018). *Refugee Livelihoods and Economic Inclusion - 2019-2023 Global Strategy Concept Note.* Retrieved from https://www.unhcr.org/publications/operations/5bc07ca94/refugee-livelihoods-economic-inclusion-2019-2023-global-strategy-concept.html

UN High Commissioner for Refugees (UNHCR). (2021). *Global Trends: Forced Displacement in 2020.* Retrieved from https://www.unhcr.org/statistics/unhcrstats/60b638e37/global-trends-forced-displacement-2020.html?query=global%20 trends

UN High Commissioner for Refugees. (2016). *Better Protecting Refugees in the EU and Globally: UNHCR's Proposals to Rebuild Trust through Better Management, Partnership, and Solidarity.* Retrieved from https://www.refworld.org/docid/58385d4e4.html

UNDP. (2017). *Social Impact Assessment of the Rohingya Refugee Crisis into Bangladesh Key Findings and Recommendations.* Retrieved 12 July 2021 from https://www.humanitarianresponse.info/sites/www.humanitarianresponse.info/files/assessments/171207_social_impact_assessment_and_rapid_host_community_impact_assessment_summary.pdf

UNDP. (2018). *Rehabilitation of Pulka dam, Gwoza LGA, Borno State.* United Nations Development Programme Office. Retrieved December from https://procurement-notices.undp.org/view_notice.cfm?notice_id=50804

UNDP. (2020). *Sustainable development goals.* Retrieved May 14, 2020, from https://www.undp.org/content/en/home/sustainable-development-goals

UNESCO. (2012). *Education for sustainable development in action. Learning and training.* UNESCO Education Sector.

UNHCR. (2016). *Lesvos Moria Campsite.* United Nations High Commissioner for Refugees. https://data2.unhcr.org/en/documents/details/46995

UNHCR. (2020). *Corona Virus Outbreak.* United Nations High Commissioner for Refugees. https://www.unhcr.org/uk/coronavirus-covid-19.html

UNHCR. (2020). *Emergency Handbook.* United Nations High Commissioner for Refugees. https://emergency.unhcr.org/entry/45581/camp-planning-standards-planned-settlements

UNHCR. (2020). *Figures at a Glance*. United Nations High Commissioner for Refugees. https://www.unhcr.org/uk/figures-at-a-glance.html

UNHCR. (2020). *Water, Sanitation, and Hygiene (WASH) Monitoring System Briefing Note 2*. https://wash.unhcr.org/download/wash-monitoring-system-briefing-note-v2/

UNHCR. (2021, July 31). *Rohingya Refugee response/Bangladesh*. Joint Government of Bangladesh. Retrieved 23 September 2021 from https://reliefweb.int/report/bangladesh/rohingya-refugee-responsebangladesh-joint-government-bangladesh-unhcr-15

UNICEF. (2017a, September 29). *Bangladesh Humanitarian Situation Report (Rohingya influx)*. Retrieved 23 September 2021 from https://reliefweb.int/report/bangladesh/bangladesh-humanitarian-situation-report-no-4-rohingya-influx-29-september-2017

UNICEF. (2017b, September 25). *Bangladesh: Humanitarian Situation Report No. 3 (Rohingya Influx)*. Retrieved 20 August 2021 from https://reliefweb.int/report/bangladesh/bangladesh-humanitarian-situation-report-no-3-rohingya-influx-25-september-2017

UNICEF. (2017c, October 15). *Bangladesh: Humanitarian Situation Report No. 6 (Rohingya Influx), 6 – 12 October 2017*. Retrieved 20 August 2021 from https://reliefweb.int/report/bangladesh/bangladesh-humanitarian-situation-report-no-6-rohingya-influx-6-12-october-2017

UNICEF. (2017d, September 12). *Bangladesh Humanitarian Situation report (Rohingya influx)*. Retrieved 12 August 2021 from https://reliefweb.int/report/bangladesh/bangladesh-humanitarian-situation-report-no-2-rohingya-influx-12-september-2017

UNICEF. (2019). *Water Under Fire: For Every Child, Water and Sanitation In Complex Emergencies*. https://www.unicef.org/reports/water-under-fire-2019

UNICEF. (2020, February 25). *UNICEF, Bangladesh Humanitarian Situation Report No. 53*. Retrieved 20 August 2021 from https://reliefweb.int/report/bangladesh/bangladesh-humanitarian-situation-report-no-53-1-january-31-december-2019

UNICEF. (2020b). *Accountability to Affected Populations: A Handbook for UNICEF and Partners*. G. United Nations Children Emergency Fund (UNICEF). https://aa9276f9-f487-45a2-a3e7-8f4a61a0745d.usrfiles.com/ugd/aa9276_0fa31f247b194585b15e98562a269e0d.pdf

UNICEF. (2021). *Water Under Fire Volume 3: Attacks on water and sanitation services in armed conflict and the impacts on children*. https://www.unicef.org/reports/water-under-fire-volume-3

UNICEF. (2021, March). *Bangladesh: Assessment of the Education Sector Response to the Rohingya Crisis – Rohingya Refugee Response*. Retrieved 20 August 2021 from https://reliefweb.int/report/bangladesh/bangladesh-assessment-education-sector-response-rohingya-crisis-rohingya-refugee

United Nations General Assembly. (2016a). *One Humanity, Shared Responsibility: Report of the Secretary-General for the World Humanitarian Summit*. Retrieved from http://sgreport.worldhumanitariansummit.org/

United Nations General Assembly. (2016b). *The Outcome of the World Humanitarian Summit: Report of the Secretary-General*. Retrieved from https://digitallibrary.un.org/record/842411#record-files-collapse-header

United Nations. (2015). *Sustainable development goals*. Retrieved from Take action for the sustainable development goals: https://www.un.org/sustainabledevelopment/sustainable-development-goals/

United Nations. (2015). *Sustainable Development Goals*. United Nations. Retrieved December from https://www.un.org/sustainabledevelopment/water-and-sanitation/

United Nations. (2021a). *The Sustainable Development Goals Report.* https://unstats.un.org/sdgs/report/2021/The-Sustainable-Development-Goals-Report-2021.pdf

United Nations. (2021b). *Neglected: Environmental Justice Impacts of Marine Litter and Plastic Pollution.* https://wedocs.unep.org/bitstream/handle/20.500.11822/35417/EJIPP.pdf

United Natons Statistics Division. (2005). *Living beyond our means: Natural assets and human well-being*. UNESCO.

Urtamo, A., Huohvanainen, E., Pitkälä, K. H., & Strandberg, T. E. (2019). Midlife predictors of active and healthy aging (AHA) among older businessmen. *Aging Clinical and Experimental Research, 31*(2), 225–231. https://doi.org/10.1007/s40520-018-1100-0

van Bergen, S. (2020). *Green chemistry in higher education.* Retrieved February 02, 2021, from https://ecology.wa.gov/waste-toxics/reducing-toxic-chemicals/greenchemistry

Van Buuren, J., & Hendriksen, A. (2010). A learning and decision methodology for drainage and sanitation improvement in developing cities. Springer.

Van Cauwenberghe, L., & Janssen, C. R. (2014). Microplastics in bivalves cultured for human consumption. *Environmental Pollution, 193*, 65–70. doi:10.1016/j.envpol.2014.06.010 PMID:25005888

van Slobbe, E., de Vriend, H. J., Aarninkhof, S., Lulofs, K., de Vries, M., & Dircke, P. (2013). Building with Nature: In search of resilient storm surge protection strategies. *Natural Hazards, 66*(3), 1461–1480. doi:10.100711069-013-0612-3

Vaudenay, S. (2020). *Centralized or Decentralized? The Contact Tracing Dilemma.* Cryptology ePrint Archive, Report 2020/531. https://eprint.iacr.org/2020/531

Visvanathan, C., & Glawe, U. (2006). *Domestic Solid Waste Management in South Asian Countries-A Comparative Analysis*. Academic Press.

Von Borgstede, C., Andersson, M., & Johnsson, F. (2013). Public attitudes to climate change and carbon mitigation—Implications for energy-associated behaviours. *Energy Policy, 57*, 182–193. doi:10.1016/j.enpol.2013.01.051

Vo, N. X., Vo, T. Q., Watanapongvanich, S., & Witvorapong, N. (2019). Measurement and Determinants of Quality of Life of Older Adults in Ho Chi Minh City, Vietnam. *Social Indicators Research, 142*(3), 1285–1303. https://doi.org/10.1007/s11205-018-1955-7

Vo, P. L., Lin, T.-Y., & Nguyen, B. G. (2018). Threat of coastal hazards to socio-economic development in the central coasts of Vietnam. *International Symposium on Lowland Technology*.

Vrdoljak, D., Marković, B., Puljak, L., Lalic, D., Kranjcevic, K., & Vucak, J. (2013). Lifestyle intervention in general practice for physical activity, smoking, alcohol consumption and diet in elderly: A randomized controlled trial. *Archives of Gerontology and Geriatrics, 58*. doi:10.1016/j.archger.2013.08.007

Wagner, M., & Lambert, S. (2018). *Freshwater Microplastics:Emerging environmental contaminants*. Springer. doi:10.1007/978-3-319-61615-5

Walker, A. (2012). *What is Boko Haram?* https://www.usip.org/publications/2012/05/what-boko-haram

Wallace, E., Salisbury, C., Guthrie, B., Lewis, C., Fahey, T., & Smith, S. M. (2015). Managing patients with multimorbidity in primary care. *BMJ (Clinical Research Ed.), 350*, h176. https://doi.org/10.1136/bmj.h176

Walters, G., Janzen, C., & Maginnis, S. (2016). *Nature-based solutions to address global societal challenges.* IUCN Report. doi:10.2305/IUCN.CH.2016.13.en

Walther, D., Curjuric, I., Dratva, J., Schaffner, E., Quinto, C., Schmidt-Trucksäss, A., Eze, I. C., Burdet, L., Pons, M., Gerbase, M. W., Imboden, M., Schindler, C., & Probst-Hensch, N. (2017). Hypertension, diabetes and lifestyle in the long-term—Results from a Swiss population-based cohort. *Preventive Medicine*, *97*, 56–61. https://doi.org/10.1016/j.ypmed.2016.12.016

Wang, C., Lu, W., Narayanan, M., Chang, D., Lord, S., & Lovell, N. (2016). Low-Power Fall Detector Using Triaxial Accelerometry and Barometric Pressure Sensing. *IEEE Transactions on Industrial Informatics*, *12*, 1–1. https://doi.org/10.1109/TII.2016.2587761

Wang, J., Lesage, A., Schmitz, N., & Drapeau, A. (2008). The relationship between work stress and mental disorders in men and women: Findings from a population-based study. *Journal of Epidemiology and Community Health*, *62*(1), 42–47. doi:10.1136/jech.2006.050591 PMID:18079332

Wang, W., & Wang, J. (2018). Investigation of microplastics in aquatic environments: An overview of the methods used, from field sampling to laboratory analysis. *Trends in Analytical Chemistry*, *108*, 195–202. doi:10.1016/j.trac.2018.08.026

Wang, Z. M., Wagner, J., Ghosal, S., Bedi, G., & Wall, S. (2017). SEM/EDS and optical microscopy analyses of microplastics in ocean trawl and fish guts. *The Science of the Total Environment*, *603–604*, 616–626. doi:10.1016/j.scitotenv.2017.06.047 PMID:28646780

Wang, Z., Lin, T., & Chen, W. (2020). Occurrence and removal of microplastics in an advanced drinking water treatment plant (ADWTP). *The Science of the Total Environment*, *700*, 134520. Advance online publication. doi:10.1016/j.scitotenv.2019.134520 PMID:31669914

Waryszak, P., Gavoille, A., Whitt, A. A., Kelvin, J., & Macreadie, P. I. (2021). Combining gray and green infrastructure to improve coastal resilience: Lessons learnt from hybrid flood defenses. *Coastal Engineering Journal*, 1–16. doi:10.1080/21664250.2021.1920278

WASH Sector Nigeria. (2021a). *WASH Sector North-East Nigeria Response Monitoring Dashboard 2021*. Retrieved October 31 from https://app.powerbi.com/view?r=eyJrIjoiNTlmZGVhZGUtYTc1ZC00OGY5LWI3ZDAtODc4N2FhYThiMjdmIiwidCI6ImY2ZjcwZjFiLTJhMmQtNGYzMC04NTJhLTY0YjhjZTBjMTlkNyIsImMiOjF9&pageName=ReportSection

WASH Sector Nigeria. (2021b). *Nigeria WASH Response Newsletter: The Time is Now*. https://drive.google.com/file/d/1W5O8vpLzP-3OVlmKmNmjDU6enM0oSzMg/view

Wassmann, R., Jagadish, S., Heuer, S., Ismail, A., Redona, E., Serraj, R., Singh, R. K., Howell, G., Pathak, H., & Sumfleth, K. (2009a). Climate change affecting rice production: The physiological and agronomic basis for possible adaptation strategies. *Advances in Agronomy*, *101*, 59–122. doi:10.1016/S0065-2113(08)00802-X

Wassmann, R., Jagadish, S., Sumfleth, K., Pathak, H., Howell, G., Ismail, A., Serraj, R., Redona, E., Singh, R. K., & Heuer, S. (2009b). Regional vulnerability of climate change impacts on Asian rice production and scope for adaptation. *Advances in Agronomy*, *102*, 91–133. doi:10.1016/S0065-2113(09)01003-7

WFP. (2017, December). *Refuge influx Emergency Vulnerability Assessment Cox's Bazar, Bangladesh*. Retrieved 20 August 2021 from https://www.wfp.org/news/ news-release/new-rohingya-arrivals-bangladesh-risk-poor-diets

Wheeler, D. (2011). *Quantifying Vulnerability to Climate Change: Implications for Adaptation Assistance*. Academic Press.

WHO. (2011). *Physical Status: The Use and Interpretation of Anthropometry Report of a WHO Expert Committee*. Technical Report Series No. 854. Retrieved 20 August 2021 from http://www.who.int/childgrowth/publications/physical_ status/en/

WHO. (2017). *World Health Organization, Bangladesh Steps Up Vaccination for New Rohingya Arrivals as Measles Cases Rise*. Retrieved 21 September 2021 from http://www.searo.who.int/mediacentre/releases/2017/1671/en/

WHO. (2021a). *International Classification of Functioning, Disability and Health (ICF)*. https://www.who.int/standards/classifications/international-classification-of-functioning-disability-and-health

WHO. (2021a, July 22). *Rohingya Crisis Situation Report #14*. Retrieved 23 September 2021 from https://reliefweb.int/report/bangladesh/rohingya-crisis-who-bangladesh-situation-report-14-22-july-2021-period-covered

WHO. (2021b). *WHO Disability Assessment Schedule (WHODAS 2.0)*. https://www.who.int/standards/classifications/international-classification-of-functioning-disability-and-health/who-disability-assessment-schedule

WHO. (2021b, August 5). *Rohingya Crisis Situation Report #15*. Retrieved 23 September 2021 from https://reliefweb.int/report/bangladesh/rohingya-crisis-who-bangladesh-situation-report-15-5-august-2021-period-covered

WHO. (2021c, August 19). *Rohingya Crisis Situation Report #16*. Retrieved 23 September 2021 from https://reliefweb.int/report/bangladesh/rohingya-influx-overview-key-changes-during-2018-monsoon-season

WHO. (2021d). *Cox's Bazar Rohinyga Camps Epidemiological Summary*. Retrieved 12 July 2021 from https://cxb-epi.netlify.app/?fbclid=IwAR3o9bRz7871J8A95a0ZgzS5rEBSNrC2FFQqAxRh3kp-aLUYmfgY9Fv9t4o

WHO/UNICEF. (2021). *Progress on Household Drinking Water, Sanitation and Hygiene 2000-2020: Five years into the SDGs*. https://www.who.int/publications/i/item/9789240030848

Wiesner, G., & Bittner, E. K. (2004). Life expectancy, potential years of life lost (PYLL), and avoidable mortality in an East/West comparison. *Bundesgesundheitsblatt, Gesundheitsforschung, Gesundheitsschutz*, *47*(3), 266–278. https://doi.org/10.1007/s00103-003-0793-0

Wilson, D. C., Araba, A. O., Chinwah, K., & Cheeseman, C. R. (2009). Building recycling rates through the informal sector. *Waste Management (New York, N.Y.)*, *29*, 629–635. PMID:18701272

Wise, R. M., Fazey, I., Smith, M. S., Park, S. E., Eakin, H. C., Van Garderen, E. A., & Campbell, B. (2013). Reconceptualising adaptation to climate change as part of pathways of change and response. *Global Environmental Change*, 10.

World Bank. (2014). *Global Value Chains*. Issue Brief. Retrieved from https://www.worldbank.org/en/topic/global-value-chains

World Bank. (2017). *The Humanitarian-Development-Peace Initiative*. World Bank Brief. Retrieved from https://www.worldbank.org/en/topic/fragilityconflictviolence/brief/the-humanitarian-development-peace-initiative

World Bank. (2017, March 27). *World Bank Approves US$100 Million in Financing to Create 100,000 Jobs for Jordanians, Syrian Refugees* [Press release]. Retrieved from https://www.worldbank.org/en/news/press-release/2016/03/27/exceptional-financing-jordan-jobs-syrian-refugees

World Habitat Org. (2018). *Plastic Bottle Houses for Sahrawi Refugees*. https://world-habitat.org/world-habitat-awards/winners-and-finalists/plastic-bottle-houses-sahrawi-refugees/#award-content

World Population Review. (2021). *Most Developed Countries in Africa 2021*. https://worldpopulationreview.com/country-rankings/most-developed-countries-in-africa

Yu, A. T. W., Wong, I., Wu, Z., & Poon, C. S. (2021). Strategies for effective waste reduction and management of building construction projects in highly urbanized cities— A case study of hong kong. *Buildings*, *11*(5). Advance online publication. doi:10.3390/buildings11050214

Zander, M., & Essife, A. (2019). *Refugees Endure Living Hell of Greece's Moria Camp on Lesbos.* https://www.dw.com/en/refugees-endure-living-hell-of-greeces-moria-camp-on-lesbos/a-51305549

Zangróniz, R., Martínez-Rodrigo, A., Pastor, J. M., López, M. T., & Fernández-Caballero, A. (2017). Electrodermal Activity Sensor for Classification of Calm/Distress Condition. *Sensors (Basel)*, *17*(10), 2324. https://doi.org/10.3390/s17102324

Zetter, R. (2019). Theorizing the Refugee Humanitarian-Development Nexus: A Political-Economy Analysis. *Journal of Refugee Studies*. Advance online publication. doi:10.1093/jrs/fez070

Zhang, Y., Diehl, A., Lewandowski, A., Gopalakrishnan, K., & Baker, T. (2020). Removal efficiency of micro- and nanoplastics (180 nm–125 µm) during drinking water treatment. *The Science of the Total Environment*, *720*, 137383. Advance online publication. doi:10.1016/j.scitotenv.2020.137383 PMID:32325555

Zheng, T., Yuan, D., & Liu, C. (2019). Molecular toxicity of nanoplastics involving in oxidative stress and desoxyribonucleic acid damage. *Journal of Molecular Recognition*, *32*(11), 3–9. doi:10.1002/jmr.2804 PMID:31373076

Zhong, L. J., & Mol, A. P. J. (2008). Participatory environmental governance in China: Public hearings on urban water tariff setting. *Journal of Environmental Management*, *88*(4), 899–913. doi:10.1016/j.jenvman.2007.04.018 PMID:17573184

Zhong, L., Mol, A. P. J., & Fu, T. (2008). Public-private partnerships in China's urban water sector. *Environmental Management*, *41*(6), 863–877. doi:10.100700267-008-9070-1 PMID:18256780

About the Contributors

Yiannis Koumpouros (BSc, MSc, MBA, PhD) is Ass. Professor in the University of West Attica, Department of Public and Community Health. Specialized in health informatics, e-health, m-health, telemedicine, assistive technologies, rehabilitation, users satisfaction assessment, and strategic management. Expert for the European Commission in the e-health and m-health sectors, and evaluator of numerous R&D projects. Responsible for the WHODAS 2.0 scale cultural adaptation for the World Health Organization in Greece. Top-level managerial experience in the private and public healthcare sector (chairman of hospitals, member of the board of directors, business development and strategic management director, etc.). Chairman of several committees in the Ministry of Development, regular member of the Consultative Technical Council (Ministry of Internal Public Administration and Decentralisation), etc. He has published more than 70 papers in reputed journals and conferences and has been serving as an editorial board member of repute. He has participated in more than 50 R&D funded projects in the field of healthcare and ICTs.

Angelos Georgoulas (MEng, PhD) is Adjunct Assistant Professor and Research Associate at the University of West Attica, teaching a number of undergraduate courses both in the Faculty of Engineering and the Faculty of Health and Caring Professions. He has completed his undergraduate, postgraduate taught and doctoral studies at the School of Electrical and Computer Engineering of the National Technical University of Athens, Greece. His related areas of expertise include mobile applications development, digital security, medical informatics, e-health and m-health Applications. Dr. Georgoulas has published several scientific papers in international journals and conferences. He is also registered expert for the European Commission, participating in several peer review panels for various programmes, proposals and projects in the field of healthcare and ICTs (e.g. FP7, HORIZON 2020, etc). He is a Chartered Engineer by the Technical Chamber of Greece (1999) and member of the IEEE, the Greek Society of Biomedical Technology, the Hellenic Association of Computer & Communication Engineers and the Hellenic Association of Electrical & Mechanical Engineers.

Georgia Kremmyda (MEng, MSc, PhD, SFHEA) is Professor and Head of Teaching at the School of Engineering, University of Warwick. Georgia completed her undergraduate, postgraduate taught and doctoral studies at the School of Civil Engineering of the National Technical University of Athens, Greece. Before joining Warwick in 2015 and for a period of 15 years, Georgia shared her time between carrying leading engineering roles in industry and delivering teaching and research in academia. Her teaching and research interests are oriented, but not limited, to the design of earthquake resilient structures, sustainable cities, infrastructure for emergencies and the delivery of interdisciplinary, problem-based education, especially in resource-constrained environments. Within her academic role, Georgia is leading initiatives related to the development of innovative programmes which are promoting equitable access to education, in Science and Engineering. Georgia leads the Warwick Humanitarian Engineering programme and Co-leads the Warwick Global Research Priority in Sustainable Cities, both aligned to the UN Sustainable Development Agenda 2030. On international level, Georgia is elected Vice President (Conferences) of the International Network of Women Engineers and Scientists for 2021-2023. Within Warwick, she is Deputy Chair of the Institutional Athena Swan Self-Assessment Team, leading the subgroup working on "Organisation and culture for inclusivity, diversity and equality." She is also Lead of the STEM Education and Student Experience workstream within the framework of the Warwick STEM Grand Challenge.

* * *

Athanasios Anastasiou received his Bachelor degree in Electrical & Electronic Engineering from the University of Newcastle Upon Tyne, in 2005. In 2006, he received his M.Sc. degree in Digital Signal & Image Processing from the University of Central Lancashire. Likewise, in 2009 he obtained his second M.Sc. degree in Biomedical Engineering held by the University of Patras and the National Technical University of Athens (N.T.U.A.). In 2015, received his Phd degree in Biomedical engineering in the field of mobile health.Since 2011, he is a member of the Biomedical Engineering Laboratory of N.T.U.A.. His current research interests comprise Bioinformatics, Biosignal Processing, Digital Image Processing and technologies for assisted living. He is a Member of the Institute of Electrical and Electronics Engineers (IEEE) - EMBS Society and Computer Society, Member of the Institution of Engineering and Technology (IET) and a member of the Technical Chamber of Greece (TEE).

Sujit Kumar Bala has been working as Professor under the Institute of Water and Flood Management (IWFM) since 2011, (BUET). He has many national, international publications to his credit. He is a member of the Institution of Engineers, Bangladesh. He has joined the IWFM, BUET as lecturer in 1997. Before joining the BUET, he worked as consultant in many national consultancy projects. He served in Surface Water Modelling Centre now Institute of Water Modelling. He was involved in the construction of the Jamuna Multipurpose Bridge. Dr. Bala had his M.Sc. in Irrigation Engineering from Tashkent Institute of Irrigation and Land Reclamation, Uzbekistan (ex-USSR) in 1982. He received Ph.D. from the same institute in 1986. His field of interests include climate change adaptation; water, gender and society; bridge hydraulics; urban roof top agriculture; Irrigation; water control structures, etc. He was involved in international and national research projects funded by EU, JICA, DFID, IFAD, etc. He participated in many national and international conferences. He served as the Director, IWFM from 2017 to 2019.

Eoin Cunningham joined the School of Mechanical and Aerospace Engineering (SMAE) in Queen's University Belfast as a lecturer in July 2013, becoming a senior lecturer in August 2020. His research focuses on the valorisation of agri-food organic waste streams via their incorporation into polymer production cycles.

Dong Doan Van, PhD (M'87), received BSc degree in electronic and telecommunication engineering from Ho Chi Minh City University of Transport, Vietnam, in 2010 and the M.S. degree in communication and information systems from Huazhong University of Science and Technology, China, in 2014, and Ph.D. degree in information communication engineering from Wuhan University of Technology, China, in 2018. He is currently Dean of the Faculty of Electrical and Electronics Engineering, Ho Chi Minh City University of Transport, Vietnam. His research interest includes Content-Centric Networking, Internet of Things, Ad-hoc Networks, Image Processing, Machine Learning, Programing, Humanitarian engineering.

Reyes Garcia is an Assistant Professor in Structural Engineering at the University of Warwick (UK). He obtained his 5-year Degree in Civil Engineering (Graduated with Honours) from the University of Michoacan (Mexico), his MSc in Earthquake Engineering and Engineering Seismology from the ROSE School (Italy) and University Joseph Fourier of Grenoble (France), and his PhD in Structural Engineering from The University of Sheffield (UK). Reyes has +20 years of research/consultancy experience in the fields of earthquake engineering, structural concrete and FRP composites for construction. In the last years, he has been also involved in multi-partner projects aimed at creating more sustainable construction in Turkey and Nepal, where he witnessed the social and shelter needs of refugees created by civil wars and earthquakes. He has published more than 60 articles, papers and book chapters. He has contributed actively to the UK TG tasks for Eurocode 8-3, and he is a member of the fib TG 5.1 "FRP Reinforcement for Concrete Structures", and of COST Action TU1207 "Next Generation Guidelines for Composite Reinforcements". He regularly reviews research grant proposals for national funding bodies in Latin America and Europe, as well as manuscripts for more than 20 international journals. He is Associate Editor of the ICE's journal Structures and Buildings. He has been member of committees and invited speaker or keynote speaker in various international conferences in Latin America and Asia. He is also Visiting Professor at Walailak University, Thailand.

Marco Geron is a Lecturer in the School of Mechanical and Aerospace Engineering at Queen's University Belfast. His research activity and interest focuses on the development and application of numerical thermos-fluid models to different disciplines. The different models utilised encompass both high fidelity methods (CFD) and low fidelity methods (state space models and POD).

Yahya Gülseven works as a senior expert for the Turkish International Cooperation and Coordination Agency. He has 16 years of experience in design and implementation of development cooperation projects in various countries. He also works as a part-time instructor in the Department of Political Science and International Relations, Başkent University, Ankara. His research interests are international aid architecture, emerging donors and the South-South Cooperation.

Khai Ha is currently a Lecturer at Faculty of Environment & Natural Resources (FENR) – Ho Chi Minh City University of Technology (HCMUT). He obtained his Bachelor degree in Hydrogeology and Geological engineering, Ha Noi University of Mining and Geology in 2007, and received Master in Water Resources Engineering, Chulalongkorn University in 2015, and PhD degrees in Environmental Engineering at Kunsan National University, South Korea in 2019. He started his academic career at HCMUT since 2021. Dr. Ha's professional experience in teaching, research and consultancy spans over 14 years in Vietnam. He teaches several undergraduate and graduate subjects, including Water quality management, Environmental Monitoring. Dr. Ha have worked as groundwater planner and researchers in Division for Water Resources Planning and Investigation for the south of Vietnam and Groundwater and soil Pollution laboratory of Kunsan National University. In these institutes, he implemented several project on groundwater quantity and quality management, modelling, and investigation.

Charles Hanson is a graduate from the Kwame Nkrumah University of Science and Technology in Kumasi. He pursued a Bachelor degree in civil engineering with a specialty in soil studies and went on to read a Master programme in Geotechnical engineering in the same institution. He has since worked with and for reputable mining and water companies across the breath and length of Ghana.

Ruby Hanson is the Dean of the Faculty of Science Education, in the University of Education, Winneba in Ghana and a full professor in Chemistry Education.

Motahar Hosen has worked as a Research Assistant at the Institute of Water and Flood Management (IWFM).

A. K. M. Saiful Islam is a Professor at the Institute of Water and Flood management of Bangladesh University of Engineering and Technology (BUET). He has received a Bachelor of Science in Civil Engineering (1989) from BUET, a Master of Science in Water Resources Engineering (1999) from BUET, and a Ph.D. in Civil Engineering (2004) from Drexel University, USA. His research interests aim at climate change impact on hydrology, water resources management, urban and coastal flood management, remote sensing for disaster risks reduction. His research adopted regional climate modeling, basin-scale hydrological modeling, river hydrodynamic modeling, urban flood modeling, coastal and storm surge modeling, weather forecast modeling, satellite remote sensing for environmental monitoring, hydro informatics for disaster risks reduction, and climate change adaption. He has published over 55 peer-reviewed journal articles, more than 100 conference papers, and contributed to 6 book chapters. Over the past decade, Dr. Islam has led several major international scientific climate change assessments and reports by the Intergovernmental Panel on Climate Change (IPCC). He contributed to the IPCC's Fifth Assessment Report and IPCC Special Report on Global Warming of 1.5°C as an Expert Reviewer. He is currently a Lead Author on the IPCC's Sixth Assessment Report.

G. M. Tarekul Islam is a Professor at the Institute of Water and Flood Management (IWFM) of Bangladesh University of Engineering and Technology (BUET). He graduated in Civil Engineering in 1994 and did his Masters in Water Resources Engineering in 1996 from BUET. He was a Lamminga Foundation Scholar before joining IWFM as a Lecturer in 1996. He received Monbusho Scholarship from the Government of Japan to pursue PhD at the Department of Civil Engineering of the University of Tokyo, Japan during 1997-2000. He was also a recipient of the prestigious JSPS fellowship to pursue post-doctoral research at the Department of Civil and Environmental Engineering of Hiroshima University, Japan during 2006-2008. He was awarded by the University Grants Commission (UGC) of Bangladesh as the best researcher in Engineering and Technology category in 2003. He has rendered many consultancies and advisory services to nationally important projects of different government and non-government organizations. He has imparted numerous talks on various issues of water and flood management as invited speaker. He has published over 130 papers in international journals and conference proceedings.

Rafiza Islam obtained her BSc and MSc degree from Department of Chemistry, National University, Bangladesh. Now she is continuing her PhD on "Investigation of antibiotic residues in meat, milk and eggs samples" at the Department of Chemistry, University of Dhaka from 2017. She got a fellowship from Bangabandhu Science and Technology Fellowship Trust, Ministry of Science and Technology, Government of Bangladesh. She attended a workshop on "Gas Chromatography Mass Spectrometry (GC-MS)" at National University Laos, Vientiane, Lao PDR, organized by ANFEC and another workshop on "Food Analysis Workshop: Proficiency Testing and Reference Materials Development" at Asia Pacific Food Analysis Network (APFAN), Bangkok, Thailand. She has four publications including an international publication entitled "Determination of chloramphenicol in meat samples using liquid chromatography–tandem mass spectrometry" in the Journal of Food Science & Nutrition. She has a particular experience in Liquid Chromatography Mass Spectrometer (LC-MS/MS), Gas Chromatography Mass Spectrometer (GC-MS), Gas Chromatography (GC-ECD), Gas Chromatography (GC-FID), High Performance Liquid Chromatography (HPLC), UV-Visible (UV-Vis) Spectrophotometer and Total Organic Carbon Analyzer (TOC).

Syed Labib Ul Islam completed his BSc in Civil Engineering from Rajshahi University of Engineering & Technology (RUET) in 2017. Professionally he is a civil engineer who has been working on a number of engineering megaprojects in Bangladesh. At present, he is working as a Railway Track Engineer on several projects of Bangladesh Railway. Concurrently he is affiliated with the Institute of Water and Flood Management (IWFM), Bangladesh University of Engineering and Technology (BUET) as a graduate student, working towards MSc in Water Resources Development. He is passionate about a diverse range of topics including long-term sustainability of infrastructure, resource management, climate change adaptation, soft engineering, and coastal zone management.

Samuel Jacob-Oricha is a seasoned Civil/WaSH, Public Infrastructures Engineering and Community mobilization professional with experiences in development, humanitarian, and fragile context in different regions of Nigeria. He has over nine (9) years' practice experience spanning across engineering planning, stakeholders/community engagement, infrastructures needs' assessment/baseline study, Programme designs, procurement planning, and construction/projects management with private firms, United Nations System, and INGOs humanitarian relief responses in emergencies, early recovery, and community stabilization. He was once a Lecturer at the Department of Water Resources and Environmental Engineering, University of Ilorin, Nigeria. His professional prowess and achievements earned him notable membership recognitions from Council for the Regulation of Engineering in Nigeria, COREN, Nigerian Society of Engineers, NSE, and Nigerian Institution of Civil Engineers, NICE. He currently works with UNDP as a Civil Engineer on the Nigeria Window of the Lake Chad Basin Regional Stabilization Facility. Samuel holds both Bachelor and Master of Engineering (B.Eng. & M.Eng.) Degrees in Civil Engineering from University of Ilorin, Nigeria. Furthermore, he holds a Professional Masters in Peace Building and Development Studies ((MPBD) at the Center for Peace Diplomatic and Development Studies, CPDDS, University of Maiduguri, Nigeria and Mini-MBA from Tekedia Institute, USA. He is widely published in reputable national & international journals with research interest in Developmental/ Humanitarian Engineering practices for Public Infrastructures, Civil Constructions, WASH, Water, and wastewater treatment & Peace Building.

Modupe Jimoh obtained her PhD from the University of Warwick as a commonwealth scholar in 2015. She is currently a Senior Teaching Fellow (Civil and Humanitarian Engineering) at the same University with experience spanning teaching, research and consulting. Her areas of interest are water and wastewater management, sanitation, climate resilience, gender inclusiveness and humanitarian contexts. She has carried out successful projects for the local communities in Nigeria and understands the challenges of water and wastewater management in both developed and developing countries. She is a member of several professional associations, some of which she has held leadership positions.

Anastasia Kadda is a Sociologist specialized in the sociology of health, illness and medicine with a historical perspective. She is also specialized in Health Units´s Management and in Educational Science. Her publications focus upon the role of wider social environment in health-illness, Social dimension of health and illness, The role of sociologists in health promotion, Stress management, Diseases and society in a historical perspective, Relation between health-medicine and society, The social role of medical manuscripts, Social utility of personalized e-health services. She is scientific associate and teaching staff in the department of Nursing of the National and Kapodistrian University of Athens and a tutor in the master of Health Services Management of the Hellenic Open University.

Ioannis Kakkos received a BSc degree in Physics (2013) from the University of Patras, a MSc degree (2015) in Biomedical Engineering from the University of Patras and the National Technical University of Athens and a PhD in cognitive neuroengineering from the National Technical University of Athens. He worked as a Research Assistant with the Singapore Institute for Neurotechnology (SINAPSE) in several research projects. He has 20 scientific publications in international scientific journals, book chapters and proceedings of international congresses and has actively participated in multiple R&D projects. He is currently working as a post-doctorate researcher in the National Technical University of Athens.

Dimitris Koutsouris was born in Serres, Greece in 1955. He received his Diploma in Electrical Engineering in 1978 (Greece), DEA in Biomechanics in 1979 (France), Doctorat in Genie Biologie Medicale (France), Doctorat d' Etat in Biomedical Engineering 1984 (France). Since 1986 he was research associate on the USC (Los Angeles), Renè Dèscartes (Paris) and Assoc. Professor at the Dept. of Electrical & Computers Engineering of National Technical University of Athens. He is currently Professor and Head of the Biomedical Engineering Laboratory. He has published over 150 research articles and book chapters and more than 350 peer reviewed conference communications. He has been the former elected president of the Hellenic Society of Biomedical Technology, HL7 Hellas and Chairman at the School of Electrical and Computer Engineering. Prof. D. Koutsouris has been principal investigator in more than 100 European and National Research programs, especially in the field of Telematics and Informatics in Healthcare. His work has received more than 1800 citations.

Tonni Kurniawan's research focuses on wastewater treatment and water purification, specifically applications of water technologies to remove heavy metals from contaminated water. His work seeks to capture water directly from non-traditional sources such as municipal wastewater and restore it to near drinking water quality, using low-cost absorbents to remove aquatic pollutants. Heavy metals are highly toxic and tend to bioaccumulate in the food chain as they are not biodegradable. Dr. Kurniawan has worked with APEC co-authors to publish over 100 scientific articles. He has been cited over 8,200 times with a Hirsch factor of 32. His most cited paper is "Low-cost absorbents for heavy metals uptake from contaminated water; a review" in the Journal of Hazardous Materials. This paper is ranked among Scopus' Top 25 Most Cited Articles in Chemical Engineering, placing Dr. Kurniawan among the top 1% of researchers in the field of Engineering according to Essential Science Indicators in 2011.

Trung Le is currently an Associate Professor and Head of Department of Resources and Environmental Informatics, Faculty of Environment & Natural Resources (FENR) – Ho Chi Minh City University of Technology (HCMUT). He obtained his Bachelor degree in Geomatics, University of Technology, Vietnam in 1981, and received Master and PhD degrees in Civil Engineering at the University of Tokyo, Japan in 1994 and 1997, respectively. He started his academic career at HCMUT since September 1981. Dr. Trung was a Research Fellow at University of Hannover – Germany in 1998; University of Hosei – Japan in 2008; Duke University, North Carolina – USA in 2011 and Maryland University - USA in 2016. He was a key member of the Union Association of Science and Technology, HCM City and Vice President of the HCM City Association of Geomatics. Dr. Trung's professional experience in teaching, research and consultancy spans over 40 years in Vietnam. He teaches several undergraduate subjects, including Surveying, Remote Sensing, Land use planning and graduate courses on Neural Network, Remote Sensing, GIS, Land Evaluation. Since 2005, he became a Director of Geomatics Center - Information Technology Park (ITP) Viet Nam National University of Ho Chi Minh City (VNU-HCM).

Alice Liddell received her BSc in Environmental Science from the University of Liverpool in 2017, and her MSc in Environmental Engineering from Queen's University in 2018. She is currently studying for her PhD at the School of Mechanical and Aerospace Engineering at Queen's University Belfast. Her research interests involve the development of novel methodologies for water sample analysis, pollutant removal by drinking water treatment technologies and particularly, microplastic behaviour and contamination.

Konstantinos Limniotis is an ICT Specialist at the Hellenic Data Protection Authority. He also holds visiting positions, as an Adjunct Faculty Member, at the Open University of Cyprus, the University of West Attica and the National and Kapodistrian University of Athens, Greece, teaching undergraduate and postgraduate courses in cryptography and network security. His main research interests include cryptography, personal data protection and information security. He has over 40 publications in these areas (book chapters, refereed international journals and conferences), whilst he is regular reviewer in international journals.

Tsung-Yi Lin achieved his Bachelor degree in 1981 and Master degree in 1990 in Taiwan. He obtained his PhD degree in Duke University (Dept. of Geology), USA in 1995. He is currently a chairman of Department of Geography, National Taiwan Normal University. Prof. Tsung-Yi Lin has intensive experience in geology, coastal geomorphology, and coastal environment and hazards over the past 26 years. He has a broad expertise in geology, oceanography, physical geography, geomorphology, coastal environments, marine geology, sedimentology, aeolian transport, natural hazards and environmental changes. He is an expert in costal dunes and barrier islands. In recent years, he paid much his research interests to coastal management, geo-hazards and sand/dust transport.

Hiep Luu obtained his Bachelor Engineer in Telecommunications in 1999 and Master degrees in GIS in 2003 in Ho Chi Minh City University of Technology (HCMUT). He is currently a senior lecturer at

Faculty of Environment and Natural Resources, Ho Chi Minh City University of Technology (HCMUT) - VNU HCM. He used to be an active member of Annual Conference Organisation Board of GISnet Conference in Vietnam from 2000 to 2005, and was member of Academy Board GIScience of HCMUT from 2000 to 2006. Mr. Luu's professional experience in teaching, technology transfer and consultancy spans over 18 years in Vietnam. He is currently a Director of Centre for IT And GIS (DITAGIS).

George K. Matsopoulos is a Professor at the School of Electrical and Computer Engineering of the National Technical University of Athens. He received the diploma in Electrical Engineering in 1988 from the National Technical University of Athens (NTUA), Athens, Greece. He received the M.Sc. degree in 1989 and the Ph.D. degree in 1993 from Bioengineering Unit at the University of Strathclyde, U.K. His interests are nonlinear image processing applied to medical applications, 2-D and 3-D registration of medical images, computer vision applications and Web-based medical systems for telemedicine application and remote image processing. He published more than 200 papers in international journals, conferences, and scientific books.

Christia Meidiana works as Associate Professor at Brawijaya University Regional and Urban Planning Department. She received her master's degree in renewable energy from Fachhochschule Koeln, Germany in 2004 and Dr.techn. degree in environmental technology from Technische Universitaet Graz, Austria in 2012. She has been involving in some projects and research projects related to environmental field such as biogas management, waste management, green house emission and renewable energy development. She is working on urban waste management project in East Java, Indonesia. She is now the Head of the Master Program for Regional and Urban Planning at the Faculty of Engineering, Brawijaya University. Beside journal papers, she writes also book chapters published worldwide. C. Meidiana is registered as a member of Asia-Pacific Chemical, Biological & Environmental Engineering Society (APCBEES).

Mithila Farzana Mumu was born in 1998 in Gazipur, Dhaka, Bangladesh. She grew up in Gazipur and was very meritorious and attentive to her study. She passed her Secondary and Higher-Secondary School Examinations with brilliant results from Dhaka Board. She received her Bachelor of Science degree from Department of Chemistry, University of Dhaka in 2019. She has expertise in analytical and environmental chemistry, and is doing MS in Chemistry from the same University.

Nilufar Nahar is a former Chairperson, Department of Chemistry, University of Dhaka. She graduated (BSc and MSc) from the University of Dhaka and did Ph.D. from the Swedish University of Agricultural Sciences, Uppsala, Sweden. She started her academic career at the Department of Chemistry as a lecturer and retired as a full Professor. She did Post-Doctoral Research at Arrhenius Laboratory, Stockholm University, Sweden, she was a Visiting Scientist at NCI, USA, Shaman Pharmaceutical Company, USA, Mahidol University, Bangkok, Thailand. Her main research areas are Natural Products Chemistry, Food and Environmental Chemistry and Carbohydrate Chemistry. Prof Nahar supervised 19 PhD and 10 MPhil students, more than 200 undergraduate students received their degree under her supervision, she published 171 papers in peer reviewed international and national journals and presented more than 165 papers on her research results in many different countries of the Asia, Europe, America and Latin America. She is co-author of a book on Co-author of a book "Use of medicinal plants for management of Diabetes" and co-author of two Chapters of two different books. Prof Nahar received research grant from ISP, IFS, OPCW and UNESCO besides national grant from University Grants Commission of

Bangladesh. She has arranged many conference/seminar/symposia in Bangladesh including farmers´ awareness programs regarding toxicity of pesticides and environmental issues. She established national, regional and international research collaboration with distinguished scientists during her tenure at the University. She is one of the founding members three Swedish supported networks; ANRAP, NITUB and ANFEC. She worked as an Executive Editor of Journal of Bangladesh Chemical Society, she was national representative IUPAC and member of Division VII (Chemistry and Human Health). She was also a Titular member of IUPAC Division VII. Professor Nahar received three distinguished Scientists awards and one gold medal for her devotion and contribution to science and chemistry.

Anh Nguyen, PhD (University of Limoges), Dean, International Education and Cooperation institute Lecturer in Bridge Engineering, Ho Chi Minh City University of Transport Research interest: Bridge inspection, Resistivity method, Heterogeneous materials, Humanitarian engineering.

Thao Nguyen obtained his doctoral degree in Civil Engineering from Yokohama National University (Japan) in 2007. Prof. Thao is working as an Associate Professor at Department of Port and Coastal Engineering, Faculty of Civil Engineering and currently a Vice President in charge of International Affairs and Communications, HCMC University of Technology (HCMUT), VNU-HCM. Prof. Thao has intensive academic and research activities in Vietnam, Asia and Africa. His research interests include Wave Forces on Coastal Structures; Port and Coastal Zone Management; Natural Disasters and Climate Change; and Change of Coastal Areas due to human's activities. He is the author and co-author of more than 60 peer-reviewed scientific papers in field of coastal engineering. Importantly, his co-research work which was entitled "Coastal Disasters and Climate Change in Vietnam" was published by Elsevier in 2014 and have a great impact among the coastal research community in Vietnam. He also cooperated with businesses to research on local issues, significantly contributed to the improvement of living conditions in coastal areas. Furthermore, Prof. Nguyen Danh Thao has been a key member who organised and chaired many conferences in local and regional areas. He bridged the academia with businesses, government, and community. Particularly, he focused on seminars which helped raise public awareness about natural disasters and climate change in the central part and southern Vietnam.

Nargis Parvin obtained Bachelor of Science (BSc) and Master of Science (MSc) degrees from Department of Chemistry, National University. At present, she is doing her PhD on "Studies of Antibiotic Residues in Beef and Chicken Meat Samples" at the Department of Chemistry (Organic), University of Dhaka which was started from 2018. She is experienced in Organic, Analytical, and Environmental Chemistry. She has particularly expertise in UV-Visible (UV-Vis) Spectrophotometer, High Performance Liquid Chromatography (HPLC), Liquid Chromatography Mass Spectrometer (LC-MS/MS), Gas Chromatography (GC-ECD), Gas Chromatography Mass Spectrometer (GC-MS) and Total Organic Carbon Analyzer (TOC). She participated in many training programs, conferences relevant to her research and attended seminars for expanding scientific knowledges for her research field and she can contribute in the field of Chemistry. She participated in a workshop on "Gas Chromatography Mass Spectrometry (GC-MS)" at The Royal University of Cambodia, Phnom Penh, Cambodia. She has two publications in Bangladesh Pharmaceutical Journal and Asian Journal of Pharmacognosy. She got a fellowship from BAN:04, ISP, Uppsala, Sweden and University of Grant Commissions of Bangladesh.

Anh Pham Thi, Director of the Institute for Environmental and Transport Studies at the Ho Chi Minh City University of Transport. Pham Thi Anh had obtained her Master degree in Urban Environmental Management in Institute for Housing and Urban Development Studies (IHS) and Wageningen University - The Netherlands; and her Doctor degree in Environmental System Analysis and Environmental Policy in Wageningen University - The Netherlands She has worked on education, research, and consultancy in environmental fields. Her research orientations are focused on urban and industrial development with specific contributions in the application of (1) Environmental System analysis; (2) industrial ecology and sustainable development; and (3) public participation and stakeholder analysis for green growth.

Mohammad Shoeb is a Professor of Chemistry at the Department of Chemistry, University of Dhaka. He obtained BSc and MSc from Department of Chemistry, University of Dhaka, and MSc from University of Aberdeen, UK. He pursued PhD in Natural Product Chemistry from the Robert Gordon, University, UK. He is leading the research group consisting of academic staff, postdoctoral, POSTGRADUATE students (PhD, MPhil, MS) and technical staff on "Analytical, Environmental and Natural Product Chemistry". He received research grants from International Science Programme (ISP), Uppsala University; International Foundation for Science (IFS), Stockholm; OPCW, Ministry of Science and Technology, Government of Bangladesh; University Grant Commissions of Bangladesh and Bose Centre. He supervised 7 PhD, 3 M.Phil. and quite a large number of MS and BS (project) students in natural products, analytical and environmental Chemistry and currently 6 PhD, 2 M.Phil. and MS and BS project students are working under his supervision. He has more than 105 research publications in peer reviewed journals with current citation: 2104, h-index: 24, i10-index: 46 and has attended 90 National and International conferences, and delivered lectures. He has collaborations with different universities of Australia, UK, Sweden, Belgium, Cambodia, Laos, Malaysia, Iran, Tanzania, Indonesia, India and Turkey.

Beatrice Smyth is a Senior Lecturer in the School of Mechanical and Aerospace Engineering, Queen's University Belfast, and is part of the Research Centre in Sustainable Energy. Her research is focused on environmental impacts, and specific interests include circular economy and life cycle analysis, optimisation of energy pathways, resource quantification, and the interface between food, energy, water and waste. She is on the research advisory panel for the Centre for Advanced Sustainable Energy and on the editorial board for UCL Open: Environment.

Thomas Toulias is an Adjunct Lecturer at the Civil Engineering Department, University of Thessaly. Fields of expertise: Information Theory, Statistics, Probability, Geometry.

Van Tran Thi is currently an Associate Professor of the Faculty of Environment & Natural Resources (FENR) at Ho Chi Minh City University of Technology (HCMUT). She obtained her Bachelor degree in Meteorology at Russian State Hydrometeorological University (former, Leningrad Hydrometeorological Institute, USSR). She then went on to pursue a Master degree in Geomatics from Ho Chi Minh City University of Technology (VNUHCM). She received a PhD degree in Using and Conserving Environmental Resources at Vietnam National University Ho Chi Minh City (VNUHCM) in 2012. She started her academic career as a Researcher at Geoscience Center, Vietnam National Academy of Sciences, in 1989. In 2005 she was elected Lecturer in Institute for Environment and Resources (VNUHCM). From 2014 up to now, she has worked for HCMUT. Dr. Van's professional experience in research and teaching over 20 years in Vietnam. She has implemented several projects and published many papers in refereed journals in the field of Geomatics, Meteo-Hydrology, Climate Change and Environment.

Viola van Onselen holds a degree in Earth Sciences from the Utrecht University in the Netherlands. In 2019 she joined the PhD program at the National Taiwan Normal University (NTNU), at the department of Geography. For her PhD project she studies how human decisions have affected the natural landscape, biodiversity and vulnerability of coastal areas in South-East Asia. She will explore Nature-based Solutions in an Eco-DRR framework and asses the vulnerability of coastal areas, to bring forward key factors that need to be addressed to enhance ecosystems and make coastal environments more resistant to natural hazards.

Phu Le Vo is an Associate Professor and Dean of the Faculty of Environment & Natural Resources (FENR) at Ho Chi Minh City University of Technology (HCMUT), VNU-HCM. He obtained his Bachelor degree in Biochemistry, Hue University, Vietnam in 1994, and received Master and PhD degrees in Environmental Studies at the University of Adelaide, South Australia in 2000 and 2008, respectively. He started his academic career at HCMUT in September 2000. Dr. Phu was a Research Fellow at Sanford School of Public Policy and Nicholas School of Environmental Science (NSOE), Duke University, North Carolina – USA in 2010 and 2011. He was also a Visiting Professor at Faculty of Life and Environmental Sciences, Tsukuba University, Japan in 2012. Dr. Phu is a key member of the Consultancy Board for developing a Master curriculum program on Water Management and Climate Change Adaptation (in VNU-HCM) in which he was a Visiting Scholar at Faculty of Environmental Sciences, Wageningen UR and UNESCO-IHE (Delft), the Netherlands in 2013. Dr. Phu's professional experience in teaching, research and consultancy spans over 20 years in Vietnam. He teaches several undergraduate subjects, including Water Resources Management, Coastal Zone Management, Clean Technology and graduate courses on Water Governance, Integrated Urban Water Management, Climate Change and Adaptation Measures. His field of expertise is Water Resources Management and Policy, IWRM, Groundwater Chemistry, Climate Change and Water Management. Since 2015, he became a national Short-term Consultant (STC) on Environmental Safeguards for various World Bank-financed projects in Vietnam in the Water and Transport Sector.

Zeba Zaidi received her MSc with Distinction in Humanitarian Engineering with Sustainability from the University of Warwick in 2021. She was amongst the first few graduates of this course. Her post graduate thesis 'Rethinking the Moria Refugee Camp: Spatial and Structural Reconsiderations' was submitted to the International Conference of Women Engineers and Scientists where it received the highest number of nominations of any paper at the conference and went on to win the Best Student Contribution Award. Zeba is currently working as a Sustainability Consultant at NUS Consulting, helping organizations to reduce their carbon emissions. She hopes to merge her two areas of interest and continue her research in the decarbonization of the humanitarian aid sector.

Index

A

B

C

D

E

F

G

H

I

L

M

N

O

P

Q

R

S

T

U

V

W

Premier Reference Source
Emerging GIS Applications for Emergency and Disaster Management

Premier Reference Source
Managerial Strategies and Green Solutions for Project Sustainability

Premier Reference Source
Comparative Approaches to Using R and Python for Statistical Data Analysis

Premier Reference Source
Solutions for High-Touch Communications in a High-Tech World

IGI Global Author Services

Providing a high-quality, affordable, and expeditious service, IGI Global's Author Services enable authors to streamline their publishing process, increase chance of acceptance, and adhere to IGI Global's publication standards.

Benefits of Author Services:

- **Professional Service:** All our editors, designers, and translators are experts in their field with years of experience and professional certifications.
- **Quality Guarantee & Certificate:** Each order is returned with a quality guarantee and certificate of professional completion.
- **Timeliness:** All editorial orders have a guaranteed return timeframe of 3-5 business days and translation orders are guaranteed in 7-10 business days.
- **Affordable Pricing:** IGI Global Author Services are competitively priced compared to other industry service providers.
- **APC Reimbursement:** IGI Global authors publishing Open Access (OA) will be able to deduct the cost of editing and other IGI Global author services from their OA APC publishing fee.

Author Services Offered:

English Language Copy Editing
Professional, native English language copy editors improve your manuscript's grammar, spelling, punctuation, terminology, semantics, consistency, flow, formatting, and more.

Scientific & Scholarly Editing
A Ph.D. level review for qualities such as originality and significance, interest to researchers, level of methodology and analysis, coverage of literature, organization, quality of writing, and strengths and weaknesses.

Figure, Table, Chart & Equation Conversions
Work with IGI Global's graphic designers before submission to enhance and design all figures and charts to IGI Global's specific standards for clarity.

Translation
Providing 70 language options, including Simplified and Traditional Chinese, Spanish, Arabic, German, French, and more.

Hear What the Experts Are Saying About IGI Global's Author Services

"Publishing with IGI Global has been ***an amazing experience*** *for me for sharing my research. The* ***strong academic production*** *support ensures quality and timely completion."* – **Prof. Margaret Niess, Oregon State University, USA**

"The service was ***very fast, very thorough, and very helpful*** *in ensuring our chapter meets the criteria and requirements of the book's editors. I was* ***quite impressed and happy*** *with your service."* – **Prof. Tom Brinthaupt, Middle Tennessee State University, USA**

Learn More or Get Started Here:

For Questions, Contact IGI Global's Customer Service Team at cust@igi-global.com or 717-533-8845

IGI Global
PUBLISHER of TIMELY KNOWLEDGE
www.igi-global.com

www.ingramcontent.com/pod-product-compliance
Lightning Source LLC
LaVergne TN
LVHW060421300926
849162LV00006B/1486
* 9 7 8 1 7 9 9 8 9 1 9 0 1 *